Brain and Being

Advances in Consciousness Research

Advances in Consciousness Research provides a forum for scholars from different scientific disciplines and fields of knowledge who study consciousness in its multifaceted aspects. Thus the Series will include (but not be limited to) the various areas of cognitive science, including cognitive psychology, linguistics, brain science and philosophy. The orientation of the Series is toward developing new interdisciplinary and integrative approaches for the investigation, description and theory of consciousness, as well as the practical consequences of this research for the individual and society.

Series C: Research profiles. Presentations of significant personal and team contributions to the empirical study of consciousness.

Volume 58

Brain and Being: At the boundary between science, philosophy, language and arts
Edited by Gordon G. Globus, Karl H. Pribram and Giuseppe Vitiello

Brain and Being

At the boundary between science, philosophy, language and arts

Edited by

Gordon G. Globus
University of California Irvine

Karl H. Pribram
Georgetown University

Giuseppe Vitiello
University of Salerno

John Benjamins Publishing Company
Amsterdam / Philadelphia

∞™ The paper used in this publication meets the minimum requirements of American National Standard for Information Sciences – Permanence of Paper for Printed Library Materials, ANSI Z39.48-1984.

Library of Congress Cataloging-in-Publication Data

Brain and being : at the boundary between science, philosophy, language and arts / edited by Gordon G. Globus, Karl H. Pribram, Giuseppe Vitiello.
p. cm. (Advances in Consciousness Research, ISSN 1381–589X ; v. 58)
Includes bibliographical references and index.
1. Consciousness. 2. Quantum field theory. 3. Brain.
4. Neuropsychology. I. Globus, Gordon G., 1934- II. Pribram, Karl H., 1919- III. Vitiello, Giuseppe. IV. Series.

QP411.B74 2004
153-dc22 2004052186
ISBN 90 272 5194 0 (Eur.) / 1 58811 550 X (US) (Hb; alk. paper)

John Benjamins Publishing Co. · P.O. Box 36224 · 1020 ME Amsterdam · The Netherlands
John Benjamins North America · P.O. Box 27519 · Philadelphia PA 19118-0519 · USA

Table of contents

Foreword

This book is the outgrowth of the meeting "Quantum brain dynamics and the humanities: A new perspective for the 21st century", which was held at the Institute for Scientific Interchange (ISI) in Torino, Italy, in November 2002. The meeting was born from an idea of Globus Gordon, to gather those involved in the formulation of the quantum model of brain, initiated by Luigi Maria Ricciardi and Hiroomi Umezawa in the middle of 1960s, to discuss in some informal but productive way the model implications for literature, philosophy, and the arts. His conviction was that the quantum model of brain could be in some sense the prototype of a new conception of making science: without loosing its characteristics and the powerfulness of the Galilean method, science must recover its merging with humanities, from which it has diverged during its development. On the other hand, humanities cannot ignore the logical and formal (mathematical and methodological) apparatus of science. Knowledge should not continue to suffer a conceptual splitting between human science and natural science. New ways of thinking are needed to effect a rapprochement.

The meeting was thus conceived to be a first limited attempt, a sort of "experiment" of "thinking together" quantum brain dynamics and humanities, organized in the frame of the ISI Project "Expanding Perception". The ISI Project aims to explore that domain of complexity theory that lies at the boundary between "hard" science, arts, linguistic and bears as well on cognitive and perception science (including the notion of mental space as induced by knowledge structuring and language).

The meeting was a successful one. After few tutorial sessions aimed at creating a common dictionary and reciprocal understanding, which by themselves were an enjoyable example of "understanding the other's reasons" – not only the other's "words"! – the participants could feel the gratifying atmosphere of a real collective thinking by a working research group. The singularity, however, of that "happening" was in the disparate cultural provenance of the participants, in the fact that the physicist's effort was to read underneath his formulas if and where the conceptual images and the logic of the philosopher could find their place. And the philosophers tried to understand where in their theoretical schemes observations in the neurophysiological laboratory could find a counterpart. The participants tried to construct links, to end up with a conceptual net. Therefore, even metaphors, when adopted, were used for what they literally are, tools to "carry over", to bridge apparently separate linguistic (conceptual) levels. To think together, indeed.

This book is the continuation of that experiment. In practically all the papers reported here there are strong echoes of the discussions among the participants at the meeting. The book, however, is far from answering questions; on the contrary, it poses them. Of course dis-homogeneity remains... thinking together science and humanities must be necessarily an interminable effort, dynamically changing in the course of time, with a somewhat dis-homogeneous appearance. This book is only a small step in such a direction. Thinking together is not a simple minded unification. The objective is on the contrary to have a coherent view of the totality of the differentiated objects of today's human sciences and natural sciences studies.

Perhaps, the secret for a true merging is in the assumption, on both sides, that humanity *is* the canon, the measure (the ratio) of everything else. It might be comforting that under such an assumption in the Italian Renaissance, painters and sculptors were listed as mathematicians in Pacioli's *Summa de Arithmetica* (1494). The first book of *De Pictura* (1435) by Leon Battista Alberti was completely devoted to "matematica". Knowledge, no matter how reached, was only one.

The meeting discussions pointed to a common direction where the participants, no matter their cultural provenance, were apparently converging: the idea that one cannot think in terms of isolated "elements" or "individuals", in biology as in physics, in psychology as in cognitive sciences. Rather one should think in terms of the cooperative dynamics among constitutive individuals which manifest themselves, at a larger scale, as a coherent system.

In many instances, the humanities, to which actually such a view is not completely extraneous, have been conditioned by the "atomistic" approach in natural sciences. During the latter part of the 19th century field theoretical concepts were regnant in the scientific community. Atomism came roaring in at the turn of the century. For the latter part of the 20th century, quantum field theory began to recognize what the earlier generation, Bohr, Heisenberg, Pauli, Dirac and Wigner had already been trying to say. But when confronted with explanations, most quantum scientists often continued to fall back on particles.

On the other hand, subsequent successes of natural sciences, which could be achieved only by letting the atomistic view evolve into the modern field theoretical view, were not so deeply influential for the conceptual apparatus of human sciences. Moreover, not only humanities did not share the shift to field theory in the scientific conception, but also large sectors of "hard" science (think, e.g., of some sectors of biology and biochemistry) are still today bound rather to the atomistic view than to the field theory conception. It is interesting to notice that the humanities, in particular, are touched by the initial revolutionary steps of the 20th century, not so much for the new results they were providing, but for the halo of mystery and unknown they were still carrying along. It is also most interesting that the conceptual content of the Maxwell equations for the electromagnetic field, which still belongs to the classical physics view, is largely ignored in very large sectors of the humanities.

Apparently, it is the concept of field which is found frightening. The ontological prejudice by which "things" are made of little beings, individuals, atoms, able to survive even in the absence of any interaction with similar beings, strongly contrasts the idea, implicit in the concept of field, of abandoning the individuals as "the actors" able to establish or not establish some sort of relation with other individuals. In the concept of field the "action" is more fundamental than the actors, and there cannot exist one isolated, single actor, but only a multitude of them. Here there is a profound shift in the conception of the Being.

This conceptual shift, embedded in the formalism of quantum field theory, is the root of the modern understanding of solid state physics, of elementary particle physics, and of cosmology. The question is, then, whether time is ripe for thinking together science and the humanities on the basis of this revolutionary shift in the conception of the Being. The quantum field model of brain and of living matter provides an appropriate venue for such a challenge. What the group of people gathered in Torino hope to offer to the 21st century is combining their transdisciplinary sensibilities with the volume of data accumulated through the 20th century revolutionary experiences.

In these dark days, their commitment is to adopt in their search the spirit of the *Tokyo '99 Declaration*:[1] Their hope is that the effort put forth in erasing "the egocentric discipline-confined approach" would be "for serving the human welfare, never warfare".

We warmly thank Professor Mario Rasetti and the ISI Director, Dr. Tiziana Bertoletti, for enthusiastically accepting the idea of our meeting and for including it in the ISI programs. The meeting could not have been realized without their help and the financial support of the ISI. We are also thankful to the ISI staff for their patience and the work done in organizing the meeting.

Finally, we are grateful to Springer-Verlag Publishing Company for giving us permission to reprint in this volume the 1967 paper by Ricciardi and Umezawa on the quantum model of brain. Our thanks also go to Mrs. Bertie Kaal of the John Benjamins Company for her patience and assistance in the editing process of this book.

Gordon Globus, Karl Pribram and Giuseppe Vitiello
Irvine/Georgetown/Salerno, November 2003

Note

1. Yasue K., M. Jibu, & T. Della Senta (2001). *No Matter, never Mind. Proceedings of Toward a Science of Consciousness: Fundamental Approaches* (*Tokyo '99*). Amsterdam: John Benjamins.

Chapter 1

Doubling the image to face the obscenity of photography

Nadia Prete
Artois University, France

The double is put in place via compositions made of photographs and photocopies. These compositions called "*photobjets*" superpose and juxtapose layers where the images thus created face "the un-possible image". This artistic questioning led us to think that if Photography can still be considered as an epiphany, it is because it unveils the obscenity it carries. It works through its two qualities: its proximity to reality and its singular fixity. Is the doubling of the image compulsory to escape its obscenity or to face it better?

Artists make objects, look at them, ask them questions.

The artist starts sounding the body of his work as a practitioner tries to understand what the symptoms have to say about his patient's problems. For our questioning of the double, symptoms are creations made from photography. They are built in various forms setting several images in a relation of duality. I named these works *photobjets* because they carry images made from traces of light and because they can be handled like objects. Their singularity lies in their composition: they are made of at least two layers, each presenting an image, that are superposed thus creating an element. Often, juxtapositions of such elements may compose a more complex *photobjet*. Studies done some ten years ago from the point of view of form will be described later on. Subsequent – and further to the realisation of these studies, I was able to start questioning myself on the subject of the photographic image and on the approach to the obscenity I feel it contains. Other works have answered this questioning but they will not be called upon here.

To photograph is to duplicate. It is an attempt to give an image of Reality – ideally speaking, for reality is moving. The photographic double of what has been there at a given moment in time – never set and already gone – will always be at a distance from the present moment which is looking at it. The difference in time and thus the distance in perception added to the photographic petrifaction imposed by exposure – between what was there in the pulse of the living object and what is now in my hand facing my eyes captured by the photo-Medusa – created an insurmountable obstacle for the photographic double, image of an image, to become the exact replica of a perceived reality. Reality cannot be trapped. Its instability forces us to accept that we have to approach it by fractions, substitutes of reality built for our understanding.

It is important not to believe in the unicity of reality and to see these part-reproductions as resemblances, both true and false, allowing us to get closer to the idea of reality. Photography is but one expression of reality, momentarily giving it an existence in which we can believe and in which we can invest. For, as Clément Rosset said, reality is insufficient to give an account of itself. This is why "what counts is that the meaning is not here but actually elsewhere – hence a duplication of the event which splits itself into two elements: its immediate expression and what this expression expresses, i.e. its meaning" (Rosset 1984: 76).

It is uncertain that photography, by doubling reality in images which delude us, is the best way to perceive reality. But, it enables the capture of a restricted reality which is then defined by the limits of the image. The advantage of these multiple limits is that they hold no illusion. They demonstrate what the act of looking really is: a mental construction of what is seen, in the form of a double. This construction is like the lining of a garment, it follows the garment's outline, fits its shapes but merely remains a light cloth reproducing it. Another self born of itself that slips into it while making a protective and discreet separation. Therefore, if all photographs put together are a thin double of the world, the photographic extract never stops recalling the limitations of the act of capturing the visual. What the eye records (just like a camera) is but a small portion brought forward for analysis, at best inviting us to stop. Paul Virillo summarizes this situation in one single formula: "To see is to be limited in one's field of vision; to see is not to see" (Virillo 1997). Thus, to see a photo and apply oneself to what it offers is evidently not seeing reality, of which, as all doubles, it merely gives the illusion, but it is to believe and through this belief, see its appearance.

"Photobjet"

I was not satisfied with the taking of one unique photographic image. The contiguous relation between two (or more) images came to me as a solution to bring me closer to what I was looking for. To come closer to this ideal image, a double was necessary. To capture reality partially one could use photography. But, to get closer to our relation with reality, a creative work in the form of a composition was necessary. I *could* not include everything in one representation; this does not mean I did not *want* to try. I could not do it from a strategic point of view, my first approach is always based on the concept of duality: bringing two units closer to find the third one, the impossible one. Two layers were necessary: the image and its double talking to each other and touching each other, thus inviting us to consider their exchange system, to question the meaning proximity generates.

Even if each unit is divided up in smaller elements (mini-units) and the construction becomes, by juxtaposing these elements, more complicated, each unit contributes to a system of proximity. The first unit, within its own divisions, is given several forms to show that one's choice among all the elements is infinite and that there are always multiple and infinite variations to be found. The unique relation between the small unit and another small unit leads to renewed waiting because each time they are representative of and question this continuity in a pair that repeats itself.

For five years, I thus made objects out of plane surfaces (cloth, tracing paper, transparencies) on which I laid images from photographs. In the beginning, these objects were made of two or three superposed and more or less opaque layers. Later on, they were made of juxtaposed elements (two to fifteen) themselves made of a transparent layer superposed on an opaque layer. First, I used canvas and picture frames that did not let through what was under the first visible image. This first layer was willingly made narrower and through the overlaps on its sides one could make out one or two underlying layers. I started using tracing paper when I abandoned the third layer. In 1991, when I made "Ecrans" using only two superposed layers, I photocopied the image on tracing paper. Whereas in 1990, all my works were made of three superposed layers and a photograph was laid on the top layer. An image was made visible under the translucence, the interrelations were then limited to the relation both images have permanently when thus trapped. The images are not separated by a supportive-layer anymore (cloth, paper, frame) and thus remain perpetually contiguous. They touch each other like a garment touches the skin, exchanging actual experiences that give meaning to their existence. When using tracing paper, a third layer was impossible since what was important was to see everything together or more precisely, to have a perception of this togetherness as a whole and to understand that it was made of two images in strong relation. Later on, the scrambling and the moiré created by the tracing paper between the images was further lightened by the use of a transparent plastic medium; it scrambled with less modesty. The two layers' proximity allows one's eye to embrace both layers at the same time, until it feels it is lost to itself. Tracing paper allows for acclimatization, for a time of latency during which the insert of the image to come does not happen with the same quality. This difference in the crossing of light, from one to the other medium, almost puts them in opposition. On the one hand, the opaqueness (of tracing paper) that retains part of the viewed object is the visible condition; on the other hand, the transparency (of transparencies) does not allow for the eye to be stopped and brings one back to the idea that there is some truth in the illusory sincerity of seeing everything. Using transparencies as photocopying material prompted me to think that with transparency one does not see anymore. One cannot stroll about the world's reality anymore when entering transparency's foldings, when moving deeper from front to rear, when exploring its contours and thus, transparency kills knowledge by erasing all mystery.

Once the two-layered structure with an image on tracing paper or transparency was fixed, I constructed *photobjets* up to 6 metres long by juxtaposing such structures (two to fifteen). In photography it is quite usual to see such horizontal juxtapositions. It is thus not the construction of the *photobjets* that intrigues and always disturbs, but the fact that the superimpositions hide the image of the lower surface. They are placed like concealing veils saying: "Look, lift me".

Photobjets, made of multiple structures, increase the number of readings. Reading is thus diachronic and synchronic. The viewer is invited to act twice: lift and move. If he does not do so he will only have a partial vision and will not see it all. By superposing images and moving them closer to one another, photography's singularity is evident: it is possible to "leaf through" it. This is what makes us use photography rapidly in everyday life for illustration and documentation (posters, newspapers, books, or for remembrance). Only afterwards is this singularity presented to us as a matter for meditation. Sur-prise[1] can thus immerge when, at the precise instant an image suddenly stands out and the eye stops, a swarm of other talkative images come to mind at this exact same moment to envelop the unexpected image.

Photobjets are not only about the relations between forms, calling to our mind some similarities in lines and volumes. They are not only about "photographicity", in the sense that a painting refers to its pictoriality as we have been taught to understand since modernism. *Photobjets* are realities perceived through a glimpse at reality, that are invited to build other images. Photography has always more to say about perception than about itself and that is what makes both its weakness and its strength. By superposing and juxtaposing, I imply that behind and between the layers, an image is always going to appear. It is this image, the image that seems veiled for eternity that I would like to catch in a constructed insert.

The insert is quite obviously the space taken, and apparently empty, that appears between photographs or photocopies. A suspension is thus created, that is favourable to the opening of various mental images, each one coming from the others which enrich the real images that provoked them. One might fear this suspension could only take us further from a truth much franker in its immediacy. On the contrary, the constructed image made of physical superimpositions and, more so, the non-image made of mental superimpositions are fleeting appearances that are closer to a moving reality. Personally, too often do I feel that the photographic image that lets itself be seen on its own is brutal. By withholding time and by keeping all things close to it, the photographic image so totally lacks self-restraint that it becomes obscene and in the end, impoverishes my reflection. Since it shows so much at once, it does not allow for distance of analysis anymore. The image is to be taken or to be left, for immediate consumption or brutal rejection that sentences it, most of the time, to indifference. Its only chance for survival is then to transform itself into a symbolic image. An image that in our view will not be more exciting since it thus becomes an emblem, the fixed totem of some fact of society.

"The un-possible image"

The execution of *photobjets* always comes back to our first concern: "Which image?" Which image should we imagine today to enable its present to face our future? Which *un-possible* image?

Un-possible could imply the possible existence of a unique image towards which all of the artist's efforts lead. An image, forever present and to be made again and again, time after time, since as it is being made, it would be thrown into the past and questioned by the present (only one possible image).

Un-possible could express the difficulty to be satisfied with executions that would express a whole, that would contain, even momentarily, what is essential to be expressed. There would be nothing new for art to draw from such an observation. If it was not for the endless quest, inhabiting all artists, to always go beyond, there would be only one work of art in each discipline, renewable in different times and different spaces, I admit to it, and it would be The representative of The Work of art (the impossible image).

Un-possible is a play on writing in French or rather, a play on sound. It is a contraction that summarizes the idea that it is *impossible* to create in *oneness*. One can thus see that, on the one hand, writing *the un-possible image* meets an inner need different to the one which would have led me to write *one impossible image*. On the other hand, the use of the definite article, *the*, implies the image is unique on universal grounds whereas the use of the indefinite article, *un*, implies the probability of several. I thus imply that no image has yet been possible and made but, that different solutions can be acceptable. "The possible," wrote Henri Bergson, "is but reality to which an act of the mind is added that projects the image in the past once it happened" (Bergson 1969: 100). The possible is a speculation that takes shape only once it has been realised. Bergson tells us in one essay entitled "The possible and reality" his answer to somebody asking him what were the possible directions for tomorrow's work of drama:

> "If I knew what tomorrow's great work of drama will be, I would do it." and adds: "however, the work of art of which you speak is not yet possible" – "It has to be possible since it will happen", insists the person. – "No, it is not, I admit at most that it will have been." (Bergson 1969: 110)

As long as the work of art is not made it remains impossible.

> As reality creates itself, unpredictable and new, reality's image reflects itself behind itself in the indefinite past; thus, it was possible at all times; but it is at that precise moment that reality starts having always been and this is why I said its possibility, which does not precede its reality, will have preceded it once reality happened. So, the possible is the illusion of the present in the past. (Bergson 1969: 111)

To think an impossible image is thus to dream it and, beyond, to work at its realisation knowing that the quest is infinite since as time goes forward, the present becomes the past; and that there always is an impossible image to create, all the way until it is made possible. *The un-possible image* is the image that would contain all: itself and everything surrounding it, time, space, reality and its imagination. The image is of course never visible. However, in no way should we become disheartened asking ourselves "what's the use" since one of art's duties is utopia.

Nadia Prete, "Vestige N° 1" (open), 1995

I use the formulation *the un-possible* because I know it is not possible to be satisfied with one photo. The mere possibility of "one" sounds impossible. My *photobjets* approach the possibility of this unique image but I am perfectly aware at the same time that its existence is only virtual. It is an image that cannot be represented. If I accept this impossibility to be represented, I have to set in place a system to show that however out of reach, I can hope to achieve it. Thus, the peculiarity of my work is not the production of the Image anymore but the image's perception that the relation between two images should create.

Everything that is produced and that tends towards it does not form a set of approaches brushing past it. Could these steps forward give us an idea of what the image would be? Does the quest make sense? Pascal Quignard wrote:

> Because the scene that is never present, that is "not-presentable", can never be "re-presented" to man, the production of that scene." "No man can hear the outcry when the semen that made him flowed out. (Quignard 1994: 146, 226)

Is it reasonable to continue searching for it, even if it is it, the only image forever present, that is of concern to us? Isn't the iconoclast right refusing all approximation? Since it seems that what makes the image impossible to produce lies there, in the observation that this scene happens in a past that our present can never find again and will never bring back. The image was, but its mystery remains untouched. All I can hope for is to suggest it, to produce a vision of it (*image of light*), to come close to what created it during the coupling of the other two. The fact is, this image cannot be pre-arranged. I try to console myself of this deprivation by producing other images, building interrelations that minimize this want, that state a more bearable human truth. Images are thus accumulating or coming back introducing with each construction a new movement. A movement which tries to approach the crack in the world; a gap that, far from leaving us mute, makes us create new chants. And here we are, suspended to these melodies; all of which seem useless and derisory and enable us to face fear and instability as well as the drama and the unbearableness of the primal scene.

The image, un-hoped for, that I call *the un-possible image* is the contrary to Roland Barthes' unary Photograph.

> The Photography is unary when, it emphatically transforms "reality" without doubling it, without making it vacillate (emphasis is a power of cohesion): no duality, no indirection, no disturbance. (Barthes 1980: 69)

This photograph does not "say" anything, it cannot be listened to and in it, there is nothing from ourselves to be heard. If the photographer is not producing this photograph, is he aware of the fact? How to be sure that the power of the photographic image is there to keep the spectator alert, wounded by the reading? One cannot ever be sure, one can only come close to it. This might only be a hope. This fear is partially in all my installations. By harmonising my images, by allocating a double to them, I secure them in the same way alpinists do when they rope themselves up to avoid falling.

I know this image is impossible to be produced. It is an image that appears and fades away during the attempts at producing it. This image does not belong to the possible of images. Let us imagine, nevertheless, that one could manage to bring it to an image's reality. Wouldn't it become dangerous for it to get too close to reality, which structure it could then take? This image does not exist and if it did, its fragility would be so great that it would disappear as soon as it is approached. I would like to achieve it though, for its inaccessibility and uncertainty are what makes its worth. The existence of this image, without believing that it can be developed, must be hoped for. And since it cannot exist, the *image of light* replaces it in the insert. This *image of light* does not pretend it will materialize; to suggest it however, there are two images. Two, very real, images that we can see, describe, analyse and touch: two wrecks gathered on the coast of reality, grounded in the desire of a photographic eye.

At my first attempt at works of art using combinations of layers and before photography even came in, three panels were organised as a polyptych. The reassuring symmetry of the construction created a hierarchy in roles. This form did not last long once photographic images were added on panels. The three panels were then superposed instead of being fixed in the form of a triptych on hinges. The narrative structure that was cluttering the discourse on form was thus questioned. The middle image was not the first one anymore, it was setting the possibility that inserts are infinite: 1+1+1... ad infinitum. I have been able to understand that my images took on their meaning in the inserts and within the complexity of the inserts' insert. And so on, until I understood that, maybe, the end-result would not be an image but rather a disturbing opening created by the arrival of another image. When two events, two objects, two people are confronted or superposed, when bonds seem to unite them, when their meeting is unexpected, we find ourselves destabilized because we have no rational explanations. That is what photography lacks most: the possibility of destabilization and uncertainty. Photography does not allow for co-presentation in one single frame of several locations, unless one traps reflections of light and makes use of projection. It does not allow either for co-representation of several configurations separated in time. Painting does not capture images but construct images from multiple visions, as cubist research showed. Photography, on the contrary, imposes a unity of time and space. Photography acts like Science – could it be because it is chemistry? It tends to unify everything in one explanation that philosophy has no obligation to accept.[2] Thus, the acceptance of a truth in all things through opposition is an attitude that goes against this unification. Dualism (if I can call so my attitude to approach truth) might go beyond my compositions. It is an attitude to fight against the reduction that is expressed within them. How long is this defensible?

The only image that could stop my quest cannot be found since it is established as impossible. We only have the imprecise mental image of the event we have constructed for ourselves. Nobody will ever possess a representation of what can be represented only at the risk of disappearing, whatever the strength of one's desire for an image is. The image will continue to be the echo of a want that Jacques Derrida calls *Différance*, meaning the operation behind delaying that cracks and delays presence.

The image and the origin of its doubles

The fact that the primal enigma is the enigma of the scene that brought us into existence does not prevent another enigma to come to us: the enigma of our death, that confronts us just as much with the mystery of selection. Our consciousness lies between these two unexplainable moments reminding us that we are made of randomness and unpredictability. This double uncertainty is our reality, pre-occupying our creative acts.

Working towards the *un-possible* image means that we want to face one explanation of this mystery, that we hope to encounter origin itself; whereas, with the search for the origin, what comes forward is the desire of an image, as if it could answer our questions. From that desire, "what matters to us is, precisely, to know the sense of existence" (Merleau-Ponty 1964: 18). My quest for an image of truth is the hope of an image that could answer the question of the meaning of our presence among things.

My approach to the common aspiration to face anxiety is singular because I search a photographic image both ideal and comforting. The sharing of the original trauma by all humans is what makes its drama so impactful. I share with the whole humanity its components, similar in their nature, but each time different in their individuality.

> Humanity never ceases to result from a scene that puts together two mammals, male and female, whose uro-genitals – if abnormality comes to them, as soon as they become distinctly deformed – fit together. (Quignard 1994: 226)

I will miss this image always. I look for substitutes that will also never fully satisfy my hope for its representation. They will be replacements for an object that cannot show its nature and they will risk the approach of other – just as worrying – compensations to aim at the impossible.

> Freud had to erect as primal scene, as radical trauma, the moment when the child discovers his genitors are two people from completely different species, who have something to do with each other, that involves violence, love and entangled bodies… from which he was born. (Sibony 1991: 342)

This image of the origin is said by Pascal Quignard to be inaccessible and thus always invented. It is to be created, to be searched for, to be produced, to be constructed with the few disconnected elements in our possession. It will be "what gives form to the formless, what provides an image of the lack of image, a representation of what is impossible to represent" (Quignard 1994: 226–227).

Rather than stating each time "this is it, this might be it", I prefer to keep it forever veiled between two imperfect images which may be acceptable in the hope that they will engender the fantasy image.

> In the end, the idea might look simple since the origin is lost (as initial limit of the human being). One can only retrieve it through its successive returns between two of its resurgences. But, – and that is what is so difficult – one has to link up together two of these resurgences confronting each other, each one being partially an image of the origin. (Sibony 1991: 337)

The image of the origin is reachable and its existence cannot be represented. However, to locate it in the insert presents the advantage of making believe that its approach depends on the thickness of the generated crack, since exploration starts with this crack.

What is even worse is that the sought-after image takes various forms to better interfere with our consciousness – sometimes these forms are softened and reassuring, sometimes they are made of visual horrors of outrageous violence. Only war suffering can go beyond this violence; it is the ultimate horror and it reflects the obsessive image of man with the truth we are trying to understand. An image that reminds us of – and insists on the "why so much suffering? why so much indifference in perpetuating it?" Should we find in these endless thoughts, the expression of the conflict inhabiting man faced with the enigma of the image that created the enigma? Man so portrayed is still made of

Nadia Prete, “Vestige Nº 1” (closed), Nº 2 (open), 1995

destructive dualities and confrontations while in search of a moment that would be the "before the first conflict".

The image of the origin carries this question of the existence. Our existence and also the existence of things.

> There is no mystery in things but there is a mystery to things. It is useless to dig into them to try and steal a secret that does not exist. Things are understandable on their surface, at the border of their existence: not that they are what they are, but simply that they are. (Rosset 1977: 40)

Photography cannot do more than present the surface of things, flattening them on its own surface. All we can expect from photography is to exhibit this cold glazing nakedness it fixes bluntly on us whereas we could still hope to catch some truth in reality's depth and thickness. Photography tells our ears, voluntarily deaf, to be without hope on the matter, and repeats it to our blinded eye; photography carries within itself the lack of meaning of things which, for it, represents a double absence: the absence of the image of the origin and the absence of origin of the image which, because of its technicality, it gives the illusion to catch.

We know that photography is not reality, that it is not even a double of reality – even if we could live with this illusion. I even wrote that to photograph is to duplicate. It is photography nevertheless that gives the image which seems to be the closest to the object it captures, the image that gets the closest to what seems to be the perception of what was originally. It is photography that makes us believe in our proximity to reality the most, because only a photograph can prove reality did exist. We know that in paintings, models were worked upon and in computer-generated images, they were fabricated, whereas with photography the subject had to be there originally. Its presence impresses upon the image while the object of the image acquires more credibility.

Photography tells us about the 'here and now' of things; it is not the "here and now". It tries to be its double but it is a fixed representation of what is merely a mirage that fades away when approached. Photography is not the image of the world it speaks of. The transpositions, concentrated on the surface, necessary for photography to exist, are different from those of the original components. And, with this weakness turned into strength and through its discrepancies, it allows us not to apprehend reality as being one and *idiotic*. Idiotic as per Clement Rosset's definition:

> Reality is what exists without double: it offers neither image nor relay, neither replica nor respite. That is why it is "idiotic", from *idiotès*, idiotic's first meaning is simple, singular, unique, not copiable. (Rosset 1977: back cover)

Photography takes us back to the origin of things but because it is not the thing itself it stimulates our desire for it. Thus, all photographs that give themselves away, most trivially, as mere doubles of reality but never really give away reality's idiocy, call upon our eye in a different way. Sometimes, it happens by stressing indifference but sometimes, by underlying some interesting aspect that our "wandering" eye would not have seen without it. Paradoxically, photography does not give us a lot with regard to its materiality and to the object's; it is, nevertheless, the technique that makes us believe the most in the existence of the object. It enables us to be particularly attentive to things. We cannot take away from photography what makes its existence, i.e. our desire for things, and what gives it its importance, i.e. our emotion towards things. Desire, in photography, is a machinery that is not only staged at each shot, but is replayed, in some obscure manner, at each visioning of each image thus produced. By choosing the word "machina" for what the French call "appareil",[3] Italian language shows the spectacular undertaking in operation for each shot, similar to the theatrical machinery. All variations are possible to stage the desires at the origin of the image: from the most detached photographic approach – left to the camera's eye and to chance –, to the most involved – constructing a "theatre" for the operation –. The spectator always tries to immerse himself in the image's "how".

The photographic obscenity

Photography, in comforting us of the loss of an ever seizable reality is fetishist, both presence and absence. In becoming a pretence of truth, it is the best man has found to fill in the want left by the unreachable reality. If photography remains what allows us to approach reality (despite cinema), if it remains indispensable to capture our interest, it is because it generates a reception only it can produce, which responsibility I place on its obscenity.

How to define obscenity? Is it what comes from the left, which is of ill omen (ob-scuevus) and what the eye should turn away from? Is it what is outside the scene (ob-scaena)? Is it what prevents all representation of the scene? Is it, in all cases, what offends modesty, what thus *must* not be shown because too real maybe, or too crude? The term would finally point out the insane pretension to say more, to go beyond all seen things, to express the naked and primary truth rather than restricting it to sex.

What makes a photograph obscene?

Its subject? as for instance a suggestive shellfish like the ones in Edward Weston's and Jim Dine's photographs. Reality itself if we can perceive it in an apparently faithful context? Or the singularity of the photographic medium that gives a singular image of the capture of reality?

"In the Photograph, Time's immobilization assumes only an excessive, monstrous mode: Time is engorged" (Barthes 1980: 142). It exhibits itself making us believe it is reality that it gives us to see. The scene to be seen is taken from reality but what walks towards us, as strongly as reality, is photography's singularity: it forced time to stop its course, it aimed at the impossible thus perverting our vision. The representation presented to us projects itself towards us to immobilize us in the net of a suspended moment.

The fixed figurative image takes the represented image away from its simple presence since what is shown is not the thing itself but the thing's force in an energetic metamorphosis. This is what Jean-Luc Nancy calls "monstrance", photography is out of the ordinary presence, bringing it closer to the nature of a monster. The image made to narrate does not draw its force from the represented thing only, but also from the violence done to the thing. The image's fixity allows us to imagine, it leaves time for the necessary work for a close prehension of things.

The image proximity (in the sense of contiguity and not faithfulness) to what is represented varies depending on technique (painting, photo, imprint). It always allows however, and whatever the image's nature, to stress its existence and thus ensures the image's power to generate apparition. The power of the photographic trace is *violent* in its ability to authenticate time and space; it is *soft* in its closeness with the thing since it happens within a protective distance thus avoiding all contact with the surface. To make a mark, teeth or steps have to come in contiguity with the material carrying the mark. Whereas with photography, there is no apparent physical contact. The mark exists only through the transportation of light (photons) from the object to the medium. I perceive this luminous flux as an exorbitance crossing space and trying to compensate for the lack of contact. To me, this exorbitance is not the monstrous immobilization Barthes talked about; on the contrary, it is the constant movement of reality calling for an image, on a continuous path going from the image to the spectator. We can thus agree with Barthes saying "The Photograph is violent: not because it shows violent things, but because on each occasion it *fills the sight by force* and because in it nothing can be refused or transformed" (Barthes 1980: 143). Photography is the witness-mark of the bodies and objects that exhibited themselves to the photographer's vision before exhibiting themselves to our vision. It has the ability to *impose*.

For me such proximity, characteristic of photography, combined with *fixity* is what makes it a carrier of obscenity. The represented scene is no more obscene than any other scene – morals or ethics are not in question here – but the way it is captured and transmitted is. Let us not be mistaken, the obscenity of the medium does not come from the subject but from a singular lighting of reality which gives it an often unacceptable indecency.[4] Because photography leaves clues, something of the object, transferred on the photographic image, always remains there; an excess of reality which is neither the object nor its envelope but which reminds us that it was captured, not so much with our hands, but with our understanding. A presence holds on to the representation, in no way embarrassed by the materiality of a medium used as a screen; the essential presence of which the thing is really made and which is why we endlessly go on its quest: the awareness of the impossibility of its presence, a rigid absence that can create anguish and rejection or cause fascination and fetishism. We accept the invention of added elements – in painting, in texts, even movies – whereas we apprehend photography only within the space of time it shared with reality. Reality comes close to our body, present today, through the violence of a projector's beam lighting too brightly what could have, or should have, remained in the shadow.

At the end of the meal with the Autodidact, Antoine, the hero of Jean-Paul Sartre's Nausea, feels the Nausea coming back. He exits the restaurant, walks, enters a tramway and jumps out because he cannot "stand things being so close anymore" (Sartre 1968: 178). "And suddenly, all at once, the veil is torn away, I have understood, I have *seen*" (Sartre 1968: 179). The existence of things appears to him. Things are not set in abstract categories anymore, they are made of the dough of things itself. We can relate this vision, maturing within him, to our consciousness of the things photographic images bring us. Something we perceive without seeing is suddenly present in a luminous trace left on a piece of paper. This is a paradoxical situation which needs a quasi immaterial image of things to give substance to their existence. This is the sign of a brutal stop that allows consciousness of "soft, monstrous masses, in disorder – naked, with a frightening and obscene nakedness" (Sartre 1968: 180). We thought the objects were harmless, Antoine said

> But as soon as you held on to them for a moment, the feeling of comfort and security gave way to a deep uneasiness: colours, tastes, smells were never

> real, never simply themselves and nothing but themselves. The simplest, most irreducible quality had a superfluity in itself, in relation to itself, in its heart. (Sartre 1968: 184)

Could obscenity be this superfluity? One can be strongly aware of it if one accepts fascination. This consciousness then turns definite as knowledge is, hence its power. And, this is why one often chooses to adopt a fleeting attitude. We accumulate shots, we multiply prints, we surround ourselves with images in all occasions and we end up so overwhelmed by these images that we do not look at them anymore. By refusing to pay them attention and to ask ourselves questions about their realisation, we turn away from their banality, we also flee from their embarrassing potential for violence.

Khmer prisoners were systematically put on files, most of the time before being conveyed from the Tuol Sleng prison to the Choeung Ek extermination camp. Photographs of these prisoners are exhibited at the "Choeung Ek Genocidal Centre". How can the exhibition of such photos violate me? If I do not know anything about why they were shot, about the conditions of detention of the featured people and the following extermination, these photographs, as all the other realistic or atrocious images that give an account of our century's conflicts, will pass and become ordinary. If, on the other hand, I know the unimaginable reality, I cannot look at them with the same indifference. The image's obscenity, what it tells me extra, that comes out of itself as a medium, comes from the knowledge I have about the conditions in which it was shot. Photography is obscene only because of the quality of disclosure of reality it gives me. It is not only photography that is obscene, it is reality itself. However, because of its proximity, it keeps some evidence of reality and because of its fixity, it confronts me with reality. Photography gives the indistinct impression that contact was made and that the definite inscription will not let me escape from it. We are caught in the means photography uses to transport reality's obscenity to us. Photography always exhibits these singular means and doing so, becomes obscene itself.

> Only photographic experience can send a glance back at itself, a *look both seeing and seen*, irremediably suspended, *fixed*, which was not the case in the specular but labile mirror's experience. Where theatre, cinema and video distil in time, disclosing their object while hiding and taking away some of its aspects, photography exhibits and gives *all at once* what it has to offer. Its nature is no longer that of a promise but that of seeing's omniscience and omnipresence. (de Meredieux 1983: 91)

The more direct, the more obscene the photography. It can be compared in this instance to sexual discourse: one request, one word is perceived as obscene and more so if they plainly designate the intention, if they offer no side issue. Fixity has it that, before and after, there is nothing around the photographic image, it is inside the image itself that the photographer can limit the eye's movement. Close-ups reinforce fixity by using the eye's fixity upon the subject. They prevent any interpretative streamline and they can be considered as obscenity's hypertrophied demonstration. The eye has no means of escaping the representation: it hits the image with the same force the image seems to use to go towards it. More than anywhere else, this is where proximity and fixity are maybe the most exhibited. The eye is blocked and the discourse presents itself as objectivity's discourse.

To counter attitudes of escape anaesthetizing the image, which most contemporary photographers have to face, I first avoided accepting photographic obscenity using coverings that at once posed the problem for the eye. I then chose to create breakaways to elude fixity and re-establish some time and a virtual image that would escape proximity, another characteristic of photography, through its total lack of contact with reality. However, I had to admit obscenity does not have the qualities of a stolen interval, it holds the evidence of confrontation. Once we have discovered the difficulty of defying obscenity, we can implement escape processes, we can wish not to have to face it. This is why we always have invented structures (myths and fantasies) to protect ourselves from reality and why we continue doing so with its faithful servant, photography. The diverted vision, at work in the *photobjets*, is the vision that bypassed and examined obscenity and its overly forceful truth. I do not want to compose from these registers anymore, they have become demonstrative insert systems. I now try to simplify vision and to look for the image that does not turn away from photographic obscenity's presence; for the image that brings out the understanding that photographic obscenity has to be dealt with and against, to better appreciate the petrifaction force coming from reality's obscenity and thus to better avoid the paralysis liable to befall all of us.

Translated by Brigitte Riera-Lund

Notes

1. In French, *Sur* means *over* and *Prise* here is used in the sense of a photo *shot*. Hence a play on word using the meaning of *surprise* and of a shot *taken over* another. (*Translator's note*)

2. However, a recent theory on the universe may put doubts in our minds with regard to monism: Jean-Pierre Petit in a book entitled "*On a volé la moitié de l'univers*" defends the idea that matter is twin.

3. An "appareil" is a device, an instrument. Here it is short for "appareil photo" meaning a camera. (*Translator's note*)

4. Beyond the value as clue or proof we give to a photograph, this could justify protecting people whose image was made visible in a photograph.

References

Barthes, Roland (1980). *La chambre claire*. Paris: Ed. de l'Etoile, Gallimard, Le Seuil.
Bergson, Henri (1969). *La pensée et le mouvant*. Paris: Ed. Presses Universitaires de France.
de Mérédieux, Florence (1983). De l'obscénité photographique. *Traverse, 29*.
Quignard, Pascal (1994). *Le sexe et l'effroi*. Paris: Ed. Gallimard, col. Folio.
Merleau-Ponty, Maurice (1964). *Le visible et l'invisible*. Paris: Ed. Gallimard.
Rosset, Clément (1977). *Le réel, traité de l'idiotie*. Paris: Ed. de Minuit.
Rosset, Clément (1984). *Le reel et son double*. Paris: Ed. Gallimard, collection Folio/essais.
Sartre, Jean-Paul (1968). *La nausée*. Paris: Ed. Gallimard, col. Le Livre de Poche.
Sibony, Daniel (1991). *L'entre-deux, l'origine en partage*. Paris: Ed. du Seuil.
Virillo, Paul (1997). Symposium at Rencontres Internationales de la Photographie d'Arles. *Image et Politique*.

Chapter 2

The self-transcendence of consciousness towards its models

Consciousness as noumenal emergence: Some philosophical remarks on the Quantum Field Theory model of Giuseppe Vitiello

Fabrizio Desideri
Università di Firenze, Italy

We deal with the problem of self-transcendence of consciousness towards its epistemic models. First, we give a short sketch of the research on consciousness developed in our book, *Listening to Consciousness: A philosophical research* (Desideri 1998), where the problem of consciousness is investigated in the context of the so-called ordinary thought. The point of view assumed in this book is that of the problematic unity of consciousness despite its physical, biological and psychological peculiarities. Then, we try to put this perspective in relation with the outlines of Vitiello's Book, *My Double Unveiled* (Vitiello 2001) and in particular with his purpose to analyze the notion of consciousness from a physical point of view: the point of view of Quantum Field Theory.

The point of departure of our discussion is given by the current characterization of consciousness as an "emergent property". Among many other philosophers and scientists this thesis is sustained by John Searle. Searle speaks of consciousness as a biological property of brain. The properties of consciousness, in the first instance intentionality, emerge – for Searle – from the brain's neural activities in a way that the former are irreducible to the latter. Therefore, the perspective of consciousness is that of the 'first person'.

In such a way, "consciousness" can also be considered as a synonym for subjectivity. The emergence of consciousness from the brain's activities means also the emergence of subjectivity. The essential subjectivity of consciousness, however, doesn't make it inaccessible to the point of view of science (to the "epistemic objectivity"). Here, Searle criticizes Dennett's perspective, developed in his book *Consciousness explained*. For Searle, Dennett's mistake is to identify the epistemic objectivity of method with the ontological objectivity of "subject matter" and so to reduce the first-person character of consciousness to the third-person point of view. His critical aim is to refuse the possibility of a science of consciousness free from the biological systems in which consciousness is embodied as an intrinsic property. Nevertheless subjectivity, the subjective phenomena i.e. what happens in the consciousness of certain beings, is objectively a part of the world.

The important distinction defined by Searle is therefore not between the mental and the physical (mind and body), but between "those real features that exist independently of observers – features such as force, mass, and gravitational attraction – and those features that are dependent on observers – such as money, property, marriage and government" (Searle 1997: 211). Searle's conclusion is that "all observer-relative properties depend on consciousness for their existence." However, consciousness itself would not be observer-relative. It is a fact of the world: "a real and intrinsic feature of certain biological systems" (human beings). As a "fact of the world" consciousness is a part of it, that will be explained when "we solve the biological problem" (Searle 1997: 201) of its being causally determined by certain characteristics of our brain activity.

In such a way, Searle refuses every dualistic perspective and maintains a naturalistic point of view. With the concept of "emergent property", he defends the ontological continuity between the phenomenon of consciousness and the phenomenon of life in the world. Its ontology, we could affirm, consists in a naturalistic ontology.

The question, which arises from Searle's theory, is whether the definition of consciousness as the emergence of a phenomenon is adequate. So characterized, consciousness is assumed in the horizon of factuality. However, here we are dealing with a singular fact, which despite all other facts of the world (despite all other phenomena) is not an object of perception.

As the philosopher Colin McGinn maintains, "conscious subjects and their mental states are not perceptual objects" (McGinn 1995: 220). First, because they do not have localization: they are not "spatially individuated". The consequence, which McGinn derives from this simple observation, is that we "have no access" (McGinn 1995: 230) to the inner constitution of consciousness. While what happens in the brain is characterized from very small spatial processes, this is not the case of mental states (thoughts, feelings, purposes and so on). If it is true that "consciousness occurs in objective reality in a perfectly naturalistic way," it is also a fact that it is "one of the more knowledge-transcendent constituents of reality" (McGinn 1995: 230). For McGinn we cannot hope to cancel "a residual sense of unintelligibility" in the "nature" of consciousness. If Searle consigns the possibility to solve the mystery of consciousness to the future of neuroscience, for McGinn the passage from the physical (and in the last instance, perceptible) properties of brain to the properties of consciousness is not cognitively transparent. Therefore, despite every possible scientific research, an inference from the physical realm to the mental is not practicable; the link between brain and consciousness remains an enigma for the possibilities of our knowledge.

It may be useful here to remember the Kantian distinction between phenomenon and noumenon and the consequent distinction between knowing and thinking. According to Kant, we cannot objectively know all ideas (ideas such as freedom, totality and so on), but we can think them. A noumenon can be thought: it is an object of a *noesis* (an act of thinking). However, consciousness is not only an object of thought; it is the possibility itself of thinking something. This possibility depends on the fact that something exists in a noumenal way: it exists in the form of "une chose qui pense" (Descartes). Consciousness is also a "strange fact" similar to the "*factum* of reason" in the Kantian sense. The singularity of the phenomenon of consciousness is therefore its noumenal character. The strangeness of this fact consists in its primary property to be self-understanding.

We could also have written self-perceptive, but self-perceptiveness is not sufficient to characterize consciousness; it is rather a necessary condition of self-understanding: the so-called first level of consciousness. Consciousness, con(cum)-scientia implies a certain form of knowing, which is not limited to the form of knowing–that (to the form of objective knowledge). We can perceive and so know the effects of consciousness (like marriage, government und so on), while the access to its noumenality is intrinsic only to the activity of thought. The transcendence of consciousness – we might affirm – is the same of thought towards the physical processes from which it arises. Agreeing with Descartes, we maintain that all the conscious mental states (thoughts, feelings, will) belong to the field of thought: according to Descartes, these acts are *cogitationes*. In his *Confessions* (X, 11.18), Augustin explains very well the sense of *cogitare* as the act of gathering (collecting) in the psyche what is perceptively scattered.

It is now necessary to define better the features of this transcendence. What does transcendence mean here? *Transcendere*, as we know, means properly to go above or beyond the limits of something. The noumenal transcendence of consciousness consists then in going above or beyond the limits of the physical and biological process, from which its existence depends. Consequently, we can also uphold that consciousness is in a relation of transcendence with the neural activity of brain and, at the same time, of immanence.

It is helpful to recall here Vitiello's distinction between structure and function (Vitiello 2001: 11–14). Consciousness – Vitiello writes – is the function of brain. Without structure – it is obvious – there is no function. Which function does consciousness carry out in respect to its structure? In what sense is this function transcendent toward its structure? Vitiello says that the function of consciousness is to elaborate the information, the external inputs, which the brain-system receives. Very appropriately, Vitiello speaks in this framework of the brain as an "open system": "the 'openness' of the brain to the external world – he writes – means that it is permanently coupled to the environment" (Vitiello 2001: 104).

By means of Quantum Field Theory Vitiello explains in a suggestive way the dynamical character of this constitutive interaction. If we have rightly understood, the dissipative nature of the interchange between brain as a system and the environment, as the "time-reversed copy" of the system, ensures the functional stability of the brain in the form of a "coherent response" to the multiplicity of external stimuli. In the connection, in the living binding between the global coherence of this response and the behaviour of "apparently separated units and physiological structures of the brain" (Vitiello 2001: 114) emerges, therefore, consciousness as a unitary response. Following a statement of Paul Valéry in his *Cahiers*, we have spoken too in our book about consciousness as a primary response, where a self forms itself (Desideri 1998: 206–207). Just because consciousness is originally a response, it is the centre of responsibility.

From the point of view of the mental field of consciousness, the structural 'openness' of the brain can be translated then in a constitutive relationship of the subject of consciousness with the otherness.

Interpreting consciousness in a radical 'privatistic' and 'internalistic' way, many contemporary philosophers, considering the subjectivity of consciousness, disregard the primary character of this relation to otherness. Without this relationship, it is impossible to think the constitution of a subject as a self-constitution. In our book, we have tried to develop this aspect making a distinction between an external sense of the relation to otherness (which implies not only the environment, but also the other minds) and an internal sense (which implies the conscious subject as a living connection between the point of view of ego and that of self). With regard to these two levels of relation to otherness, consciousness can be thought as an active threshold between the external world and the internal field of its paradoxical noumenal nature. Consciousness appears also as a threshold between its biophysical origin and its noumenal contingence (its emerging as a transcendence). We could also define this threshold-character of consciousness as the threshold of otherness (Desideri 1998: 110–119).

This last concept concerns also the relation between the noumenal emergence of consciousness as a whole (a global field) and the global field of quantum brain dynamics. The self-reproducing *automaton*, of which Mari Jibu and Kunio Yasue speak in their very interesting book (Jibu & Yasue 1995), is not the same as consciousness. *Automaton* in ancient Greek means chance, a spontaneous contingency. Perhaps it is the same, but only on the physical-naturalistic side of the threshold. On this side, we can observe only a set of unconscious sub-processes. This set can also be considered as a whole (as something, that works as an orchestra) only from the point of view of the emergent consciousness, only in virtue of what Samuel Coleridge defines as the "power of initiative" of intellect (the engine itself of consciousness).

In investigating consciousness, however necessary is the circularity between understanding-processes top down and bottom up, we think that the point of departure is always top down. This possibility (as potentiality) is internal to essential actuality of consciousness itself. We could also define in terms of a supervenience the transcendence of consciousness as the global noumenal field, in which the self-reflexive feature of thought unfolds.

For at least twenty years, the concept of supervenience has been object of an intensive debate among philosophers of mind, who intend to avoid either a reductionist or a dualistic solution of the so-called mind-body problem. As Jaegwon Kim points out, just the concept of emergence combines the three components of supervenience. These are: (1) property covariation ("if two things are indiscernible in base properties, they must be indiscernible in supervenient properties"); (2) dependence (supervenient properties (mental acts) are dependent on, or determined by, their subvenient bases (neural processes in the brain); (3) non-reducibility (supervenient properties are not reducible to their base properties) (Kim 1994: 576).

Without going into a very sophisticated debate, we would like to observe that the notion of "supervenience" appears for the first time in a *Commentary on Aristotle's De Anima* by Themistius, a Greek Philosopher of IV century after Christ. Themistius speaks here of the relationship between *intellectus agens* and *intellectus in potentia* inferring, according to the first Latin translation of his Commentary by Wilhelm von Moerbeke, that the *intellectus agens,* the intellect in act, is *superveniens* on the *intellectus in potentia* (an expression, which can be translated with the brain dynamics) and becomes one thing with it. The actuality of intellect shows here itself as the logical presupposition of its biophysical condition (the *intellectus in potentia*). According to the actuality of this necessary presupposition, we could also reverse the relationship between nature and intellect. With Themistius'words in their Latin translation, we could then affirm:

> "et usque ad hunc natura cessavit tanquam nihil alterum habens honoratius, cui utique ipsum faceret subiectum. Nos igitur sumus intellectus activus" (the nature stops here, as if it did not have anything noblest to offer itself as a subject. We therefore are active intellect). (Themistius, In Libr. Arist., 185 1–10)

What can this phrase for our problem mean? The question is, if it is possible to reverse also the relationship between structure and function and then, if it is possible to consider brain as a function of consciousness. In this way, the dynamical interchange would be considered in a true circular way. In order to understand better this theoretical hypothesis, we must reconsider the function of consciousness with respect to its structure. Briefly, we could uphold that the function of consciousness is a strange function that unfolds itself bearing (producing) a self-theory.

We therefore think that to define consciousness as an intelligent system in sense of A.I. (Artificial Intelligence), as a system that has the function to elaborate information, is inadequate and neither the self-reference is sufficient to signify the features of the singular (strange) function of consciousness. It is necessary to specify too that consciousness, in its ability to refer the elaborated information to itself, works in a reflexive way and so it produces the autonomy of self. We could also affirm that consciousness implies always, in a tacit form too, self-thinking. This question, as we know, constitutes the centre of gravity in Kant's *Critic of Pure Reason*. It is just in self-thinking that consciousness, the point of view of first person, appears – in Vitiello's terms – in a nonlinear coupling or dialogue with the inseparable own self (Vitiello 2001: 141). It is in virtue of this difference between self and consciousness that the unity of consciousness is dynamically constituted by its intrinsic property to investigate oneself: to question oneself. In the light of this self-questioning property, consciousness displays its power to generate models of itself. This supposes a distance between consciousness and its models, but does not exonerate us from proving the goodness and the fitness of each model. A criterion to estimate the goodness and the fitness of a model of consciousness may be its capability to consider the self-implication of the model in the field of consciousness and therefore the fact that consciousness is self-transcendent toward its models too. Without going into details for lack of competence in physics, we think that Vitiello's model takes a meaningful step toward an understanding of consciousness from a scientific point of view, which avoids either reductionism or dualism.

As Stanley Klein remarks in a Review, which appeared in Psyche, of *Shadows of the Mind* by Roger Penrose, it is extremely meaningful from an epistemic point of view the fact that in quantum mechanics the split between the observer and the observed is moveable. Mentioning von Neumann, Wigner and Stapp, Klein points out that quantum theory, with its flexible placement, allows the neural correlates of awareness to be above the split (the neural correlates of awareness become the observer) while the unconscious neural activity remains below the split. The challenge is – as Klein writes – "to find a satisfactory way to associate the 'observer' of subjective awareness with the observer of quantum mechanics" (Klein 1995).[1]

Vitiello answers to this challenge searching for a solution of the problem in the scientific sphere of Quantum Field Theory. Following the theses sustained by Susan Greenfield, Vitiello analyses three essential properties of consciousness: (1) its non-locality (its being "spatially multiple" and "temporally unitary"; (2) its continuity (its growing as the brain develops); (3) its being "conscious of something, never of nothing and not of everything at once" (Vitiello 2001: 131–134). In relation to these three aspects or properties of consciousness, Vitiello explains with strict arguments how the dissipative model (the quantum model) can provide the dynamical ground of the emergence of a simultaneous coherent cooperativity among neurons. In relation to this subject, we have found especially interesting the way in which it is developed the argument of a recording process (the forming of memory) as necessary for the identity of self-consciousness. Opposite features, such as "non locality" from one side and space-temporal "localization" from the other side, continuously merge one into the other. From this dynamics – we could affirm – emerges the activity of consciousness as a categorization-activity that presupposes a permanent non-symmetric exchange or "trade" (commercium) between mind and world. In this categorization-process, at which origin there is the growing of a memory-ability, how can we think the existent idea of a self?

Our thesis is that the "self" is not a category among the others, but a supercategory. In the three features of consciousness recalled above we did not find that of the self-relation, which we must suppose as a implicit condition of being conscious of something. Here we can touch the problematic nucleus of Vitiello's book: that of the relationship between the dynamical identity of the self and the dynamical structure of the universe. If the possibility of a consciousness is given only by a self-relation, which arises as a response to the "informational inputs from the external world", can we also suppose a self-relational character in the constitutive processes of the universe? In other terms, the "permanent trade between subjective and objective" as a "breakdown of symmetry" is the last word for the relationship between self and universe (observer and observed)? How can we think further on the unity of this relationship? Only in terms of contingence, of emergent property, or otherwise? In front of this problematic or better *aporetic* nucleus, physics and metaphysics are indissolubly intertwined. The focus of their common researches is the possibility of symmetry beyond and in the inner of every breakdown and of every dissipative interchange. From a philosophical point of view the alternative is here that between Kant and Spinoza.

Note

1. We want to recall here that in Latin the term *consciens* means witness: the *consciens* is necessarily an observer.

References

Desideri, Fabrizio (1998). *L'ascolto della coscienza. Una ricerca filosofica*. Milano: Feltrinelli.

Jibu, Mari & Yasue, Kunio (1995). *Quantum Brain Dynamics and Consciousness*. Amsterdam & Philadelphia: John Benjamins.

Kim, Jaegwon (1994). Supervenience. In S. Guttenplan (Ed.), *A Companion to the Philosophy of Mind* (pp. 575–583). Oxford & Cambridge: Blackwell.

Klein, Stanley A. (1995). Is Quantum Mechanics Relevant To Understanding Consciousness? A Review of *Shadows of the Mind* by Roger Penrose. *Psyche, 2*(3).

McGinn, Colin (1995). Consciousness and space. *JCS, 2*(3), 220–230.

Searle, John R. (1997). *The Mystery of Consciousness*. New York: The New York Review of Book.

Vitiello, Giuseppe (2001). *My Double Unveiled*. Amsterdam & Philadelphia: John Benjamins.

Chapter 3

The unthinkable

Nonclassical theory, the unconscious mind and the quantum brain

Arkady Plotnitsky
Purdue University, USA

Extending the epistemology of Bohr's interpretation of quantum mechanics as complementarity, the article introduces the concept of nonclassical theory. A nonclassical theory places the ultimate objects it considers not only beyond its own reach or the reach of all knowledge, but also beyond any possible conception, thus making these objects literally un-thinkable. The article, then, considers the possibility of the nonclassical theorizing of the human mind, specifically the unconscious, on the one hand, and of the human brain, considered as a quantum system, on the other. The article also discusses the nature of consciousness and its relation to the unconscious.

1. An outline of nonclassical theory

This article offers two main arguments. The first concerns the human *mind*, considered primarily as a *mental entity* (i.e. apart from material systems, such as the brain, that may be responsible for its functioning), and the relationships among thinking, consciousness and the unconscious. The second concerns the human *brain*, considered as a *material entity*, and the physics of its functioning. This argument is conditional in one important respect. It proceeds under the (speculative) assumption, which governs several recent investigations, that the brain processes responsible for consciousness and thinking are fundamentally quantum rather than classical in nature.[1]

Both arguments concern the same type of theoretical thinking and the epistemology arising from it, which I shall call nonclassical. The present conception of nonclassical theory owes most to a particular interpretation of quantum mechanics, which follows the ideas of Werner Heisenberg and Niels Bohr, developed by Bohr into his interpretation of quantum mechanics as complementarity. I argue that understanding both the workings of the mind and the dynamics of the brain responsible for them (if this dynamics is quantum) may equally require nonclassical theory. For the reasons explained below, I shall not be concerned with relating these two arguments, in particular with whether the nonclassical character of the brain is responsible for the nonclassical character of the mind. The mind may need to be seen nonclassically even if the physics of the brain responsible for it is classical, and conversely, a physically nonclassical brain may be responsible for a classically describable mind.

My argument will be framed, historically and conceptually, by Kant's *proto*-nonclassical philosophy, at one end, and, at the other, by Sigmund Freud's, also *proto*-nonclassical, psychoanalytic theory and by *nonclassical* philosophy. The latter extends from both Kant's and Freud's work, along with that of (closer to Kant's) G. W. F. Hegel and (closer to Freud) that of Friedrich Nietzsche and Martin Heidegger, and is developed in, among others, the work of Bohr, now seen as a philosopher, Georges Bataille, Jacques Lacan, Jacques Derrida. I shall, however, only address Bohr's ideas here.

The nonclassical theoretical-epistemological *situation* proceeds from a given theory, such as quantum mechanics (the mathematical formalism cum the experimental data treated by this formalism). The *nonclassical* character of the theory (or, as in the case of quantum mechanics, the theory in a certain interpretation) is defined by the fact that this theory *rigorously*, by means of an argument given from within the theory, places certain objects it considers not only beyond the reach of the theory or knowledge but also beyond any possible conception. Such objects (I shall call them nonclassical in turn) are made unthinkable, in the literal sense of un-thinkable, as being beyond thinking, ultimately including as objects in any conceivable sense. It is essential that unthinkable entities are rigorously defined by means of (from within) this theory, rather than are merely postulated as such. The unthinkable is placed *inside* and is made a constitutive part of this theory, rather than positioned beyond the purview of or otherwise *outside* the theory. By the same token, the presence of unthinkable objects and the fact that they are unthinkable are essential to what the theory can do in terms of knowledge, explanation, prediction, and so forth. It also follows that, while always unthinkable, the field of the unthinkable is different each time, depending upon the theory in which it is established as unthinkable. Such a theory may further differentiate the unthinkable entities it considers, in the way quantum theory differentiates quantum objects into photons, electrons, quarks, and so forth, or various composite quantum systems. Quantum mechanics in a nonclassical interpretation does not say that one is not concerned with knowing or thinking about the nature of quantum objects, but that, thus interpreted, the theory, in principle, *precludes* the possibility of knowing, saying, or thinking about their nature. All it can conclude about quantum objects is that they exist. This "existence" itself, however, may not be conceived of in any specific form available to our thinking, beginning with those attributes of (wave or particle) motion that define classical physics, but ultimately extending to all conceivable attributes. Accordingly, the term "existence" or any other term referring to quantum objects ("quantum" and "object" included) is ultimately inapplicable.

Nonclassical epistemology moves us beyond the limits defined by Kant's conception of things in themselves. While unknowable, the latter are still thinkable, and, thus, are theorized as *classical* in the present view. In Kant's words, "even if we cannot *know* these same objects as things in themselves, we at least must be able to *think* [of] them as things in themselves" (Kant 1997: 115; emphasis added). One may, accordingly, define as *classical* an account that would, at least in principle, determine all of its objects (which I shall call classical) as knowable or, analogously to Kant's things in themselves, at least as thinkable. By contrast, the objects of nonclassical theories are *irreducibly* unthinkable, even as objects or things in themselves.

Accordingly, nonclassical theories suspend realism at the ultimate level they consider, if we understand by realism the possibility, at least in principle, of mapping or conceiving of the properties and behavior of a given entity or system. (Thus, one might try to conceive of quantum objects and behavior on the model of classical physics.) The suspension of causality, at the same level, is an automatic consequence: if one cannot conceive of nonclassical objects and their behavior at all, one cannot claim this behavior to be causal. Quantum theory and other nonclassical theories, however, suspend causality at other levels as well. Classical theories are, by definition, realist in the sense just explained, which allows for the possibility of knowledge or at least conception concerning all of their objects. Many are also causal.

Nonclassical theories, too, contain classical and even strictly knowable strata. Indeed they must do so, given that the existence of unthinkable objects are rigorously derived by a given nonclassical theory, as opposed to being merely postulated. For such a rigorous derivation is not possible otherwise than on the basis of something that could be and is known, even though it must also be seen as impacted by what is not and cannot be known or thought of. We *know* of the existence of nonclassical objects and *know* (rather than only think) them to be unthinkable through their effects upon the knowable, and only through these effects. Accordingly, nonclassical knowledge only concerns effects produced by the nonclassical objects in question in a given nonclassical theory upon other, knowable and hence classical, objects. At the level of such effects realism and classical thinking would apply. Causality appears difficult to claim even at the level of effects in both quantum theory and in the nonclassical view of the mind by virtue of nonclassical processes intervening between and affecting classical events in question.

Nonclassical knowledge and thinking are, however, hardly less rich or deep than those of classical theories, which are part of nonclassical theories in any event, and as such, or in their own right, could be as rich, deep, and important as nonclassical theories. It is not a matter of epistemological preference (or prejudice) but of theoretical necessity that may compel us to classical approaches in some cases and nonclassical in others. Nonclassical theories do expand our understanding of the nature of fundamental explanation in science, philosophy, and other fields. Indeed, when one says that nonclassical theories place their ultimate objects beyond any knowledge or even conception available to us, the terms knowledge and conception are used as classically conceived. One may, however, expand the conception of knowledge to include the nonclassical unknowable and to allow for knowledge to be conceived in terms of effects of this unknowable upon what is knowable. This conception itself is still classical as a *conception*, and there is, by definition, no other way for us to conceive of anything rather than classically. What is different is the *character* of knowledge and thinking.

It is primarily nonclassical theory that brings together the two main arguments of this article. These are, however, two separate arguments, offered here as independent of each other and operative in their own spheres, defined by the functioning of the mind and by the physical, material constitution and dynamics of the brain, respectively. From the disciplinary viewpoint, the first is the domain of philosophy and the second is that of the natural sciences, such as physiology, biology, chemistry, and physics. Establishing rigorous efficacious connections between the material dynamics of the brain and the workings of the mind is difficult, especially given the history of modern, post-Galilean physics, as an experimental-mathematical science of matter, on the one hand, and the history of philosophy, as a science of the mind, on the other. This difficulty is mitigated little by the fact (taken for granted here) that the materiality of the brain or the body is irreducible in philosophy and could, accordingly, be a subject of a proper disciplinary investigation there. Nor is the mind reducible in physics, at least in quantum physics, or elsewhere in science. Ultimately, the relationships between the brain and the mind may prove to be beyond our capacity of properly theorizing them, even though we may be able to develop nonclassical (or classical) theories for each separately. These relationships are bound to be nonclassical if the workings of the brain responsible for consciousness and thinking are fundamentally quantum in nature and if quantum theory is nonclassical. That we may eventually know. But, even if such proves to be the case, this is not the same as to have a rigorous theory of these *relationships* themselves. I make no claims concerning bringing philosophy and physics together, and am inclined to remain skeptical as to how far one can travel on this road. Using these disciplines to help each other or using findings from still other disciplines in either is a different matter. One always can and often must borrow specific concepts, tools of argumentation, and so forth, or, more generally, deploy differences and affinities between different disciplines, thus bringing together what seems heterogeneous and distant. This traffic has been and is likely continue to be instrumental to both philosophy and physics, or mathematics and science in general.

2. Kant, Freud, and the nonclassical epistemology of the mind

It would not be possible to trace here the history of nonclassical thinking, which reaches as far back as the pre-Socratics. I would argue that one encounters nonclassical theorizing in the present sense for the first time in Nietzsche's work in philosophy and in Bohr's work in science. Kant's work, however, appears unavoidable in considering the modern, post-Cartesian, part of this history. According to Kant:

> We have no concepts of the understanding and hence no elements for the cognition of things except insofar as an intuition can be given corresponding to these concepts, consequently ... we have cognition of no object as a *thing in itself*, but only insofar as it is an object of sensible intuition, i.e. as an appearance [phenomenon]; from which follows the limitation of all even possible speculative cognition of reason [*Vernunft*] to mere objects of *experience*. Yet the reservation must also be noted, that even if we cannot *cognize* [*kennen*] these same objects as things in themselves, we at least must be able to *think* [*denken*] [of] them as things in themselves. To *cognize* an object, it is required that I be able to prove its possibility (whether by the testimony of experience from its actuality or *a priori* through reason). But I can *think* whatever I like, as long as I do not contradict myself, i.e., as long as my concept is a possible thought, even if I cannot give any assurance whether or not there is a corresponding object somewhere within the sum total of all possibilities. But in order to ascribe objective validity to such a concept (real possibility, for the first sort of possibility was merely logical) something more is required. This

> "more," however, need not be sought in theoretical sources of cognition; it may also lie in practical ones. (Kant 1997: 115)

Kant proceeds next to an example of the freedom of the human soul. For my purposes, this example is most significant insofar as it refers to *mental*, rather than material, things in themselves. While we may think more readily of things in themselves as material *objects* (also in Kant's sense), for Kant the concept equally refers to mental objects and equally distinguishes them from appearances or phenomena, although in this case both the objects and the phenomena are mental. This view has significant implications for our understanding of the nature of thinking, specifically understanding, logical or other, and reason, also in Kant's sense of *Vernunft*. Kant, I would argue, ultimately assigns reason to the unconscious, even if, to put it in deconstructive terms, without quite saying so or against himself, and against the history of philosophy, which has nearly always associated reason, especially ethical reason, with consciousness and self-consciousness. Building on Kant's, perhaps unwilling, insight and on Freud's and Lacan's psychoanalytic theories, I see such activities as logical thinking, understanding, argumentation, reasoning and so forth, as fundamentally *unconscious* and thus unknowable and possibly (nonclassically) unthinkable, as opposed to those mental formations that are or could be made present to our *consciousness*. Such formations may also include certain *products* of the unconscious activities just mentioned and might require the unconscious to be processed. Among such products are those effects of the workings of the unconscious that compel us speak of it in terms of things in themselves or in terms of the nonclassical unthinkable, or to infer the unconscious, to begin with.[2]

Freud used Kant's argument for unknowable but thinkable *mental* objects, things in themselves pertaining strictly to the human mind, in his analysis and his very conception of the unconscious. As he writes in "The Unconscious" (1915):

> In psychoanalysis there is no choice for us but to declare mental processes to be in themselves unconscious, and to compare the perception of them by consciousness with the perception of the outside world through the sense-organs; we even hope to extract some fresh knowledge from the comparison. The psychoanalytic assumption of unconscious mental activity appears to us, on the one hand, a further development of that primitive animism which caused our own consciousness to be reflected in all around us, and, on the other hand, it seems to be an extension of the correction begun by Kant in regard to our views of external perception. Just as Kant warned us not to overlook the fact that our perception is subjectively conditioned and must not be regarded as identical with the phenomena perceived but never really discerned, so psycho-

> analysis bids us not to set conscious perception in the place of the unconscious mental process which is its object. The mental, like the physical, is not necessarily in reality just what it appears to us to be. It is, however, satisfactory to find that the correction of the inner perception does not present difficulties so great as that of outer perception – that the inner object is less hard to discern truly than is the outside world. (Freud 1963: 121)

It follows that all evidence concerning these unconscious dynamics is irreducibly indirect, which, as Freud (rightly) argued, does not prevent the possibility of rigorous and even scientific investigation. Indirect evidence is often used by science, for example, quantum physics, where indeed, in a nonclassical view, all evidence concerning quantum objects themselves is irreducibly indirect. Nonclassical epistemology may be described as the epistemology of irreducibly indirect evidence, extended to the point of the impossibility of knowing or even conceiving of the ultimate dynamics behind the evidence and yet deriving this inconceivability from this evidence. I take the nonclassical view concerning the *unconscious* character of thinking, at least certain aspects of thinking, which places their ultimate nature beyond knowledge and leaves it open, at most, to thinking about, rather than knowing, and *possibly* makes it unavailable even to thinking about it. I say "possibly" because a rigorous theory will not be offered here, thus making my argument for the nonclassical nature of thinking conjectural.

In this view, consciousness has primarily to do with the *presence* of phenomena, including of itself as a phenomenon (the phenomenon of self-consciousness), and far less to do with *thinking*, at least as logic, understanding, reason, and so forth. Far less, but not altogether nothing! This type of unconditional separation, without mutual interaction and inhibition, may not be possible, as Freud tells us, even if, again, against himself. Viewed nonclassically, the unconscious is not some exterior reservoir that is fully outside consciousness and that may or may not, in part or as a whole, become available to consciousness, although this type of traffic between both domains plays a role. Instead, the unconscious refers to the nonclassical dynamics that continuously involves the reciprocal and mutually inhibiting interactions with consciousness upon which it produces certain effects. These interactions are analogous to those between quantum objects and measuring instruments or the classical macro world (or what we perceive as such) in the nonclassical view of quantum physics. Some among these effects compel us to theorize such unconscious processes nonclassically by placing the ultimate character of these dynamics beyond all knowledge and thought, just as certain experimental data compels us to take the nonclassical view of quantum physics, where we place beyond knowledge and thought quantum objects and processes.

While, in the present view, ultimately conforming to the same type of mutually inhabiting dynamics, the interaction between mental and material unthinkable objects or even (thinkable) things in themselves in Kant's sense is a more complex matter. For, once the ultimate character of certain mental processes is that of things in themselves, something to be thought of but not to be known, we may also think of them as material (say, in terms of the brain rather than the mind), and they may in fact be material. But then, this materiality, or any ultimate (such as quantum) materiality, may be equally or even further removed from our knowledge and thought. In other words, we may be able to *think* the mental things in themselves, the mental unconscious, but not the material one – physiological, biological, chemical, or physical. Indeed Freud believed the material unconscious to be further away from our knowledge, if not thinking, than the mental one, and, on these grounds, suspended the material dynamics responsible for mental processes from the field of psychoanalysis, perhaps wisely, at least at the time (Freud 1963: 118). Here, I take a more symmetrical view, in part by virtue of taking a more nonclassical view of both the material and mental unconscious. In this view, neither one would be any more (or less) known or knowable or thinkable than the other.

On the other hand, the actual material dynamics responsible for mental processes (conscious or unconscious) remains a formidable problem, in spite of major advances of the last fifty years in several fields. Accordingly, it may be prudent to exercise maximal caution in trying to bring them together. To argue for a particular form of brain dynamics, say, classical vs. quantum, as responsible for consciousness (or the unconscious), is a difficult task already. A demonstration that this dynamics is quantum would be an extraordinary achievement. Still, this is not the same as to link the actual logical, epistemological, or other architecture and dynamics of mental processes and those of the physical processes in the brain, responsible for the mental ones. A very limited set of links may be sufficient to demonstrate that our mental life is the product of a particular, say, quantum, brain dynamics, since to do so one might only need to establish a limited set of effects relating both without linking their architecture. In other words, the ensuing physics may conform to nonclassical epistemology, but might not be able to account rigorously (classically or nonclassically) for the connections between the architectures of both domains, beyond certain minimal links or correlations. This is why I prefer to make *strong* claims only in separate domains of the brain and of the mind, and rigorously respect the disciplinary boundaries involved. Both domains are decoupled analytically and disciplinarily, even though they may be ultimately connected materially.

By contrast, nonclassical epistemology can be argued to be consistent with other disciplinary requirements of these fields. Indeed, in quantum theory, in contrast to philosophy or other nonscientific disciplines, the deployment of this epistemology is even facilitated by these disciplinary requirements, specifically by the mathematical-experimental character of the theory and of modern physics in general. We may use the mathematics of quantum theory to make excellent *predictions* of the outcomes of the experiment, without making any claims concerning the *description* of the quantum physical processes involved. We may even rigorously argue that such a description or, again, even a conception of such processes is ultimately impossible, which is what nonclassical interpretations of quantum theory do. Once we move to philosophical or psychological theories of human nature, this type of approach and attitude, which is not always easily and sometimes not at all accepted even in science, encounters a much greater resistance and may indeed be less effective. The disciplinarity of philosophy tends to demand an epistemologically classical explanation, at least by way of thinkable, even if not ultimately knowable, things, in other words, Kantian things in themselves.

As the passage cited above suggests, Freud appears to be more optimistic than Kant (let alone than a nonclassical theorist of the unconscious) would be as concerns a possible access to unconscious mental processes, even in terms of knowledge rather than only things in themselves, for example, in terms of their Oedipal dynamics. Freud even seems to argue primarily that what we actually think (in our unconscious) is not what we (consciously) think we think, rather than, as Kant does, for the ultimate unknowability but possible thinkability of the unconscious, or, as I argue, for its ultimate unthinkability. Freud *thought* about the possibly unknowable dynamics of the unknowable and gave it specificity. He then argued that this dynamics could be ultimately made nearly empirically knowable on the basis of the irreducibly indirect evidence, that is, evidence manifest only in certain (more) ascertainable effects. Nonclassically, while such effects are equally indirect, some of them also compel us to infer the workings of not only the irreducibly unknowable but of the irreducibly unthinkable behind these effects.

But then, one can never be sufficiently cautious in making claims concerning Freud's thinking. Freud never stopped stratifying his pictures or his un-pictures, his visual and unvisualizable models, of consciousness and the unconscious, of the knowable and the unknowable, of the thinkable and the unthinkable, the material and the mental, and of the interactions between and among them. Freud, even if against himself, may have been closer to nonclassical theory than it might appear.

He certainly said on several occasions that consciousness may well be the ultimate enigma of human nature, which would make the program announced in the passages cited here more difficult to fulfill, as ultimately proved to be the case. Lacan, who entertained a nonclassical view of the unconscious and thinking, said: "Freud has told us often enough that he would have to go back to the function of consciousness, but he never did" (Lacan 1981: 57).

As I said, I primarily associate consciousness with phenomenological presence and hence with what is knowable in the sense of that which can be made present to consciousness, appearing, as it were, on the mental terminal screen of consciousness, rather than with the processes related to logic, understanding, reason, and so forth. I see the actual dynamics, mental or physical, governing these processes as fundamentally unconscious, hidden in the black box of our mental software and hardware, and thus also ultimately beyond our knowledge and possibly conception, and, if the latter is the case, as subject to nonclassical treatment. Any articulation resulting from our (unconscious) thinking concerning such processes, including in terms of logic, understanding, reason, or whatever, can be made available to knowledge, in the same way one can learn and know quantum theory or philosophy. In other words, our ability to think of these things, including as concerns the unknowable or unthinkable, may result in learning or knowledge. But that does not mean that we can *know* or even *think* how matter or mind actually works, including, in the case of mind, how it ultimately enables us to know these or other things, in other words, how it is that we can think about them. Rather than phenomena, which we can know, these may be Kantian things in themselves of which we can only think or, at the nonclassical limit, ultimately something of which we cannot even think, as Gödel's theorems perhaps tell us in the case of mathematical thinking. In short, we might not be able to think how we think or how it is that we are capable of thinking.

This view may be seen as extending, if, again, nonclassically radicalizing Freud, via Lacan. As Lacan says, crediting Freud with "truly unprecedented boldness":

> When Freud realized that it was in the field of the dream that he had to find confirmation of what he had learned from his experience of the hysteric, he began to move forward with truly unprecedented boldness. What does he tell us now about the unconscious? He declares that it is constituted essentially, not by what consciousness may evoke, extend, locate, bring out of the subliminal, but by that which is, essentially, refused. And how does Freud call this?

> He calls it by the same term by which Descartes designates what I just called his point of application – *Gedanken*, thought.
>
> There are thoughts in this field of the beyond of consciousness, and it is impossible to represent these thoughts other than in the same homology of determination in which the subject of the *I think* finds himself in relation to the articulation of the *I doubt*. (Lacan 1981:43–44)

Thus, the *psychoanalytic* unconscious, the unconscious as *theoretically* defined in the field of psychoanalysis, is primarily thinking – *Gedanken* – and, according to the present view, thinking is primarily unconscious. This unconscious, moreover, may need to be theorized nonclassically, as something that is ultimately beyond our ability to think about it, except for its actually or potentially manifest effects, which also make us infer this unthinkability.

This view does not imply any lesser significance of consciousness or self-consciousness than that assigned to them by classical theories. Quite the contrary, the role of consciousness is decisive in the human mind and indeed in human (and perhaps animal) life, from perception to theoretical, including scientific, knowledge. Consciousness and conscious knowledge are the necessary, inevitable starting point of any investigation of any perception and thinking (conscious or unconscious), as both Kant and Hegel, or most major figures before and after them, from Plato and Aristotle to Nietzsche and Freud, to Lacan and Derrida, indeed *knew*. For, again, how else could we think and theorize anything, classically or, and especially, nonclassically, except by starting with and investigating one or another type of manifest conscious effects? One might, accordingly, appreciate Giuseppe Vitiello's invocation of "*Now* you know it! …" as a necessary aspect of quantum (specifically quantum-field-theoretical) approach to consciousness and thinking in *My Double Unveiled* (Vitiello 2001:106), as well as his ellipsis, extending what you know beyond knowledge, into the unconscious. We may neither appreciate nor approach the work of thinking, from dreams to quantum theory (and the dreams of quantum theory), without this "now you know it! …", even though and because most of this work is beyond the domain of consciousness and is possibly unthinkable. It takes place in the region designated by Vitiello's ellipsis "… ".[3]

3. Quantum theory as nonclassical theory

I shall now outline the key features of quantum theory from the nonclassical perspective, specifically insofar as these features make nonclassical interpretations of quantum theory possible and possibly necessary. While I shall focus primarily on quantum mechanics, I shall also address the epistemology of quantum field theory.[4]

It is worthwhile to begin by considering the double-slit experiment, a kind of archetypal quantum-mechanical experiment. The well-known arrangement consists of a source; a diaphragm with a slit (A); at a sufficient distance from it a second diaphragm with two slits (B and C), widely separated; and finally, at a sufficient distance from the second diaphragm a screen, say, a silver bromide photographic plate. A sufficient number (say, a million) of quantum objects, such as electrons or photons, emitted from a source, are allowed to pass through both diaphragms and leave their traces on the screen. Two set-ups are considered. In the first, with both slits open, we cannot, even in principle, know through which slit each quantum object passes. In the second we can, either in practice or, importantly, in principle.

In the case of the first set-up, a "wave-like" interference pattern will emerge on the screen, in principle regardless of the distance between slits or the time interval between the emissions of the particles. The traces, once a sufficiently large number of them are accumulated, will "arrange" themselves in a pattern, even when the next emission occurs after the preceding particle is destroyed after colliding with the screen. This pattern is the actual manifestation and, according to, at least, nonclassical interpretations, the only possible physical manifestation of quantum-mechanical "waves."

If, however, in the second set-up, we install counters or other devices that would allow us to check through which slit particles pass, the interference pattern inevitably disappears. Merely setting up the apparatus in a way that such knowledge would in principle be possible would suffice. The fact that even the possibility in principle of knowing through which slit the particles pass would inevitably destroy the interference pattern may be shown to be equivalent to uncertainty relations.

These facts are extraordinary and difficult to confront, even though or because quantum mechanics rigorously predicts them, including the particular distribution of traces on the screen defined by the distances between different parts of the arrangement, and by the specific placement of the slits.[5] Accordingly, such locutions as strange, mysterious, incomprehensible, or paradoxical are not surprising. Attempts to conceive of the situation in terms of physical attributes of quantum objects themselves appears to lead to unacceptable or at least highly undesirable consequences. Among such consequences are logical contradictions; difficult assumptions, such as attributing volition or personification to nature in allowing particles individual or collective "choices" (e.g. quantum objects appear to "know" whether both slits are open, or whether counting devices are installed); or, as Einstein was first to note, nonlocality of the situation, making it incompatible with relativity.

Bohr, by contrast, saw the situation, in nonclassical terms, as indicating the "essential ambiguity" and ultimately impossibility of ascribing any physical attributes to quantum objects themselves or to their behavior. In this view, in considering individual marks on the screen we may rigorously speak of them only as *particle-like effects* or, in certain circumstances, as *wave-like effects*, and *not* as traces left by collisions with classical-like *particle or wave objects*. He writes: "To my mind, there is no other alternative than to admit that, in this field of experience, we are [rather than with properties of quantum objects] dealing with individual phenomena [in Bohr's sense] and that our possibilities of handling the measuring instruments allow us only to make a choice between the different complementary phenomena we want to study" (Bohr 1987: v. 2, 51). In other words, we are dealing with two different and mutually exclusive types of *effects* of the interaction between quantum objects and measuring instruments upon those instruments under specific physical conditions.[6] This, apparently irreducible, mutual exclusivity of certain types of arrangements (they can never be used at the same time) and yet equally the necessity of using them all for a comprehensive (consistent and complete) description is the proper physical meaning of what Bohr calls complementary physical descriptions. It is also Bohr's interpretation of Heisenberg's uncertainty relations. Measuring and even defining the variables involved in them, such as position and momentum, or time and energy, require mutually exclusive arrangements as well. These variables can now only apply to the classical physics of measuring instruments, and not to quantum objects themselves, for which no such variables can ever be ascribed, even each by itself, let alone jointly. By contrast, we can ascribe either one such variable or another, but never both together, to the measuring instruments involved under the impact of their interaction with quantum objects. This situation is the origin of Bohr's "complementarity" terminology. Ultimately the term came to designate his overall interpretation of quantum mechanics, the nonclassical nature of which is indeed correlative to the unavoidability of such complementary situations of measurement.

What we see on the screen is now assumed to be the manifest effects of the interactions between quantum objects and measuring instruments upon the latter. Such effects are classical physically, insofar as they (but not their emergence) are described in terms of classical physics, and classical epistemologically insofar as they (but, again, not their emergence) could be manifest, present to our consciousness as phenomena, including in Kant's sense (as different from things in themselves). Each of these effects or marks is a discrete, particle-effect, entity and an individual phenomenon in Bohr's sense. Bohr defines *quantum-mechanical* phenomena in terms of individual effects of this type, which, accordingly, also makes the corresponding material entities (i.e. the effects in question) available to our consciousness in terms of phenomena in Kant's sense, while anything at the quantum level is not available to such a representation.

It is crucial that in this interpretation an unambiguous reference to quantum objects and processes would remain impossible even when one speaks of single such attributes, rather than in the case of a simultaneous attribution of joint properties involved in uncertainty relations, and even at the time when the measurement takes place. In other words, *neither one nor the other* complementary variable could be assigned or even defined for quantum objects themselves, rather than only one or the other, say, a position or momentum. Accordingly, Bohr argues that "in quantum mechanics [at least in this interpretation] we are not dealing with an arbitrary renunciation of a more detailed analysis of atomic phenomena, but with a recognition that such an analysis is, *in principle*, excluded" (Bohr 1987: v. 2, p. 62). As a result, quantum objects are placed in the position of nonclassically unthinkable entities.

It is sometimes argued that higher-level quantum theories, such as quantum field theory, are likely to remove the nonclassical features of quantum mechanics, which are unacceptable to some, beginning with Einstein. Such higher-level theories are also seen as more suitable for quantum approaches to the brain. One might more readily agree with this latter argument.[7] By contrast, the contention concerning the possibly classical-like character of such theories is less compelling. Given the complexities of quantum field theory, one might plausibly anticipate that the movement from quantum mechanics to it may proceed not toward epistemological classical-like theories but toward as yet more complex, at least mathematically and experimentally, if not epistemologically, forms of theoretical nonclassicality. Let me briefly indicate why such may be the case.

Suppose (speaking provisionally in terms of quantum objects themselves) that one arranges for an emission of an electron from a source and then performs a measurement at a certain distance from that source. Merely placing a photographic plate would do, and the corresponding traces could be properly treated nonclassically. According to classical physics, one would encounter at this point the same electron, and its position could be predicted exactly (within the capacities of the measuring instruments used) by classical mechanics. In quantum mechanics, by contrast, one would encounter either an electron or nothing, and the alternative probabilities are properly predicted by quantum mechanics. Once the situation involves higher energies, however, and is governed by quantum electrodynamics (a form of quantum field theory), one may find an electron, nothing, a positron, a photon, an electron-positron pair, or still something else. The probabilities for the alternatives are, again, properly predicted by quantum electrodynamics. One can also formulate the situation in terms of the ground state difference between quantum mechanics, in which one can assign a single ground state, and quantum field theory, in which one cannot do so. The upshot is that in quantum field theory, an investigation of a particular type of quantum objects (say, electrons) not only irreducibly involves other particles of the same type but also other types of particles, conceivably all existing types of particles or their combinations.[8] It is crucial that the situation involves *different types* of particles, since one cannot distinguish different particles of the same type, such as electrons, and, accordingly, one can never be certain that one encounters the same electron in the experiment just described even in the quantum-mechanical situation.

The situation appears difficult to negotiate classically. The epistemology of quantum field theory is, however, a complex and little developed subject, which cannot be addressed within the scope of this essay. Accordingly, I shall not make a strong claim either way. This is, moreover, not essential for my argument, which concerns only nonclassical theories of the brain and the mind. This argument, however, allows me to make the following claim, which is rigorous. If the dynamics of the brain responsible for consciousness and thinking is quantum and if quantum theory, including quantum field theory, is epistemologically nonclassical, we may, beyond the fact that this physics is quantum, never know or even be able to think (in the classical sense of both terms) how thinking is ultimately possible physically.

4. Conclusion

From the quantum perspective, the brain or the body may be seen as a kind of measuring machinery, a conglomerate of measuring instruments, suited and developed for both classical and quantum measurement. This machinery enables both our unconscious thinking and our consciousness. It is through consciousness and only through consciousness, and the classical domains it gives us, that we infer the workings of the unconscious, possibly itself unthinkable even if considered in terms of its mental dynamics. This is why we must begin with consciousness. There is no other place to begin, including in the case of quantum physics, whose data, too, is ultimately given, as data, only to our consciousness. For one reason or another, our bodies appear to enable us to "see," to consciously experience, only a classical world. Our minds, however, through mostly unconscious thinking, can, by using some of the data of consciousness, conceive and even rigorously, including, in the case of quantum theory, mathematically, derive that which defies knowledge and even classical thinking. To think, let alone to know how our brains or our bodies enable our minds to do so is a formidable task indeed, the hard problem of the unconscious. This problem may have no solution short of a complete unification of many disciplines, which does not appear likely. All the more credit then is due even to the smallest steps forward. Greater things are not impossible either, within and between different disciplines. Nonclassical thinking, thinking with the unthinkable, may help our thought here, even though and because it will also reshape how we define and approach our theoretical problems.

Notes

1. I refer primarily to the work of Hiroomi Umezawa and his followers, especially that of Giuseppe Vitiello (Vitiello 2001), based on quantum field theory. Neither Bohmian approaches nor those of Roger Penrose and his followers will be considered here.

2. These workings are not restricted to those of understanding or reason. The present argument does not claim to encompass all aspects of our mental constitution, such as (beyond thinking) feelings, emotions, and so forth, all subjects of complex discussions and debates in several fields.

3. The question of the relationships between consciousness and temporality is crucially implicated here, but it cannot be addressed within my scope.

4. I can only offer an outline here and refer to Plotnitsky (2002) and Plotnitsky (2003) for further details.

5. The situation can also be given a statistical interpretation indicating that the statistical nature of quantum mechanics may be and, in the present view, is irreducible.

6. The overall situation also involves the so-called quantum-mechanical entanglement between the quantum object considered and certain parts of the overall measuring arrangement. This circumstance is pertinent to my argument, but the subject cannot be addressed here.

7. See Vitiello's discussion (Vitiello 2001), based on such features as an open (dissipative) character of the brain as a physical system, the inter-connective, network-like architecture of this system, and, implicitly, the quantum-field-theoretical rather than quantum-mechanical nature of the quantum entanglements involved.

8. For important features of this situation in terms of quantum-field-theoretical (vs. quantum-mechanical) entanglement and, mathematically, the so-called Hopf's algebras, see

(A. Iorio, G. Lambiase, & G. Vitiello 2002), which offers a profound but inevitably technical analysis.

References

Bohr, Niels (1987). *The philosophical writings of Niels Bohr*, 3 vols. Woodbridge, CT: Ox Bow.

Freud, Sigmund (1963). *General psychological theory: Papers on metapsychology*. New York: Collier.

Iorio, A., G. Lambiase, & G. Vitiello (2002). Hopf's Algebras, Thermodynamics and Entanglement in Quantum Field Theory. quant-ph/020740, 8 July.

Kant, Immanuel (1997). *Critique of pure reason*. Paul Guyer & Allen W. Wood (Trans., 1997). Cambridge: Cambridge University Press.

Lacan, Jacques (1981). *The four fundamental concepts of psychoanalysis*. Alan Sheridan (Trans.). New York: W. W. Norton.

Plotnitsky, Arkady (2002). *The knowable and the unknowable: Modern science, nonclassical thought and the "two cultures"*. Ann Arbor, MI: University of Michigan Press.

Plotnitsky, Arkady (2003). Mysteries without Mysticism and Correlations without Correlata: On Quantum Knowledge and Knowledge in General. *Foundations of Physics*, *33*(11).

Vitiello, Giuseppe (2001). *My double unveiled: The dissipative quantum model of the brain*. Amsterdam/Philadelphia: John Benjamin.

Chapter 4

Mental presence and the temporal present

Georg Franck
Technische Universität Wien, Austria

This contribution ventures a look at quantum brain dynamics (QBD) through the glasses of phenomenology. In this view, QBD is about perception and recollection. Perception implies mental presence. Recollection makes sense only in a context in which present and past denote distinguished modes of existing. In physical theory, both mental presence and the temporal present are supposed to be conscious phenomena. QBD thus is confronted with the question of how the physical and the phenomenal are interrelated. So far, the difference between the physical and the phenomenal aspect of the brain has been predominantly discussed in terms of the third-person and first-person perspective. In the following, an alternative approach is put forward. The perspective of the first person and the perspective of the third person share a common viewpoint: the temporal present. In the perspective of the first person, the temporal present is indistinguishable from mental presence. In the perspective of the third person, the present is the viewpoint in time shared by all persons. The paper asks how this communality can be made productive for mediating the ontological difference between phenomenal consciousness and the reality described by physics.

How does it feel to be a brain?

Conscious brains are the strangest of objects. They present themselves in completely different ways depending on whether they are investigated from outside or sensed from within. In the perspective of the outside observer, the brain is an anatomical structure with physiological functions, a conglomerate of chemical and physical processes, whose prime ability is the processing of information. In the perspective of the person who *is* the brain, a world endowed with sense qualities, value and meaning appears. From the third-person point of view, the brain is a net of neurons and glia cells swimming in a water bath. From the first-person point of view, our brain is what provides us with the phenomenal world: the world that is present when we are in the state of mental presence, and absent when we are unconscious.

Mental presence is the feeling of being a conscious mind. Is it the feeling also of being a conscious brain? The question may seem nonsensical since the brain has no organ for sensing itself. On the other hand, does the very fact of one's feeling not imply feeling to be a brain? The difficulty of taking one's mental presence as the way it feels to be a brain is this: Even though our being in the state of mental presence is something we cannot help to be acquainted with most intimately, it is, at the same time, something completely alien to us. The state of mental presence is what we know best of all because it is what every act of experiencing is in. Yet, it is completely alien to us because we cannot grasp it in its own reality. Mental presence is a byword for concreteness. Still, it is not a thing we can experience with our senses. Nor is it accessible by abstract thought. It ceases to be what we are trying to grasp as soon as turned into an object of thought. Mental presence eludes our grasp since it is neither a thing nor an idea. The only way it can be experienced lies in its taking cognizance of itself. In order to reach this state of *self-awareness* we must rupture, however, the everyday intimacy with our being aware *of* something. Self-awareness means being aware without being aware *of* something, an attitude for which there are no words.

An account of what it feels to be a brain that deals with this difficulty is Martin Heidegger's ontology of *Dasein*. "Da sein" literally translates as "to be there". When related to the mode we exist in as conscious beings, "to be there" assumes a double meaning. "Da sein" can mean to exist as a living organism, and it can mean to be present in the sense of mental presence. It is this double meaning of "being there" that is characteristic of the existence of a conscious brain. The organism is an entity. Mental presence is not an entity, but a mode of existing. Dasein covers this intrinsic difference. According to Heidegger, Dasein is the entity (*Seiendes*) that is aware of itself and cares about its Being (*Sein*). Being, thus understood, means presence. Being, as distinct from entities, means presence, as distinct from the things and events presencing. The conscious brain is that one distinguished entity that itself performs the ontological differentiation between Being and entities.

From the dualism of views to the difference in ontology

Presence, in the sense of Being, is not restricted to the perspective of the first person. Presence, when tied to the perspective of the third person, is called the temporal present. In the perspective of the first person, mental presence and the temporal present are one. In the perspective of the third person, the temporal present is as objective, however, as is the fact that the reality given to perception consists of things and events that are separated from their temporal environment. It is the *cut of nowness* that singles out the collection of "res" that the concept of reality classically refers to. Part of this collection is the brain as appearing in the perspective of the third person. Accordingly, the ontological differentiation between Being and entities is neutral with regard to the dualism of the first-person and third-person perspective.

The temporal present, though implied in the classical concept of reality, does not come forth in post-classical physics. The present explicitly is expelled from spacetime, it has no place in the universe described by the wavefunction. In relativity as well as in quantum theory, there is no distinguished section of the universe that is raised to presence while the presentification of the rest is suppressed. From a physical point of view it is not the dualism of views, thus, that is relevant for demarcating the physical from the phenomenal. Both the perspective of the first person and the perspective of the third person are bound to the now. They both presuppose the present. When it comes to the question how the physical and the phenomenal are interrelated, Heidegger's account of *Dasein* is more to the point than the dualism of personal perspectives.

On the other hand, Heidegger's philosophy epitomizes a way of thinking that scientific thought progressively has distanced itself from. Scientifically, entities are all there is. For Heidegger, presence is what deep thinking is all about. Scientifically, even the temporal present is denied objective existence. For Heidegger, modern science is the upshot of *Seinsvergessenheit*: of forgetfulness of Being. In fact, ignoring presence means to ignore the very existence of consciousness. For us, as conscious beings, everything that is supposed to exist does so by making appearance in conscious awareness. The only way we have access to the reality deemed to exist independently of being experienced is conscious thought.

Accounting for the existence of conscious brains thus means to think together totalities that seem to be most intimately related and, at the same time, to be separated by an abyss. It is as if we had to do with different universes, a subjective one that is present and an objective one that is real. It is as if these universes had to come into contact before entities can manifest. Hence, is there any account of the functioning of the brain that accounts for this dualism of universes?

Remarkably, there might be one. Even more remarkably, the account coming into question is a physical one. In Hiroomi Umezawa's quantum theory of thermodynamics, a dualism of universes emerges from the formalism. In thermofield dynamics "every dynamical degree of freedom is doubled; to any operator A is associated its tilde conjugate A~" (Umezawa 1993: 144). It is an open question what this doubling means beyond the job it does in the formalism. However, thermofield dynamics has developed into dissipative quantum brain dynamics (Vitiello 1995, 2001). Quantum brain dynamics (QBD) is an approach to the brain's capability to memorize (Jibu & Yasue 1995). Memory is that one capability of the brain that has most immediately to do with the differences engendered by presence. Memory lies at the base of time perception and of the brain's way of dealing with temporal change. Temporal change means that world states having been future become present, only to vanish into the past. Without memory, there is no past and no future. Memory, however, has so far escaped explanation by non-quantum approaches to the functioning of the brain. In the brain, memory storage is not localised (see Pribram 1991). QBD is that one approach that accounts for both the enormous capacity and the non-local way of storage. Dissipative QBD, in addition, accounts for the self-organisation of the brain as a system that maintains itself through energy exchange with its environment (Vitiello 2001).

The brain, as a dissipative system, is an open system. Its environment, in terms of thermodynamics, is a heat bath that is an open system as well. In order to say something specific of the brain embedded in its environment, the system brain plus environment has to be closed somehow. Due to the holistic character of quantum theory, this closure is a delicate operation. The device accomplishing this closure is the thermofield dynamical doubling of the system. By this doubling the system is closed and kept open as well. Due to the holistic character of quantum theory, the doubling of every dynamical degree of freedom amounts to a doubling of "worlds". The description of the brain as a dissipative system is thus accomplished by relating two universes, the (nontilde) universe A and its (tilde) double ~A. Umezawa (1993: 34) speaks of "the presence of other universes which are totally dissociated from our world, though they share the vacuum with our world."

The vacuum state is the state of minimum energy. It is the key to the explanation of the enormous capacity and the nonlocal way of memory storage. We shall come back to it in a moment. Since we are looking for an account of brain functioning that accounts for the production of mental presence, Umezawa's talking of the "presence of other universes" looks promising. This promise, however, relies on a misunderstanding. The descriptions of both the nontilde and the tilde universe are wavefunctions. The states that wavefunctions give expression to are possible states. Possible states differ from actual states in that they lack presence. Wavefunctions can be interpreted as relating to presence in a way only that Umezawa certainly had not in mind (see below). Most probably, Umezawa's talking of presence is just a way of dealing with the hard to express existence of wavefunctions in ordinary language.

It was first Gordon Globus who explicitly addressed the question of how the universes addressed by Umezawa manifest. Since manifestation means presentification, he asks how the dualism of un-present universes might relate to the ontological difference between Being and entities. The wavefunctions of the nontilde and tilde universe have the same form, but differ in the sign of the imaginary unit number **i**. Tilde is the complex conjugate of nontilde. Observables, i.e. descriptions of entities that can be made to manifest, result from a conjugate match of nontilde and tilde. When being made to manifest, the observable turns into an actual happening. An actual happening is what happens when entities present themselves in the present. Since it is fundamentally unclear what the temporal present might be in the absence of mental presence, we should be entitled to say that entities surface in the present when entering some Dasein – or, rather, when getting involved into the happening of Dasein. Globus now embarks on thinking together the conjugate match of the universes and the happening of Dasein. He illustrates his approach by referring to Descartes' dualism of *res cogitans* and *res extensa*. "Descartes' dualism of incompatible yet interacting substances is succeeded by a thermofield dualism in which an interaction takes place in the vacuum states upheld by the living brain. This interaction is ... a lighting process in which *res extensa* is disclosed in virtue of a ~conjugate match. In the case of match, the physics equations show real numbers, which are associated with observables" (Globus 2003: 81). *Dasein* is what results from the conjugate match between nonpresent tilde situatedness and nontilde potentiality.

Quantum memory

Physics, to repeat, does not thematize presence. Hence, the conjugate match is definitely not a self-contained account of Dasein. Dasein means to live mentally present in the present. The only way of relating the ontological difference to physical theories of the brain lies in interpreting these theories in the light of the accounts we have of our existence as conscious minds. The account Globus makes use of is the most fundamental notion of *Dasein:* of Heideggerian *Existenz.* Globus deems the notion of phenomenal consciousness to be inappropriate since too closely related to that of classical reality. This is a point with which I disagree. I think phenomenal consciousness to be indispensable for accounting for the brain's way of dealing with presence. Presence, for the conscious brain, is not an all-or-nothing mode of existing. In experience, presence varies in two different, if not independent, respects. Presence can vary in the sense of the varying degree of concreteness to which things are present; and it can vary in the sense of the varying degrees to which we are mentally present. The varying degrees of concretness are epitomized by the soft transition from future to present and then past (as described by Husserl in terms of protention and retention). The varying degrees of mental presence are epitomized by the spectrum of arousal that ranges from highest alertness or even shock to the verge of sleep and further from dreaming sleep to dreamless sleep down to the definite loss of consciousness.

Compared to the very existence as a conscious mind, these varying degrees of presence are subtleties. These subtleties, however, are of prime interest when understanding QBD as a theory of the brain's capability to memorize. Memory lies at the base of time perception and of the differentiation between perceptive and reflective consciousness. Time perception is the awareness of the involuntary change of the degree to which things and events are present. The differentiation of perceptive and reflective consciousness relies on the capability of the brain to uncouple the presence controlled by temporal change from the presence controllable by the brain itself. When interpreting dissipative QBD[1] in the light of these capabilities, the following features of the memory system appear in the foreground. (1) Everything lived through consciously is automatically printed to memory, (2) total memory is constantly held accessible and (3) past experience, though having lost its original presence, can be made to reappear in mental presence. Feature (1) relies an enormous capacity of information storage, feature (2) relies on the conservation of printed memory and its protection against overwriting, (3) means that the experience passed is reproduced by recombining the information stored with mental presence.

Ad (1) According to QBD, memory is printed to vacuum, i.e. minimum energy, states of quantum fields that extend over macroscopic distances in the brain. The main burden of information storage lies on fields that are generated dynamically by the exchange of quanta that correlate the rotational and vibrational dynamics of water molecules in a coherent manner. The submicroscopic constituents of the water electric dipole field are vibrating in phase to the effect, that the field behaves as if it were one molecule that assumes macroscopic size. The quanta exchanged by way of these long-range correlations are massless, which means that they do not add to the energy of the system. What the number of dipole wave quanta condensed in the system does change is the phase that the system is in. Quantum field theory allows the electric dipole field of water to assume an infinity of phases. The idea behind memory printed into minimum energy states of this field is as follows. The equations controlling the time evolution of the field are invariant under some (group-theoretic) groups of continuous transformation. This symmetry spontaneously breaks down when the system reaches its minimum energy state. The vacuum states are no longer invariant under the full group. Nevertheless, there are as many vacuum states as there are phases potentially assumed by the system. The vacuum states are capable carriers of memory if it is possible to stabilize the imprints and to protect them against overwriting by subsequent states of minimum energy.

Ad (2) Vacuum states, by virtue of being states of minimum energy, are stable in principle. The electric dipole field of the brain water is a system, however, that is constantly undergoing phase transitions. The possibility of stabilizing vacuum states of a system that is thus "living over many ground states" (Del Giudice et al. 1988) crucially depends on two conditions. The first is that the system is allowed to be in an open number of states at the same time. In quantum theory this condition is allowed by the so-called superposition principle. Superposition means that the state of a system can be described as the sum of a set of independent (orthogonal) states. The first condition for the possibility of stabilizing vacuum states of the electric dipole field is that the brain is a quantum system that is, and continues to be, in stabilized superposition. This condition is necessary but not sufficient for protecting the vacuum states against overprinting. The sufficient condition is that the system is allowed to collect physically inequivalent vacuum states without limit. This condition can only be fulfilled by open systems, i.e. by systems that are connected to their environment. The brain is such a system. As a dissipative system, it is in constant exchange of energy with its environment. By virtue of the brain's being a dissipative system, an overprinting of the vacuum states of the water electric dipole fields can be prevented. By thus being protected, an unlimited number of ground states are allowed to co-exist in stabilized superposition. It is this feature that might explain the enormous capacity of the brain to memorize.

Ad (3) Superposition is the mode in which quantum theory allows sums of orthogonal states of a system to exist or, rather, to sub-sist without being manifest. These superposed states are what the so-called state vector of the system is made of. The states entering the state vector are no actual states. No system whatsoever can actually be in orthogonal states at the same time. Conventionally, the state vector is interpreted as an expression of the states that the system possibly is found to be in when subject to a measurement. In the context of quantum theory, measurement means that one of the potential states of a system is turned into an actual state. An actualization of this kind must take place also when a memory state is selected for making appearance in mental presence. The states having subsisted in superposition before being actualised in an act of recollection cannot have been just potential states, however. They must have subsisted *really* without existing *actually*. This difference is inconceivable in classical theories of the brain. It is not before quantum degrees of freedom come into play that states are allowed to be real without being manifest. In a sense, thus, the states entering the state vector of the QBD system are sub-present. They do not come forth but by a so-called reduction of the state vector. Reduction of the state vector means that one of the many sub-present states is selected and raised to full presence. This many-to-one projection happens both when a measurement, as understood quantum theoretically, or when a recollection in the sense of the re-actualization of a printed vacuum state take place.

Actualization and temporality

The memory system theorized by QBD is a macroscopic quantum system. It is macroscopic in two regards. It is macroscopic regarding the coherence lengths of the quantum fields it consists of. And it is macroscopic regarding the perceptions it conserves. The macroscopic coherence lengths of the fields whose vacuum states wear the information could be the explanation of the non-local way memory is stored in the brain. A major part of the information these states wear represents the environment objectified. Perceptions are more than just impressions. The environment objectified consists of the collection of "res" that the concept of reality classically refers to. In terms of quantum theory, this concept of reality is derived. It depends on the process of actualization, i.e. on the process by which one of the alternatives entering the state vector of the system is selected for being turned into an actual happening.

The environment objectified in perception, even though presenting itself in only ever a single state at a time, is actually never in the same state. Actualization means that a state singled out of the state vector is temporarily raised to full presence. That the state is only temporarily raised to full presence implies that each moment another state of the world is made to manifest. Temporality, thus understood, means that each moment a state having been future is made to appear in the present only to vanish into the past. In order to survive the moment of its first presentification, a state surfacing in the present needs to be refreshed, i.e. made to reappear in the present.

It is not before perception is augmented by recollection that this kind of repetition can become effective in perception. Moreover, it is not before the states actualised are recorded and kept presentified that change becomes susceptible to perception. It is only through re-actualization of the states printed to memory that the reality given to perception extends beyond the immediate present. Or, to put it differently, past and future do not come forth but by re-presenting or pre-presenting, respectively, states that are not immediately present. Hence, it is the interplay of actualization and re-actualization that presents us with a classical reality.

One of the revolutionary novelties hypothesised by QBD is that the states that give rise to the world we perceive are macroscopic quantum states. These quantum states, taken together, contain the trajectories of the objects that make up our classical environment. The space of the memory states are classical trajectories (Vitiello 2003). The states themselves exist in the mode of superposition as long as they are not re-actualised in an act of recollection. This, however, means that the states underlying our feeling of living in a real world are not just sub-conscious when not being in the foreground of attention, but *sub-present*. They are real without being manifest. In order to manifest they have to be (re-) actualised.

The process by which states having subsisted in superposition are turned into actual states is not yet finally understood. One of the unanswered questions is how actualization and mental presence are interrelated. Since not even QBD addresses this question immediately, let us put it for the moment aside. Let us observe, instead, that two further conditions have to be met in order to bring forth a perception. The first is that the brain is coupled to its environment by exchanging energy in a highly specific way. The energy input that results in perception does so by assuming the character of sensory input. The second condition to be fulfilled is that the causal chain connected with the energy input is intentionally inverted in such a way that the object perceived is located out there. Only a tiny fraction of the input processed in the perceptive brain results in conscious perception. In conscious perception, the brain exchanges energy not only, but gets correlated with its environment in a highly specific way. Locating objects out there means that the activity of perceiving reaches out and does not stop short of the object objectified. It makes no sense to say that the perception happens in the brain. The act of perceiving happens in the brain as well as in the place that the object perceived occupies. The relation thus established is non-local in that the object is not just a representation, but identical with the content that the consciousness is conscious *of*. The *intentionality* of consciousness, thus understood, assumes the form that quantum theory describes as *entanglement*.

When recollecting perceptions, these entangled states are re-actualised. In re-actualization, the brain state performing the actualization and the state that is actualised are separated in time. However, how do we get a notion of this? How is it that we distinguish perception from recollection? Both the state initially actualised and the state re-actualised are states presenting themselves in mental presence. What has changed is the source of information. It is not an awareness of the source of information, though, by which we distinguish perception from recollection. It is much more our sense of concreteness that we make use of. The things perceived are more concrete that the things recollected. This difference is even characteristic of the states printed to memory. When being aware of some part of our biography we know, in a how-it-feels sense, whether we recollect something immediately perceived or something we had only indirect notion of. Memory printing depends on energy input. Hence, it should be the measure of energy exchanged between brain and environment by which we distinguish perceptions from recollections. In recollection, the energy exchange with the environment is interrupted. The result is the how-it-feels difference between perceptive and reflective consciousness.

Interestingly, the brain's capability of re-actualization is not restricted to an either full or nil presentification. We find ourselves capable of interlacing recollections into the ongoing stream of perceptions. We are able, that is, to divide attention. While being aware of what happens before our eyes, an inner eye may attend to happenings quite different from those in the foreground. Moreover, we feel free to switch between foreground and background. The scene perceived in the foreground and the episode recollected in the background are easily made to change position in attention space. When the scene in the foreground of our attention is past or future we just seem absent-minded to those observing our behaviour. In fact, however, it is quite normal that we switch between past, present and future by manipulating the weights of presence of the states that are manifesting at the same time. In order to allow a state to come forth in mental presence it is not necessary, thus, to fully reduce the state vector of the QBD system. We seem capable, rather, to manipulate the weights of presence in such a way that an 'actual' superposition of the states coming forth results.[2]

We are back to the process of actualization. Actualization means that one of the alternatives entering the state vector of the system is selected for being turned into an actual happening. In the context of perception or, for that matter, of measurement, the states entering the state vector are interpreted as possible states. Actualization, thus interpreted, means that out of the cloud of previously possible states an actual state precipitates. In the context of recollection, a slightly different interpretation may be more appropriate. The vacuum states of the QBD system are not just potential states. They are real states, tracing facts. The states wearing the information of veridical memories are real states that exist in a less-than-full degree of presence as long as they are not selected for re-actualization.

By this interpretation, a distance in time gets involved that separates the initial actualization of states from subsequent re-actualization. At first glance this time seems to be just the distance measured by clocks. On closer inspection we see, however, that distance, i.e. difference in date, is not the only difference. There is a difference in presence as well. Our sense of concreteness does not need a clock for distinguishing perceptions from recollections. It just discriminates degrees of presence. It is a progress hard to be overrated that QBD is capable of translating this difference into terms of energy. By this translation, however, the interpretation of the weights – i.e. the complex terms – of the wavefunction changes. Each state entering the state vector of a system contributes with a definite weight to the superposition. In the context of measurement, these weights (the square moduli of the complex terms) are interpreted as the probability of obtaining the alternative in question when the system is measured. Actualization, thus understood, means that the probability of one of the alternatives shifts from a value less than unity to the value unity (see Stapp 1993 for this formulation). As soon as the states waiting for actualization are not just potential states, the weights with a value less than unity turn into measures of a less than full degree of presence. Accordingly, actualization turns into the process in which sub-present states are temporarily raised to full presence.

Epistemologically, this 'temporalistic' interpretation is equivalent to the probabilistic one. The closer the degree of presence of a state is to full presence, the higher is the probability of meeting the system in this state when a measurement is performed. The difference lies in the concept of time involved. Presence explicitly refers to *temporal* change. Temporal change has to be clearly distinguished from *real* change. Temporal change, to repeat it, means that world states having been future become present and then past. Real change means that states differing in date also differ in structure or function. Temporal change and real change are independent of one another. We can abstract from temporal change by leaving real change perfectly intact. Accordingly, when consistently disregarding the processes of real change, we are left with nowness as such.

The system described by QBD either presupposes of engenders temporal change. The manifestation of perceptions and recollections is bound to temporal change. Dealing with temporal change is what recollection and anticipation are good for. It is not before perception is distinguished from recollection that we realize that time goes by. In the domain deemed to be physically fundamental time does not pass. The question thus is how the passage of time is accounted for in dissipative QBD.

The passage of time and the mirror image in time

Asking thus may seem to be besides the mark. As a physical theory, QBD does not account for the process we experience as temporal change. Temporal change means that the state manifesting in the present never is the same, whereas the present itself persists. The states, including the 'inner' states of the consciousness, come and go. The Now just *is*. It has been now since time began to pass. As soon as the temporal present is treated as a permanent Now, maintaining its identity while moving relative to the states that make appearance in it, temporal change appears as a relative kind of motion. In QBD, there is no account for permanent presence, nor is there one for relative motion.

Nevertheless, dissipativity allows a singularity on the axis of time to come forth whose emergence involves a symmetry break between the directions of time (Vitiello 2001: 107). The singularity means that there is a distinguished place in time. Breaking the symmetry means that there is a preferred direction of time. The distinguished place in time and the irreversibility are necessary, though not sufficient, conditions for temporal change. What is lacking, still, are the differences in presence and the spontaneous movement we experience as passage. It may be that these latter ingredients are tied to presence as a mode of existence for its own. Still, there is the strange doubling of universes in dissipative QBD waiting to be considered in the light of the experience we have of time. The reason is that the doubling of the system brain plus environment is accomplished by way of mirroring the system in time. The tilde universe ~A is the *time-reversed mirror image* of the nontilde universe A.

Is there a way of interpreting this mirroring in time in the light of the experience we have of time? As Globus (2003: 138) makes clear, it is pointless to think of a movie that runs forward and backward at the same time. There may be an interpretation, however, when we take seriously the description of passage as a relative kind of motion. Relative motion is a concept that includes, or gives rise to, a mirror image. By definition, relative motion can be looked at in two ways. Each of the relata can be looked at as being at rest while the other one is in motion. This applies to the travel of the Now as well. The Now can be looked at as being at rest while the sum total of the states having passed through or being destined to pass are in motion. Or the Now can be seen as being in motion while the states are at rest. In order to switch between these views, a transformation is needed in which the direction of time is reversed. In the first view, the Now travels forward in time, in the second view, the states travel backward in time (Franck 2000, 2003).

Both the relative motion and the mirror image in time deal with the relation between the perceiving self and the world perceived. It is thus tempting to assume that the experience of the moving Now is the phenomenal correlate of the dynamics described by dissipative QBD. This interpretation, however, faces difficulties. Even though the dynamics of dissipation breaks the time-reversal symmetry, it does not give rise to a constant shift of an entity relative to another. Since energy is not conserved in dissipation, there seems to be no base for the translational invariance implied in relative motion.

Nevertheless, the brain plus environment is an entangled whole. In dissipative QBD, the brain is constantly entangled with its environment. It is entangled not just in the way that gives rise to a non-local correlation in space. It is entangled in a way also that maintains the unified whole in time. This ongoing entanglement leaves as its trace the perceptions printed to memory. The perceptions printed to memory are processed into the *cognitive map* we construct and maintain of our environment. It is this mental map that appears to be in relative motion when we have the impression that time goes by.

From a physical point of view, this map is not extended in time. It is built up from information that shares the date of its actual use. It is only by interpreting this information in a certain particular way that it turns into a re-presentation of experiences undergone. In order to turn information available at time t_0 into a re-presentation of an event having taken place at time t_{-1} an attitude is needed that deploys perspectival depth behind the pattern presented. This *intentional* attitude has to perform in time what the perception of objects out there performs in space.

Intentionality is a property of phenomenal consciousness. Accordingly, the mental map, as a map representing temporal regions extending beyond the present, does not come forth but in mental presence. The kinematics of temporal change may thus be purely phenomenal. The relative motion may be absent in the absence of cognitive maps. Hence, we should be careful of not committing a category error when looking for a physical correlate of the impression that the Now travels through time. Before asking what physical process gives rise to the impression that time goes by we should further go into the analysis of the impression itself. How is it that we have the impression of living in a Now that maintains its identity while constantly changing its location in time? What precisely are the relata the combination of which gives rise to the impression of relative motion?

The 'paradox' of temporal change

The only descriptions we have of time's flow are phenomenological. Remarkably, however, the description of this basic experience proves to be frustratingly hard. Since Henri Bergson there is a noted incompatibility between physical time and the time containing the Now. For Bergson (1889), the difference is that between the distance measured by clocks and the duration experienced subjectively. In order to account for the aspect of time that escapes measurement by clocks, Bergson introduced the concept of 'durée'. 'Durée' gives expression of the fact that the Now endures. Bergson did not consider the possibility, however, that duration may assume two totally different meanings when applied to the Now. The meaning of duration suggesting itself when applied to the Now is the eigentime of the present, i.e. the interval in clock time covered by mental presence (on this point see Pöppel 1997 for an overview). This interval however, as extended as it may be, has to be distinguished from the permanence of the Now. This latter kind of duration is different not only, but even independent of the interval spanned by the so-called specious present. Since Bergson has never been clear about this point, his concept of durée became a notorious source of confusion.

The operation of singling out the span covered by mental presence from the duration lived through is delicate when to be performed on the level of phenomenology. On the level of phenomenology, the Now is indistinguishable from the presence of phenomenal consciousness. It is one of the tenets of phenomenology that consciousness is intentional as such. That consciousness is essentially intentional means that being conscious always means to be conscious *of* something. By virtue of its essential intentionality, the presence of consciousness cannot be separated from the contents presenting themselves in this presence. Hence, the problem of drawing the line between the different meanings of duration implied in the concept of durée lies in demarcating the pure 'form' of mental presence from the things and events surfacing in it.

The problem consists, to put it differently, in distinguishing mental presence from the stream of conscious phenomena. It is this distinction that Edmund Husserl never stopped tackling. Husserl realized that there is a kind of relative motion relating the presence of consciousness to the contents consciousness is conscious of. He clearly saw that there are two views of passage, the one showing the Now as passing while the moments passed through are at rest, the other showing the Now at rest while the moments passing through are in motion. Husserl called the experience of this ambiguity the '*arch-impression* of the both standing and flowing Now' ("die Urimpression der stehend-strömenden Gegenwart"). However, Husserl was prevented from taking the relata of this relative motion apart by the dogma of the essential intentionality of consciousness. Taking these relata apart means separating phenomenal consciousness from its intentional content. Instead of daring the cut, Husserl felt driven to treat the ambiguity of the both immobile and flowing Now as a kind of paradox. He was downright haunted by this 'paradox'. He never stopped fighting with the problem. From his lectures on "The Phenomenology of Internal Time Consciousness" in 1905 (Husserl 1966/1996) until the manuscripts dating from the year of his death, 1938 (see Held 1966), he grappled with the problem again and again. The tension remained unresolved.

It was first Heidegger who left the dogma of the essential intentionality of consciousness behind. For Heidegger, it is the very distinction between *presence* and the things and events *presencing* that becomes essential. It is this distinction that is drawn when *Being* (Sein) is distinguished from the *being-there* of things and events (Seiendes). Being, as distinct from the things being, means being aware in the sense that a world of experience is present at all. Being, as distinct from the events being in time, is the time being. We have to leave subtleties aside in order to translate "Sein" as *presence*.[3] It is this translation, however, by which Being is related to time. Presence, when related to time, is nowness as such. Nowness is neither a thing nor is it an event. The things and events presented come and go; presence just *is*.

The reason why it seems so strange and outlandish to draw the distinction circumvented by both Bergson and Husserl becomes clearer when we consider the novelty that the advent of Heidegger's philosophy meant in the tradition of western thought. The distinction between Being and entities is a differentiation requiring an attitude that never has been cultivated in western philosophy. It requires awareness to get rid of intentional content. As long as awareness is assumed to mean being aware of some *thing*, abstraction from the contents surfacing in awareness is tantamount to abstracting from awareness as such. Only by forgetting or annihilating the objects awareness habitually is concerned with, presence as such comes to the fore. Or, put the other way around, it is only in *conceptless self-awareness* that phenomenal consciousness comes to its own. Only through the exercise of disregarding anything surfacing in awareness, the differentiation between Being and entities becomes intelligible. The way in which presence as such can be experienced consists in avoiding any distraction from one's own state of being aware.

The novelty of Heidegger's philosophy is that it bridges western and eastern thought in an unprecedented way. Being, as distinct from the being-there of things and events, is what eastern tradition calls the *empty Being* or *filled Nothing*. Being, as distinct from any thing there is, is presence void of any specific content. Being, as distinct from any event, is nowness experiencing itself and nothing but itself. Being is empty in that it is without inner structure and outer circumscription. It is Nothing in that it is the fulfilment of the absence of anything detracting from presence as such. In this equivalence of Being and Nothing, consciousness escapes the determination of being consciousness *of*. In conceptless contemplation, the unity of the phenomenality and the intentionality of consciousness breaks.

Considering the deep-rooted habits of thought that need to be overcome when drawing the ontological differentiation performed by Heidegger, we understand better why Bergson and Husserl did not succeed in separating the relata that need to be separated in order to be truned into constituents of the relative motion we perceive as time's flow. On the other hand, not even Heidegger asks how the dimension **t** is turned into the process of temporal change. Instead of putting the relata separated by the ontological differentiation together again, he treats these relata as if one had to decide oneself which one to be concerned with in the first place. He belittles the being-there of entities by emphasising, again and again, the overriding dignity of Being. Instead of developing a philosophy of time that reconstructs in depth the interplay between presence and reality, he emphatically focuses on the depth of Being.

Ontological complementarity

Nevertheless, Heidegger cuts through a Gordian knot. Distinguishing Being from entities shakes the deep-rooted habit of thinking the real to be present as such. Being, as distinct from entities, is the mode in which anything appearing in the light of an awareness manifests. This mode may or may not be included in the description of reality. It may be excluded in the name of objectivity. This is what contemporary physics does. Manifestation means presentification. Presence may ultimately be synonymous with mental presence. However, we do not know whether the advent of presence had to await the evolution of conscious brains. It may be as well that presence is as primordial as is material reality (i.e. matter and physical energy). Heidegger leaves it open how, and whether, Being relates to subjectivity.

Presence, when distinguished from the entities presencing, can be purified from anything particular that manifests. Conversely, material reality can be purified conceptually from presentification. Presence, when purified from any thing and event manifesting, is the empty Being or filled Nothing. When purifying physical reality from presentification we end up with the universal wavefunction. Remarkably, both the elimination of entities from presence and the elimination of presence from reality result in fathomless wholeness. Both concept-less awareness and the totally entangled quantum whole are backgrounds of existence that cannot be transgressed. They are both fundamental and extremely symmetrical. The emergence of anything particular presupposes that symmetries are broken. Symmetries break as soon as the pure extremes are left. Concept-less awareness turns into intentional consciousness as soon as the awareness becomes selective. The totally entangled quantum whole turns into a selection of possible states as soon as observables are identified. In both cases, something particular emerges by way of reducing the one extreme and allowing the other mode of existence to enter. The categories that enter when concept-less awareness turns into focussed attention are the forms that entities assume when becoming objects of thought. The observables that enter when probabilities are calculated are forms that entities assume when conceived as objects located in space and time. It is an open question how the selection of states described by the wavefunction and their transformation into local states is related to presence. In the case, however, that it is legitimate to talk of sub-present states whose presence is amplified when an actual happening takes place, the constitution of macroscopic reality implies presence. On the other hand, conceptual awareness ranges from abstract thought to concrete perception. Abstract thought is distanced from concrete perception by steps of reflection whose effect is that the materiality represented is diminished. Most importantly, there is a middle between the extremes where presence and materiality match in a highly particular way. The matching case is the perception whose description renders the collection of "res" that the concept of reality classically refers to. Classical reality is what results when physical reality is reduced to an only ever single and manifest state of the world, and when presence is reduced to the social average of the temporal present.

What we are facing is an ontological complementarity of presence and materiality. We can have presence up to the extreme of fathomless wholeness, and we can have materiality up to the extreme of fathomless wholeness. We can have the maximum of either, but we cannot have both. We can have each to the extent only that the other is reduced. The more materiality, the less presence, and vice versa. This ontological complementarity is one of the fundamental conditions of existing as a conscious brain: of *Dasein*. Dasein participates in both materiality and manifestation. The degrees of participation are not fixed, but subject to change. Dasein implies the feeling of being free to combine materiality and presence at will.

The ontological complementarity of presence and materiality is a topic also of the philosophical reflection of quantum theory. It was Wolfgang Pauli who suggested that phenomenality and physicality should be thought together in terms of complementarity. "The general problem of the relation between psyche and physis, between the inner and the outer, can ... hardly be said to have been solved by the concept of 'psychophysical parallelism' which was advanced in the last century. Yet modern science may have brought us closer to a more satisfying conception of this relationship by setting up, within the field of physics, the concept of complementarity. It would be the most satisfactory of all if physis and psyche could be seen as complementary aspects of the same reality" (Pauli 1994: 260). Recently, Hans Primas took up the idea, starting from an ontologically undifferentiated whole, an *unus mundus*, in which "[n]either time, nor mind, nor matter and energy are taken to be *a priori* concepts. Rather, it is assumed that these concepts emerge by a contextual breaking of the unitarian symmetry of the unus mundus" (Primas 2003: §2.3). In Primas' concept of primordial symmetry, not even the symmetries to be broken are predetermined. Rather, there may be different separations, leading to complementarity descriptions of the unus mundus different from the one using the concepts of presence and materiality. Hence, the ontological differentiations characterising specific kinds of existence are assumed to be contingent on the evolution of the universe. Primas comes very close thus to what Heidegger conceives as the *history of Being*.

Primas refrains from speculating about how the breaking of the symmetry between presence and reality might relate to the evolution of conscious brains.[4] Primas also refrains from asking how the breaking of the unitarian symmetry relates to the phenomenology of time. This restraint is wise insofar as temporal change still awaits consistent phenomenological description. The question becomes crucial, however, when we ask how the kind of actualization that QBD involuntarily is concerned with relates to the ontological complementarity. In order to go into this question we have to look for the symmetry that needs to be broken for conceiving time as a process instead of as a dimension only. In order to look for this symmetry, let us try to reverse the abstraction that is performed when the physical concept of time is purified from nowness.

When abstracting from the Now, a two-fold abstraction has to be performed. On the one hand, we have to abstract from the spontaneous movement that the Now is in relative to the world states that pass. On the other hand, we have to abstract from presence as a mode of existing. In order to reverse this two-fold abstraction, let us first ask what it means to recombine presence with dimension **t**. Since it is unclear what presence means in total absence of mental presence, let us start from the specious present. In terms of presence, the specious present denotes an atomic chunk of nowness, i.e. a minimal duration. This duration appears as the eigentime of mental presence when the temporal resolution of our sense of time is measured in clock time. Conversely, to the interval that mental presence covers in clock time a quantum of nowness belongs. These individual nows can be ordered in the same way as the eigentimes can. When ordering the eigentimes chronologically, we obtain dimension **t**. When ordering the individual nows accordingly, we obtain a present that extends over all time (Franck 2003). When thus co-ordinating distance in time and nowness, we face a perfect symmetry: The diameter of the present is co-extensive with **t**. Hence, the breaking of the symmetry must have to do with the emergence of the spontaneous movement we experience when having the impression that time goes by. In fact, the synthesis of the individual nows into the permanent Now fundamentally differs from the synthesis of the eigentimes into time **t**. The synthesis rendering the permanent Now renders a Now whose lifetime is (or may be) co-extensive with **t**. This (nearly) unlimited lifetime does not mean, however, that the span covered by the Now extends without limit. Rather, the permanent Now has the same diameter as have the individual nows. The 'paradox' of a Now that lasts 30 milliseconds and forever at the same time is resolved by breaking the symmetry between presence and distance in time. The symmetry broken is that between the lifetime and the diameter of the Now. The Now is allowed to last 30 ms and forever at the same time by being put into motion relative to the eigentimes adding up to **t**.

From symptoms of amnesia we know that the synthesis of the individual nows into the permanent Now has to be actively performed in order to engender the impression that time goes by. Without this synthesis, there is just a sequence of unconnected atoms of presence. However, the synthesis that is needed is of a highly particular kind. What is needed is a synthesis that establishes *identity in time*. Such a kind of synthesis is performed when the re-actualization of states that have passed is processed into the experience of recollection. In order to turn re-actualization into a recollection, the state re-actualised has to be identified as a state of this same consciousness. The unity of consciousness lies in the self-sameness of the mental presence on which it relies. This self-sameness does not mean that the 'inner' states of the consciousness are prevented from changing. Nor does it mean that the intensity of the mental presence must not change. It means, rather, that presence as such perdures while the states presencing come and go.

It is not before the unity of consciousness is thus established that the impression of temporal change comes forth. However, the emergence of this impression is tantamount to the ontological differentiation between Being and entities. In the impression that the Now endures while the states come and go, presence as such is distinguished from the entities passing through. At the same time, the ambiguity characteristic of Bergsonian durée is imminent. In order to avoid this ambiguity we have to clearly distinguish the distance denoted by **t** and the duration, let it be $\boldsymbol{\tau}$, implied in the awareness that the Now endures. The Now extends in both $\boldsymbol{\tau}$ and **t**. Moreover, the extension of the Now in **t** is independent of its extension in $\boldsymbol{\tau}$. In order to resolve the paradox that haunted both Bergson and Husserl, distance **t** and duration $\boldsymbol{\tau}$ have to be assumed to be orthogonal. If **t** and $\boldsymbol{\tau}$ are orthogonal, we are left with **t** when the Now is abstracted away. Accordingly, we are left with $\boldsymbol{\tau}$ when presence is purified from entities.[5]

We are back to the question of how the time-reversal mirror image in the formalism of thermofield dynamics relates to temporal change. We have seen that the relative motion we experience as time's flow gives rise to a dualism of views that can be interpreted as a mirroring in time. This mirroring, however, is not the only one that is characteristic of the experience we have of time. With the passage of time we mean the Now travels along **t**. Travel is a process that combines way and time. The "way" travelled by the now lies in **t**. The "time" this travel takes is denoted by $\boldsymbol{\tau}$. The travelling Now is thus a process that may be inseparably tied to the existence of the self that undergoes the experience. The experience of the travelling Now is the self-experience of an I that maintains its identity while suffering incessant change. The impression of relative motion relies on self-identification. Self-identification is a feat of the intentionality of consciousness. Since intentionality may ultimately rely in the phenomenality of consciousness we have to proceed very carefully when relating the process of temporal change to the processes described by thermofield dynamics. Nevertheless, self-identification presupposes a kind of mirroring for its own. In every act of recollection a reflective doubling of the self takes place:

A state other than the one that the self is feeling to be in is identified as a state of this same self. "Thus the overall mathematical structure of the model and in particular the specific dissipative character of the dynamics strongly point to consciousness as a 'time mirror', as a 'reflection in time' which manifests as a nonlinear coupling ... with the inseparable own Double" (Vitiello 2001: 141). It is this doubling of the self that Giuseppe Vitiello associates with the doubling of universes.

Acknowledgements

The paper is the outcome of a discussion I have had with Gordon Globus for years. His comments as well as his constant encouragement have been main inputs to this contribution. Basil Hiley and Teruaki Nakagomi contributed by generously providing support in interpreting physics from the viewpoint of presence. Giuseppe Vitiello's hints and warnings were more than inspiring and helpful. He saved me from much more serious errors than those that are still to be found in the paper. Thanks to the Institute for Scientific Interchange Foundation (ISI) for the wonderful Torino workshop in November 2002.

Notes

1. In the following I most heavily rely on Vitiello (2001, 2002).
2. For the formalism of such a superposition see Jibu and Yasue (1995), Appendix A.
3. The subtleties disregarded are not negligible. The translation holds, however, as far as "Being and Time" (Heidegger 1927/1962) is concerned.
4. As an approach to this question see Teruaki Nakagomi's quantum monadology (Nakagomi 2003a, 2003b).
5. For further reasons for assuming that t and τ are orthogonal see Franck (2001, 2003).

References

Bergson, Henri (1889/1960). *Time and free will. An essay on the immediate data of consciousness* (F. L. Pogson, Trans.). New York: Harper.

Del Giudice, E., R. Manka, M. Milani, & G. Vitiello (1988). Non-constant order parameters and vacuum evolution. *Physics Letters, B 206*, 661–664.

Franck, Georg (2000). Time and presence. In R. L. Amoroso, R. Antunes, C. Coelho, M. Farias, A. Leite, & P. Soares (Eds.), *Science and the primacy of consciousness* (pp. 68–72). Orinda, CA: The Noetic Press.

Franck, Georg (2001). Time, actuality, novelty and history. In A. U. Frank, J. Raper, & J.-P. Cheylan (Eds.), *Life and motion of socio-economic units* (pp. 111–123). London: Taylor & Francis.

Franck, Georg (2003). How time passes: On conceiving time as a process. In R. Buccheri, M. Saniga, & W. M. Stuckey (Eds.), *The nature of time: Geometry, physics and perception* (pp. 91–103). Dodrecht, Boston & London: Kluwer Academic Publishers.

Globus, Gordon G. (2003). *Quantum closures and disclosures: Thinking-together postphenomenology and quantum brain dynamics*. Amsterdam & Philadelphia: John Benjamins.

Heidegger, Martin (1927/1962). *Being and time* (J. Macquarrie & E. Robinson, Trans.). New York: Harper & Row.

Held, Klaus (1966). *Lebendige Gegenwart*. The Hague: M. Nijhoff.

Husserl, Edmund (1966/1996). *The phenomenology of internal time consciousness*. Bloomington: Indiana UP.

Jibu, Mari & Kunio Yasue (1995). *Quantum brain dynamics and consciousness: An introduction*. Amsterdam & Philadelphia: John Benjamins.

Nakagomi, Teruaki (2003a). Mathematical formulation of Leibnizian world: A theory of individual-whole or interior-exterior reflective systems. *BioSystems, 69*, 15–26.

Nakagomi, Teruaki (2003b). Quantum monadology: A consistent world model for consciousness and physics. *BioSystems, 69*, 27–38.

Pauli, Wolfgang (1994). *Writings on physics and philosophy* (C. P. Enz & K. von Meyenn, Eds.). Berlin: Springer.

Pöppel, Ernst (1997). The brain's way to create "nowness". In H. Atmanspacher & E. Ruhnau (Eds.), *Time, temporality, now* (pp. 107–120). Berlin: Springer.

Pribram, Karl (1991). *Brain and perception*. Hillsdale, NJ: Lawrence Erlbaum.

Primas, Hans (2003). Time-entanglement between mind and matter. *Mind and Matter, 1*, 81–119.

Stapp, Henry P. (1993). *Mind, matter, and quantum mechanics*. Berlin: Springer.

Umezawa, Hiroomi (1993). *Advanced field theory: Micro, macro, and thermal physics*. New York: American Institute of Physics.

Vitiello, Giuseppe (1995). Dissipation and memory capacity in the quantum brain model. *International Journal of Modern Physics, 9*, 973–989.

Vitiello, Giuseppe (2001). *My double unveiled: The dissipative quantum model of the brain*. Amsterdam & Philadelphia: John Benjamins.

Vitiello, Giuseppe (2002). Dissipative quantum brain dynamics. In Kunio Yasue, Mari Jibu, & Tarcisio Della Senta (Eds.), *No matter, never mind* (pp. 43–61). Amsterdam & Philadelphia: John Benjamins.

Vitiello, Giuseppe (2003). Classical chaotic trajectories in quantum field theory. hep-th/0309197.

Chapter 5

The psycho-emotional-physical unity of living organisms as an outcome of quantum physics

Emilio Del Giudice
Istituto Nazionale di Fisica Nucleare, Milano, Italy

The concept of psycho-emotional-physical unity of a human (and, more in general, a living organism) has been a consequence of the Freudian revolution. As recognised by Freud and the other founding fathers of "psychodynamics", classical physics and the other "hard sciences" based on it as modern molecular biology are inadequate to deal with this fundamental notion. I discuss in this contribution how modern Quantum Field Theory (QFT) could open the way to an understanding of a psyche deeply rooted in the body.

Introduction

A living organism is simultaneously an ensemble of molecules governed by the laws of physics and chemistry and a subject with a psyche, an emotional world and, at least in the case of humans, a mind. How the component molecules know that they must reproduce when they are an aggregate, Jack or Ursula, which are at different times happy or somber, extroverted or withdrawn persons?

The bridge between the molecular model and the psycho-emotional model of humans and, more in general, living organisms is still missing. One century ago, at the dawn of the 20th century, a number of conceptual revolutions occurred both in the physical sciences and in the understanding of psychic phenomena. Sigmund Freud (1925) has been able to sketch the foundations of the dynamics of the emotional sphere, that originates from the dark outcome of the molecule reactions, termed by Freud (1925) and Groddeck (1923) the "Es" or the "Id", namely "the Thing". It evolves toward higher structures such as the unconscious, the preconscious and so on, until reaching the conditions that allow the formation of an Ego. It has been recognised more and more in subsequent times that mind and body cannot coexist as a unity without the psyche, the cauldron where emotions and passions are boiling, the hot source of every cool mind; see, for instance, the interesting book of Antonio Damasio (1994): *Descartes' Error: Emotion, Reason and the Human Brain.* Some centuries earlier, Spinoza (1955) has recognised that every cool (geometrical) mind emerged actually from a passionate soul and Vico (1999) distinguished the three levels of understanding, by stating that humans firstly sense, then feel with a moved and perturbed soul, eventually think with a pure mind.

Actually mind is quite unable to govern the body apart from a few voluntary functions, whereas the psyche plays the role of governor of the vast ensemble of bodily chemical reactions, as the existence of psychosomatic phenomena shows with an ever growing evidence. On the contrary the Freudian tradition has accumulated a vast evidence about the dependence of the rational sphere upon the emotional level.

Freud believed that psyche emerged from the obscure world of the chemical reactions, but he was also convinced that the hard sciences of his time – physics, chemistry and, consequently, molecular biology – were utterly unable to account for the basic concepts of the psychoanalytic revolution, so that he cautioned his disciples from falling in love with the hard sciences and their arrogance ("what we don't understand does not exist") up to the point of preferring the contribution of humanists to the high brow opinion of hard scientists. Actually the opinion of Freud was not at all a submission to an irrationalistic preconception, but, more simply and correctly, the recognition of an historical inadequacy of physics and chemistry of his time to decipher a mystery much too complex for them.

Let me now address from a completely different point of view, namely the point of view of the molecules and their interactions, the problem of the inadequacy of hard sciences to understand life. Consider two opposite chemical systems:

a. The system of molecules enclosed in a chemical reactor in the absence of catalysts; these molecules react through the mechanism of random collisions, whose rate is controlled by temperature and pressure. In this case the outcome of the full ensemble of chemical reactions occurring in the reactor is quite frequently a huge number of reaction products (that give rise in the industrial reality to a huge number of chemical wastes to be disposed of) and, moreover, wide fluctuations of the physical parameters (temperature and pressure) inside the reaction chamber and a large demand of energy in the form of low-grade heat.

b. The system of the component molecules of a living organism. In this case chemical reactions occur as sequences, ordered in space and time, of encounters of selected molecules able to find each other in a very short time and without (almost) any mistakes. Everything occurs as specific codes of molecule mutual recognition and recall were at work; the genetic code is a well-known example. What is the dynamical origin of these codes? It is impossible that each pair of molecules required to interact at the nth step of a selected sequence be already at the very close distance demanded by the short range (a couple of molecule diameters) of chemical interactions. Thus the existence of codes of molecule mutual recognition and recall together with the short time of reaction imply that long range (at least several tens of Angstroms) interactions should play a role in biochemistry.

Unfortunately modern molecular biology is construed as an ensemble of strictly local encounters of selected molecules occurring according to codes whose existence is not the consequence of a rational dynamics, but is just a dogma. The missing element is just the network of long range interactions among molecules that are then no longer independent, as the dogma of local causality actually demands, but become the components of an extended object and behave according to the signals circulating in the network of long range interactions, that appears as the inner "soul" of the ensemble of molecules. The system acquires a meaning just from the ensemble of signals of the network, which in turn does not fall from the blue but emerges from the energetics of the molecular level. In the following we will explore whether this point of view could receive support from the ideas produced by the scientific revolutions of the 20th century and could also produce the foundation of a physics more acceptable to Freud than the atomistic physics underlying molecular biology:

1. The revolutions in the physics of the 20th century

The end of the 19th century has seen physics, what we term now "classical physics", as a monumental construction, which, according to some of its founders such as Lord Kelvin, Ostwald, Poincaré and other prominent figures, was bound to be the "ultimate explanation of Nature," the "frame of the system of the world". Physical reality was conceived as the interplay of two basic entities, particles of matter and fields of forces. Matter was conceived as separable into independent particles, each one to be isolated from the rest of the world and moreover, inert, namely unable to change by itself its state of uniform motion (principle of inertia). Each particle interacted with the other particles through forces filling the whole space (fields), as for instance the gravitational and electromagnetic forces; such fields were the means through which energy and momentum were exchanged among particles. Moreover, particles were able to exchange energy and momentum through collisions.

By using the above theoretical concepts physicists were able to explain the whole body of thermodynamics founded on the two celebrated principles in terms of the "classical" behaviour of the component atoms, at least in the case of gases. This "reduction" (hence the term reductionism given to the complete explanation of macroscopic matter in terms of microscopic components) of the macroscopic behaviour of gases to the microscopic "classical" behaviour of atoms was a triumph of scientific rationalism and was made possible by the introduction of the so called "phase space" by Boltzmann and Gibbs.

Consider a physical system made up of N atoms. According to classical physics the trajectory of each atom is completely specified by giving the position and the momentum, namely the velocity times the mass, of the atom at a given time. Both position and momentum are each specified by three numbers (the components along the three axis oriented as the three directions of the space), so that the whole system of N atoms is described by $6N$ numbers at every instant of time. This set of $6N$ numbers could be the set of coordinates of a point in an abstract space of $6N$ dimensions, which is just the "phase space" of the given system of N atoms. Suppose that during the time the trajectories of all the N atoms fill the subspace Γ of the full phase space corresponding to a given macroscopic state (namely the state specified by macroscopic variables as temperature, pressure and so on); Boltzmann was able to show that the entropy S of the system of N atoms was connected to the volume of the subspace Γ by the formula:

$$S = k \ln \Gamma \tag{1}$$

where k is the Boltzmann constant and the function ln is the logarithm. According to eq. (1) the entropy of a macroscopic state is an index of the complexity of the ensemble of microscopic states corresponding to the macroscopic state. The successful reduction of the thermodynamics of gases to the classical mechanics of a multitude of atoms was made possible by the bridge provided by eq. (1). At the turn of the 20th century, however, physicists and chemists were suddenly plunged from the heaven of triumph into the hell of conceptual failure. On the ground of a wide body of phenomenological evidence, the

German chemist Walter Nernst (1969) was able to state the Third Principle of Thermodynamics: *The entropy of any physical system vanishes when the absolute temperature vanishes.*

Classically all atoms are at rest, namely all momenta are zero. When atoms are at rest, their positions too are completely fixed by the condition of minimum energy. So the subspace Γ corresponding to the state of zero absolute temperature of a physical system shrinks to a single point in the phase space. Here lies the tragedy that suddenly brought classical physics to an inescapable deadly catastrophe. The volume of a single point is zero and it is well known that

$$|\ln 0| = \infty \quad ! \tag{2}$$

When temperature vanishes, classical physics predicts – and it is impossible to circumvent this conclusion – that entropy must diverge. That implies universal death in a hot furnace created by the ultracold! Complete nonsense.

The way out suggested by Max Planck was the seemingly technical trick that points in the phase space should be replaced by tiny regions whose size was controlled by an universal constant of Nature, the Planck constant h, that was strictly forbidden to be made vanishing. By taking this naturally defined fundamental volume as the unit volume, then the volume of the subspace Γ of each physical system is a multiple integer of a fundamental constant. So in the case of a macroscopic state corresponding to a single microscopic state we get:

$$\ln 1 = 0 \tag{3}$$

and Walter Nernst could finally smile.

This detailed exposé shows clearly how deep and inescapable has been the crisis that undermined classical physics. This crisis is usually misrepresented in the textbooks since the general argument about the Third Principle of Thermodynamics is specialised to the particular case of the black-body radiation, whose spectral distribution depends upon the frequency ν and the temperature T through the ratio ν/T (Wien's law); the limit $T \to 0$ can thus be mimicked by the limit $\nu \to \infty$, so that the "cold catastrophe" could be misrepresented as an "ultraviolet catastrophe". Since, moreover, $\nu \to \infty$ implies that the wavelength $\lambda \to 0$, finally the crisis of classical physics, which is actually a catastrophe occurring at low temperature, quite near to the daily life (room temperature is not so far from $T = 0$), has been presented as a catastrophe occurring at very small distances, in the remote underworld of subatomic physics, very far from the daily life where classical physics and the related hard sciences pretend to survive the catastrophe of the Third Principle of Thermodynamics.

Let us now ask ourselves whether the seemingly innocent trick of Planck implies some fundamental consequence. For the sake of simplicity, let us concentrate on the simplest system: one particle ($N = 1$) moving just in one dimension. The phase space of this system is thus a two-dimensional one, a plane having as coordinate variables the position q and the momentum p. Planck tells us that the physical states of the single particle are represented by small regions, whose sides are δp and δq and whose area $\delta p\, \delta q$ cannot be smaller than the fundamental value h:

$$\delta p\, \delta q \geq h \tag{4}$$

It is easy to realise that inequality (4) is nothing else than the celebrated Heisenberg uncertainty's principle, but in our frame its meaning is quite transparent. Inequality (4) is not a free choice of our intellect or the consequence of the inability of humans to know in a complete way the "real world". Inequality (4) is an ontological property of the real world and states that, contrary to the principles of classical physics and to the philosophy of Democritus, matter and its component atoms cannot be conceived as inert; otherwise entropy would destroy the whole world, should an observer be present or absent. The meaning of inequality (4) is *not* our inability of measuring simultaneously position and momentum of a particle, but that these variables do not exist in the real world except in the sense of average values. What does that mean? It means that in the new physics governed by the law of Max Planck, which is the only way out from the entropy catastrophe, matter should be conceived as intrinsically fluctuating without external perturbations. There are then two types of movement in matter:

a. The movement from outside produced by externally applied forces or collisions with other particles. Externally induced fluctuations of a system contribute to its entropy.
b. The movement from inside, the self-movement produced by the spontaneous "quantum" fluctuations of the components. These fluctuations do not contribute to the entropy.

We can conclude, in full agreement with the philosophy of Epicurus, that, in order to prevent the entropy catastrophe, which emerges as a consequence of the very principles of classical physics, we must recognise the existence of a deep "*horror quietis*", fear of rest, in matter. Consequently a new fundamental physical variable appears in the physics of particles alongside energy and momentum; this variable has been present so far in the physics of waves and is termed the phase. It describes the rhythm of the spontaneous oscillations that each particle cannot avoid performing.

After the above discussion on the emergence of quantum physics from the deep contradiction between classical mechanics and thermodynamics, let us address the other revolution which occurred in the same years: the relativistic one.

Consider an isolated electron. Is it a pointlike object or an extended one? In the latter case the electron might not be stable since it should be torn apart by the repulsive forces mutually exerted among the different parts of this tiny ball. The former assumption of a pointlike object implies that the energy of the associated electrostatic field is infinite. A new catastrophe for the unfortunate classical physicist!

Making a long story short, relativity heals this new deadly disease of classical physics by investing the excess energy into the production of new particles (the mutual transformation of energy and mass is the main feature of relativity). Classically the energy of the electron field increases when the distance from the particle decreases and then diverges when the distance vanishes. In the relativistic framework, at the distance r_0 where the energy matches the mass energy $2mc^2$ of a pair electron-positron, the pair is actually produced. When r decreases further below r_0 and approaches zero, an infinite number of electron-positron pairs are produced. The negatively charged pointlike electron breaks these pairs by attracting the positively charged positrons and repelling afar the negatively charged electrons. So in a sphere of radius

$$r_0 = e^2/2mc^2 \text{ (e is the charge of the electron)} \qquad (5)$$

around the original "bare" electron a "dressing" of positrons is built, giving rise to a "dressed" electron having a charge

$$q = e - \text{sum of the charge of positrons} \qquad (6)$$

Since the last sum is infinite, e too should be infinite in order to get a chance to have a finite q. Luckily the mathematical structure of relativistic quantum electrodynamics is such that the difference of the above infinities is just a finite number, namely, as said in the jargon of theorists, Quantum Electro Dynamics (QED) is a renormalisable theory. Renormalisability is an essential requirement for every acceptable theory.

The classical "elementary" electron becomes relativistically a highly complex object; a pointlike "seed" surrounded by a cloud of seeds with opposite charge. However, there is more than that; there are also an infinite number of pointlike electrons created for ensuring the stability of the original single electron and pushed far away by its field. These pointlike seeds become in turn "dressed", i.e. physical, electrons, so that we realise that a single isolated electron cannot exist without being a component of a vast ensemble of electrons, of an electron species, of an electron field, a field whose quanta are the electrons.

After the collapse of the first pillar of classical physics, the inert matter, we watch the collapse of an additional pillar, the possibility of an isolated independent body. Just as Robinson Crusoe is an abstract entity, the real thing being the whole human species, so single particles become metaphysical entities, the only real thing being the quantum field having particles as its quanta.

The classical dualism between matter and light is eliminated by the holy alliance of the revolutions of the 20th century. Matter is a field just as light and light has quanta that look like particles. The classical dualism gets replaced by the quantum monism where the basic physical objects are quantum fields (Bjorken 1965; Preparata 2002). This point of view is the full embodiment of the scientific revolutions of the last century and has been kept confined by hard "scientists" in realms (elementary particle physics, cosmology) quite remote from daily life and in particular from biology where this point of view could give rise to a tremendous potential innovation.

An important consequence of the assumption that Quantum Field Theory (QFT) is the conceptual frame of Nature is the existence of an additional uncertainty's principle, like the Heisenberg principle described by the formula (4). The number N of quanta of a field and the phase φ of the same field – namely, as stated above, its rhythm of oscillation – obey the inequality:

$$\delta N \, \delta\varphi \geq h \tag{7}$$

where δN and $\delta\varphi$ are the fluctuations of N and φ.

From inequality (7) we derive that it is impossible to prescribe a definite phase φ to a system having a definite number of particles ($\delta N = 0$). Then Quantum Mechanics (QM) where a wave function, and thus a phase, is prescribed to ensembles of a definite number of particles $(1, 2 \ldots, N)$, should be regarded still as a semiclassical theory violating the uncertainty's principle given by formula (7). The existence of paradoxes in QM should then be regarded not as a real phenomenon, but merely as a reflex of the semiclassical approximation underlying QM. Inequality (7) suggests two extreme cases:

a. $\delta N = 0$ so that N is a definite number and φ is totally indefinite. This case is liked by hard sciences, which concentrate on the atomic (or molecular) composition of the system and neglect every wavelike feature

b. $\delta\varphi = 0$ so that the system has a definite rhythm of oscillation, whereas the number of component particles is indefinite. Such states are named in QFT "coherent states" and should play an essential role in living matter (Preparata 2002).

2. Coherence in condensed and living matter

Every quantum object (particles and interaction fields, such as the electromagnetic (e.m.) fields) cannot but fluctuate. Every physical field, in particular, cannot ever be absolutely vanishing, since an absolute vacuum is a non-fluctuating entity; zero as well as infinite are physically not acceptable quantities. Thus, in the empty space the statement: "the e.m. field is vanishing" cannot be understood in the sense that there is an absolutely dark night, but in the milder sense that this night is punctuated by sudden short-lived flashes, by weak lights switched on and off, by the "fluctuations" of the e.m. field. The vacuum is no longer, as in classical physics, the absolute nothing in an empty space, but is the ensemble of all the quantum fluctuations of all physical fields; in other words what exists is already implicit in the vacuum. A classical "god" does not need to create the vacuum; a quantum "god", before creating the existing things, has also the burden of creating the vacuum (Preparata 2002).

The proof that what we have just said is rigorous physics and not cheap metaphysics is the existence of many physical effects. Let us briefly describe two of them: the Lamb shift and the Casimir effect.

A. Lamb shift. An atom of hydrogen is made up by a proton and an electron orbiting around. Proton and electron interact through electrostatic and magnetostatic forces only, provided that e.m. fields are absent. Since theorists are able to solve exactly, without approximations, the two-body problem, the energy of the hydrogen atom kept together by electrostatic and magnetostatic forces has been calculated with great accuracy, providing a predicted value having many digits.

Technology too, has become very accurate and in 1947 Willis Lamb was able to measure this same value with an adequate number of digits. A discrepancy appeared! The measured value was shifted with respect to the calculated value, the Lamb shift, a tiny discrepancy, but in any case a non-negligible one.

This discrepancy is completely accounted for in the frame of QED by the energy of the interaction between the electric current produced by the orbiting electron and the fluctuations of the e.m. field in the vacuum. Consequently vacuum fluctuations must exist!

B. Casimir effect. E.m. fluctuations give rise to a distribution of energy in the empty space. This distribution can be affected by material bodies present in the space. Take for instance a pair of two parallel metal plates separated by a distance d. In the space enclosed by the two conducting plates the only e.m. modes allowed to exist are those whose wavelength is smaller than d and, consequently, whose frequency is larger than c/d, where c is the speed of light. The modes having a frequency smaller than c/d are then excluded, including their fluctuations. As a consequence of the truncated distribution of energy between the plates, there is more energy per unit volume in the open space around the plates than in the space enclosed between them. This mismatch of energy gives rise to an attractive force between the plates, a small force indeed, that has been revealed and measured. The vacuum fluctuations thus must exist!

In 1916 Walter Nernst (Preparata 2002) proposed that a complex body could emerge from its component atoms just when they become able to tune together their fluctuations, producing a common phase of oscillation that could be regarded as the "identity" of that complex object. Conventional forces, electrostatic or chemical, produce a further binding in the basic structure created by the tuning of quantum oscillations.

This pioneering proposal was left unanswered for a long time, until, starting in the fifties, a sequence of theorists, from Robert Dicke (1954) to Klaus Hepp and Elliott Lieb (1973) to Herbert Fröhlich (1968) and finally to Giuliano Preparata (1995) in the have worked out a new vision of the dynamics of the transition from gaseous matter, that is an ensemble of basically independent particles, to condensed matter (liquids and solids).

Leaving aside the technical derivations, let us concentrate on the essential result that can be stated as follows. An ensemble of a very large number N of particles, which are able to assume two different configurations and are coupled with the e.m. field (as every particle made up of charged components, for instance atoms and molecules, does), enters a coherent state when its density exceeds a threshold and its temperature lies below a critical value. This coherent state has an energy lower than the original gas-like state; in the coherent state the particles oscillate between the two configurations in unison, in tune with an e.m. field grown up from the vacuum fluctuations and trapped within the assembly of coresonating particles. This coherent regime of both matter and e.m. fields holds within a region of space whose size is the wavelength of the e.m. oscillations; this region is called the "coherence domain" (CD) and its size ranges from a fraction of a micron for liquids and metal electrons to some microns for solids to several tens of microns for ions in solution. When the CD is open to an environment at a temperature T the thermal collisions of the external molecules against the coherent fabric of the CD put some of the coherent molecules out of tune with the common coherent field, so that a non-coherent fraction is produced. The total chunk of matter is then the sum of two mesoscopic components, a coherent fraction made up by a large number of CD's and a gas-like non-coherent fraction made up by particles filling the interstices among CD's. CD's are able to establish bounds among themselves through the tails (evanescent fields) of the quite intense e.m. fields existing inside the domains. So there is a large cage of the CD's within which there is a definite phase and zero entropy, trapping a "gas" of non-coherent particles, whose pressure increases with temperature; the non-coherent "gas" accounts for the total entropy of the global physical system. When temperature is large enough, the pressure of the non-coherent fraction breaks the cage of the CD's and boiling occurs, so that the full ensemble of particles reverts to the gaseous state. In the above picture, the electromagnetically induced coherence compels the particles to contract into a stable piece of matter, where as temperature extracts a fraction of molecules from the dynamics of contraction and compels them to try to expand. It is the similarity of the above conclusion of rigorous quantum field theory to the Chinese vision of matter as the union of the two opposite and inseparable agents of the yin and the yang.

Let us add a few words on the dynamics of the growth of a coherent e.m. field within a CD from the vacuum fluctuations. In the original gas of particles whose two internal configurations have a difference of energy E, a vacuum e.m. fluctuation of frequency $\nu = E/h$ (according to the Einstein formula) pushes the particles from the first to the second configuration; since the fluctuation is short-lived, the same happens for the particle excitation. However, during the particle excitation, the charged components get in motion, giving rise, as in the case of the Lamb shift, to an electric current, which interacts with the e.m. field of the fluctuation and produces a (negative) attraction energy which decreases the energy brought about by the fluctuation. If we imagine the fluctuation as a loan of energy from the bank of the vacuum to the physical system, that must be returned in the time of duration of the fluctuation, then the above "Lamb-shift" effect reduces the amount of the debt. The very important feature of QED is that, while the energy required to excite the particles increases as N, the (negative) energy of the "Lamb-shift" attraction increases as $N\sqrt{N}$, so that the fraction of the debt cancelled by this effect increases as $\sqrt{N}$. There is consequently always a critical value of N (and then of the density of particles) where the energy debt of the physical system toward the vacuum is completely cancelled out and the oscillatory regime involving particles and e.m. field demands no expense of energy, so that it is allowed to keep on indefinitely. More, when *N* exceeds the critical threshold, the total energy of the fluctuation becomes negative, so that the physical system gets an energetic advantage when it assumes the new oscillatory configuration; so it is encouraged to increase *N* as much as it can in order to maximise the energy gain. This is precisely what happens in Nature when a gas becomes a liquid! There is a huge increase of density (in the case of water by a factor 1600) that occurs in a very short time when the density of the gas reaches the critical value.

Many condensed systems (liquid water, electrolytes, crystals) have been so far analyzed in this new framework and a good agreement between theoretical prediction and experimental finding has been found. Unfortunately this new conception is still confined within a small circle of physicists, since the scientific community at large, whose non technical understanding of QFT is poor, whose intuition is still largely modelled by classical physics and its tenet of matter as a collection of inert tiny balls, remains to the conventional pre-QED concepts.

In the QED frame matter appears then as an array of correlated mesoscopic objects characterised by a phase and behaving as natural microlasers, which enclose coherent e.m. fields. It is a well-established result of electrodynamics that, where an e.m. field falls off deeply enough (namely where the gradient of the field is large enough), as occurs on the boundaries of CD's, large selective long-range forces develop which depend on the oscillation frequencies of the interacting particles and on the frequency of the e.m. field. The range of these forces is the full distance that can be travelled, by remaining on the surfaces of the correlate CD's, that can reach values as large as several hundreds of Angstroms. The intensity of these forces, confined to the CD's surfaces, is negligible when the oscillation frequencies of interacting particles are very different, but reaches a very large value where their frequencies approach each other. The coherent picture of condensed matter provides a very flexible and selective code of interaction; molecules able to co-resonate among them and with the CD's of the containing medium develop a highly selective long range attraction, whose vehicle is the e.m. field present in the CD's.

The above result could be the key element for the still missing bridge between physics and biology (Del Giudice 1988, 1989, 1995, 2002). In non-living matter the components are quite well defined and the CD's structure is quite stable. In living matter the huge number of molecule species able to interact chemically and dissolved in a medium, water, having a large number of CD's surfaces, make possible a very rich scenario of events that we try to sketch.

On the surfaces of CD's "protomembranes" can be formed (in the sense of Oparin) made up of the molecules coresonating with the CD's of water. When such protomembranes reach their critical density they become CD's of their own constituent molecules and are able to attract new molecule species that in turn. . .and so on. Moreover these molecules are able to give rise to chemical reactions, that by changing the nature of the molecules change also the e.m. structure of their CD's. The energy output of the chemical reactions is released to a coherent ensemble of molecules, so that it is not dissipated as heat and then does not induce changes of temperature. Like the energy supplied to a string of a violin, the energy released to a coherent region produces a coherent excitation of the structure that modulates its e.m. field.

In conclusion living matter appears as the outcome of the interplay of a chemical level and of a self-produced electromagnetic level. Molecules do not move at random, but follow tracks on the surfaces of the coherent regions under the influence of e.m. fields according to a code of mutual recognition and recall based on matching frequencies. Vice versa chemical reactions, through their energy output, change the patterns of the e.m. structure and consequently the pattern of the e.m. assisted further chemical reactions. We face the possibility of sketching a time dependent non-random biochemistry free from the conceptual difficulties outlined in the Introduction. The fundamental physical variable in this picture is the phase, which on one hand organises selective encounters of molecules and on the other hand is varied by the energetic outcome of their interactions.

Let us conclude this Section by quoting a recently discovered phenomenon (the Zhadin effect) (Zhadin 1998) that allows to organise the long range traffic of ions and the crossing of cell membranes. When a current of ions is subjected to two parallel magnetic fields, one static, the other alternating, it exhibits a strong increase of the current when the frequency of the alternating field matches a characteristic value, typical of the ions species. This phenomenon displays its maximum effectiveness within a coherent structure and makes it possible in such structures the existence of magnetically driven electric currents (Del Giudice 2002).

3. The interaction through the phase: Emotions and perception of the emotions; the dual system

The coherent picture of living matter cancels out the strictly local, atomistic, random features of conventional molecular biology that leave no room for the appearance of those excitations involving the full body, that are the emotions. The coherent picture opens the way to the existence of a network of long range flows of particles, ordered in space and time, driven by the phase as it is determined within the organism by the interplay of the two levels of organisation of living matter, chemical and electromagnetic. Consider a pair of two neighbouring CD's and suppose that there is a difference of phase between the two; then in order to re-establish equilibrium a flow of ions is induced along the gradient of the phase in order to suppress it; if some agent keeps the gradient non vanishing a permanent current is developed (Josephson effect) (Josephson 1962). In living matter the phase of CD's can be affected by energy supplies (chemical, hydrodynamic, electromagnetic); any non-utilised amount of energy is spent to increase the phase of CD's. When the energy production is confined at definite sites, the phase becomes a decreasing function of the distance from such sites and then a phase-driven long-range current develops. However, if along the current some CD's get energy non otherwise dissipated from the surroundings, their phase gets enhanced, the gradient of the phase cancelled and the ion current stopped.

In the first, healthy, case all the cells downstream form the flow of ions get supplied and biochemistry goes on. In the second, pathological, case, all the cells downstream from the region where the jam of energy has stopped the flow of ions are no longer supplied and biochemistry is in trouble. This very rough picture resembles in a way to the picture worked out in the thirties by Wilhelm Reich (1948) and makes it possible to identify the emotions flowing in an organism with the organised system of long-range flows of ions driven by the phase.

Quantum systems having a definite phase can establish among them long-range interactions which do not imply exchange of energy and momentum and thus are not bound by the limit of the speed of light, which applies only to the interactions mediated by exchanges of particles. This is the content of the Böhm-Aharonov effect (Bohm 1959), formulated on theoretical grounds in the fifties and confirmed experimentally in the eighties. The interaction through the phase does not carry energy and momentum, does not exert forces, but limits itself to tune together the phases of the partners provided that, of course, they have a definite phase. So coherent systems only (and living organisms are coherent systems) can interact through the phase, at a speed that can well exceed the speed of light. Here is the possible rationale for the so-called subtle influences so despised by the no-nonsense hard scientists, or for the cosmic superimposition suggested by Wilhelm Reich (1948).

Physiological evidence suggests that the phase driven long-range currents of ions should flow along tracks contained in the connective tissue of the organism. It is tempting to identify the above tracks with the Chinese meridians.

Suppose now that near the system of "Chinese meridians" a brother network exist, i.e. the nervous system. The switching on and off of the currents in the first network produces spikes of voltages in the second network, starting a system of currents that are in this case electrically driven. We are in the presence of a pair of dual systems, where one "observes" the other. The phase-driven ion currents, produced by the ordered pattern of chemical reactions (the Freudian "Es") carry all along the body the feeling of oneness through the long-range correlations they maintain (the emotional system); the electrically driven electric currents in the nervous system are induced by the first system, they perceive it, and in a limited extent can provide a feedback to the first system by affecting it with its fields. As a matter of fact, the existence of two dialoguing systems is the precondition for the development of a language.[1]

At this stage of intellectual work, may we hope that Freud could remove the ban on the dialogue between psychoanalysis and a "hard science" at last softened?

Conclusions

The picture presented above is based on rigorous results, but also, mainly in the above Section, includes dreams. However, progress in science cannot occur in an atmosphere of absolute and unmitigated rigor. One needs to dream; not, of course, wild dreams, unrelated to Nature and to the previous knowledge, but possible dreams that do not clash trivially with what is well known. One needs to have dreams that oscillate in phase not only with ideas, but also with the emotions of the human species, not only the colleagues, but all the people which are all interacting through the phase among them and with the cosmos.

From these dreams driven by a hot emotional content, cool and pure rigorous thinking will flow out eventually.

Acknowledgements

I am grateful to Nicola Del Giudice, Antonella De Ninno, Martin Fleischmann, Giuliano Mengoli, Claudio Poggi, Getullio Talpo, Margherita Tosi, Giuseppe Vitiello for the ideas and opinions exchanged in many discussions.

Note

1. I owe this observation to Getullio Talpo.

References

Bjorken, James & Sidney Drell (1965). *Relativistic Quantum Field Theory*. New York: McGraw-Hill.

Böhm, David & Yakir Aharonov (1959). Significance of Electromagnetic Potentials in the Quantum Theor. *Physical Review, 115*, 485–491.

Damasio, Antonio (1994). *Descartes' Error: Emotion, Reason and the Human Brain*. New York: Putnam.

Del Giudice, Emilio, Silvia Doglia, Marziale Milani, & Giuseppe Vitiello (1988). Structures, Correlations and Electromagnetic Interactions in Living Matter: Theory and Applications. In Fröhlich Herbert (Ed.), *Biological Coherence and Response to External Stimuli* (pp. 49–64). Heidelberg: Springer.

Del Giudice, Emilio, Silvia Doglia, Marziale Milani, Giuseppe Vitiello, & Cyril Smith (1989). Magnetic flux quantisation and Josephson behaviour in living systems. *Physica Scripta, 40*, 786–791.

Del Giudice, Emilio & Giuliano Preparata (1995). Coherent dynamics in water as a possible explanation of biological membranes. *Journal of Biological Physics, 20*, 105–116.

Del Giudice, Emilio, Martin Fleischmann, Giuliano Preparata, & Getullio Talpo (2002). On the "unreasonable" effects of weak ELF magnetic fields upon a system of ions. *Bioelectromagnetics, 23*, 522–530.

Dicke, Robert (1954). Coherence in Spontaneous Radiation Processes. *Physical Review, 93*, 99–110.

Fröhlich, Herbert (1968). Long Range Coherence and Energy Storage in Biological Systems. *International Journal of Quantum Chemistry, 11*, 641–645.

Groddeck, Georg (1923). *Das Buch vom Es*. Wien: Internationaler Psychoanalytischer Verlag; reprinted (1961), Wiesbaden: Limes Verlag.

Hepp, Klaus & Elliott Lieb (1973). On the Superradiant Phase Transition for Molecules in a Quantized Radiation Field: the Dicke Maser Field. *Annals of Physics, 76*, 360–404.

Hepp, Klaus & Elliott Lieb (1973). Equilibrium Statistical Mechanics of Matter Interacting with the Quantized Radiation Field. *Physical Review, A8*, 2517–2525.

Josephson, Brian (1962). Possible New Effects in Superconductive Tunnelling. *Physics Letters, 1*, 251–253.

Nernst, Walter (1969). *The New Heat Theorem*. New York: Dover.

Preparata, Giuliano (1995). *QED Coherence in Matter*. Singapore, London, New York: World Scientific.

Preparata, Giuliano (2002). *An introduction to a realistic quantum physics*. Singapore, London, New York: World Scientific.

Reich, Wilhelm (1948). *The discovery of the Orgone*, Vol. 1 *The function of the orgasm, sex-economics problems of biological energy*. New York: Orgone Institute Press.

Spinoza, Baruch (1955). *Works of Spinoza*. New York: Dover.
Vico, Giovanbattista (1999). *A New Science*. New York: Penguin Books.
Zhadin, Mikhail, Valentin Novikov, Frank Barnes, & Nicholas La Pergola (1998). Combined action of static and alternating magnetic field on ionic current in aqueous glutamic acid solution. *Bioelectromagnetics, 19*, 41–45.

Chapter 6

Dual mode ontology and its application to the Riemann Hypothesis*

Gordon Globus
University of California Irvine, USA

Dual mode quantum brain dynamics (QBD) is reviewed and examined ontologically, with special attention to consciousness, subjectivity and existence. The provenance of dual mode QBD is both ontological and epistemological: presence and trace. Quotidian and monadological ontological interpretations of dual mode QBD are compared. The monadological approach is applied to the Riemann Hypothesis (RH) regarding prime numbers. The prime numbers, Riemann's nontrivial zeros of the zeta function and Being itself are closely connected.

Foreword

The idea that we are fundamentally quantum creatures in our functioning is one of the most intriguing ideas of our time. Not that the brain can be described more finely at the quantum level – a rock or a computer can be so described – but that quantum degrees of freedom are actually integral to brain functioning. For this idea, the immensely successful contemporary brain science has been "barking up the wrong tree." Meanwhile the crucial quantum operations of a "cryptic brain," as Jibu and Yasue (1995) say, have gone undiscovered, at least up until the first glimmerings in the late sixties. (See Fröhlich 1968; Ricciardi & Umezawa 1967; Stuart, Takahashi, & Umezawa 1978, 1979.) We perhaps feel Kuhned out on "scientific revolution" but a quantum brain would be a paradigm shift if there ever was one!

If brain functioning has quantum degrees of freedom, then it ought to be able to do very different things than a brain conceived as wet computer might do. Analogous to the way that 20th century quantum physics was a revolution for common sense, quantum brain theory threatens revolution in the 21st century against the common sense that the brain is a wet computer. (I am not talking about a fancier form of computation with qubits, currently under great development, but something amachinative and alogical. Cf. Penrose 1994.) Dual mode quantum brain dynamics (Vitiello 2001) opens up a way of understanding brain functioning which frees us from the computational *Zeitgeist* of contemporary technoscience, overthrowing Turing hegemony. If we are fundamentally quantum creatures, then our very existence might at last be consistent with, rather than divided from, the world of science.

Introduction

Umezawa's (1993) proposal of an alter quantum universe has deep ontological implications. (See Umezawa 1995; Jibu & Takahashi this volume; and Jibu & Yasue this volume for the historical development.) The quantum universe previously described by physics, in the first quantization of quantum mechanics (QM) and the second quantization of quantum field theory (QFT), is quite sufficient for dynamics. Umezawa enhances QFT with the addition of thermodynamical degrees of freedom, made possible by admitting an alter universe – a "∼universe" that serves as heat bath for our non∼ universe. As Vitiello emphasizes (this volume), the two modes should not be considered independent since they are entangled in the common least-energy vacuum state. Thermodynamics and the open, dissipative, far-from-equilibrium living system are beyond traditional QFT. Umezawa's ∼universe gifts QFT with thermodynamical degrees of freedom, however, and so his quantum thermofield dynamics is applicable to the living brain. Whereas Umezawa posited the alter universe, it appears naturally in the dissipative quantum brain dynamics developed by Vitiello (1995, 2001).

Umezawa (Chapter 6) also achieved quantum description of macroscopic quantum objects (as envelope structures of coherent quanta), which is of great ontological importance. The macroscopic is classically present, unlike the unperceivable microscopic ("microscopic" in the quantum physics, not the biology, sense). So the scale microscopic-macroscopic tacitly contains an ontological disjuncture, between unpresent microscopic and present macroscopic. Umezawa's theory opens a way to heal this break by making the microscopic-macroscopic on the same (unpresent) dimension – ontologically unbroken – which leaves the problematic of presencing open to new interpretation.

Jibu and Yasue (1995, this volume) developed Umezawa and coworkers' earlier theoretical formulation of memory traces into a full-fledged quantum brain dynamics (QBD) associated with consciousness. Vitiello (1995, 2001) extended Umezawa's thermofield dynamics to a dual mode QBD, whose ontology is discussed here. I have shown elsewhere (Globus 2003) that a form of dual mode QBD can be "thought together" with the postphenomenological philosophies of Heidegger and Derrida.

Section 1 gives an overview of quantum brain dynamics. Section 2 considers consciousness and subjectivity within the framework of QBD. Section 3 compares quotidian and monadological interpretations of dual mode ontology. Section 4 makes the surprising suggestion that this formulation can be applied to the famously unproved Riemann Hypothesis.

1. Dual mode Quantum Brain Dynamics (QBD)

Alter universe as default

The alter universe is strange indeed. Umezawa denotes it with the symbol of negation, '∼'. This '∼' is not the negation we understand as no-thing, which already assumes objectuality in the reference to "thing." Of course, *our* quantum universe also lacks objectuality but it readily collapses to the objectual on measurement, whereas the alter universe is unmeasurable, indeed, unknowable. This '∼' does not negate some-thing but annihilates objectuality altogether. The alter universe *defaults* our quantum universe, at addresses in relation to which "negation" and "thing" can never in principle be applied, even as a potential. Derrida (1981) calls such a default a "re-mark." A re-mark marks only an "address," where the directly and indirectly knowable both default to an altereity ever beyond our ken.

This unknowable is more radically unknowable than Plotnitsky's (2002, this volume), for whom the unknowable exists, though not in a form that is conceivable. Nonetheless, the unknowable can be conceived via its "efficacity," its effects upon the knowable. This is an indirect knowledge of the unknowable through its effects, in the time-honored fashion of physics. "Unknowable" in the present sense defaults our quantum universe to an alter universe forbidden to us in principle, only re-markable as the address of default. What exists is the re-mark, the address of singularity. The default as such does not even exist, only its address does.

Between-two

The only interchange between these dual universes of Umezawa is in the least energy vacuum state common to both of them, which functions as an ontological *between-two*, the vacuum state between of dual universes. Dual universes *participate* in the between. Umezawa develops "~conjugation rules" (7.2.2) that govern relations between dual universes in the vacuum state they share. One of the dual universes is our conventional quantum universe whereas the other universe is radically alter. Umezawa gifts ontology a wondrous between-two in the vacuum state. The most splendid of between-two states is when the participants belong-together, a peace in which the *lumen naturale* is lit and we find ourselves, ungrounded, thrown-ready-to-act, amidst a presencing world of affordances.

Creation and annihilation dynamics

Let's follow the creation and annihilation dynamics of a shower of quanta losing their energy and settling into the non~ mode of the vacuum state (remembering that creation and annihilation dynamics are restricted only by the law of energy conservation). "The" vacuum state is misleading here ... there are infinitely many interpenetrated (superposed) "theta-vacua," which turn out capable of recording memory traces.

We consider the case where there are already dual mode memory traces of past input invariances, called "symmetrons" since they preserve symmetry broken by past inputs. Dual mode symmetrons can be traced in the various theta-vacua. Let the input shower of quanta be a "reference signal" which repeats (or has been contingently associated with) a former signal that has been dual mode symmetron-encoded. Vitiello (2001) writes of this case,

> Then the excitation of a [non~] quantum A *from* the vacuum (its annihilation *in* the vacuum) corresponds to the creation of its "hole" *in* the vacuum, namely to the creation of the corresponding ~A mode, *which may indeed occur under the external replication signal.* (113, brackets and ending italics added)

Under energy conservation law, when the non~ mode gains some quanta, the ~mode must lose an equal number of quanta, and vice versa. The creation of non~ quanta corresponds to ~mode annihilation of quanta and an equal number of ~mode defaults. Alternatively, the showering of quanta into the non~ mode may, in the case of recognition, excite A-mode quanta already there to higher energy levels, effectively annihilating them from the vacuum.

This must be accompanied by creation of an equal number of quanta in the ~mode of the vacuum.

Following out the story, when the excited quanta dissipate their energy and sink back again into the non~ mode of the vacuum state, then an equal number of quanta would be annihilated ~mode. Everything would be as it was before, before the recognition of the reference signal. No trace would remain of the re-cognition. The two modes are symmetrical.

But the quanta excited out of the non~ theta-vacuum mode don't settle into the same vacuum state as before. There is a new input shower of quanta that the just excited quanta settle with, falling into a new non~ mode of the theta-vacuum. This means that an equal number of quanta must be annihilated from the ~mode of that theta-vacuum. "Locally," in a particular theta-vacuum there may be more quanta in one mode than the other, but globally, across all theta-vacua, the modes have equal numbers of quanta. (The Hamiltonian energy difference, H-hat, is zero when the dual modes match, and must always be zero globally.)

Vitiello adopts the meta-physical stance that characterizes physical science. Here subjectivity lies outside of the dual universes with their creation and annihilation dynamics, so one can reflect upon both. Standing extrinsically one can comprehend dual universes and that the two universes are mirror-images. But metaphysics should not be allowed in this dual mode case. One quantum universe is *ours* – nontilde – and the alter ~universe is unknowable. To stand meta- to both universes with "God's eye" violates the nomological framework. We can never transcend our non~ quantum universe, only note its defaults that re-mark an unknowable alter universe.

Dual mode and unimode memory traces

One form of memory trace in the dual mode model – the symmetron trace – is itself dual mode. Input coming into the brain breaks vacuum state symmetry – imposes difference on the undifferentiated – but under fundamental energy conservation law, the broken symmetry is conserved, in the form of massless dual mode Nambu-Goldstone bosons. The dual mode N-G trace preserves the broken symmetry specific to the input invariants, indeed, preserves over time a "total memory" (Jibu & Yasue 1995) of all symmetry-breaking input. (Since there are an infinite variety of theta-vacuum states, the vacuum capacity to form traces of unique orders is infinite. However, the traces deteriorate over time due to quantum tunneling (Jibu & Yasue this volume). Vacuum states upheld by the living brain carry dual mode traces of past invariants in the input flux. Symmetry-breaking invariance within the input flux from the surrounding reality is encoded in a vacuum state symmetron condensate of dual mode N-G bosons, where the condensate density is the memory code.

There is a second and more functional form of memory trace than the dual mode N-G quanta that indiscriminatingly trace *everything*. This second form of memory is a *unimode* recognition trace, made when a new input to our non~ quantum brain universe is coherent with the dual mode N-G trace. (This includes contingent recognitions, where traces of "accidental" co-occurences become coherent in their contiguity, without the glue of invariances in common.) Traces of recognition are ~unimode. Our quantum universe is restricted to re-marking these recognition traces.

Unimode tilde recognition traces provide an attunement for the vacuum state encounter with the non~ input flux. The more a re-cognition is repeated, the more heavily weighted is its trace in the attunement. The non~ default re-marks the ~trace of re-cognition. So dual mode quantum brain dynamics can also inscribe *unimode* recognition traces in the ~mode, which are only re-marked by the non~ mode as addressible self-default.

In Vitiello's formulation of dual mode QBD, there are no unimode recognition traces, only dual mode total memory traces. The density of the boson condensates is the same in each mode. (In terms of the Hamiltonian, the energy difference between the dual modes, H-hat, is zero.) Density difference is allowed between the dual modes in particular theta-vacua, however, so long as across all theta-vacua, H-hat = 0.

Time

Vitiello emphasizes that the succession of vacuum states, where symmetry is broken by input invariances and traced dual mode, cannot be reversed to the original state of symmetry. This gives time its arrow. The dual quantum modes are mirror-images, the alter mode ~conjugate to our mode, symmetrical under ~conjugation. The theta-vacuum state encodes a "total memory" of all past inputs + present input, a total memory that is continually updated, while retaining the previous total memory in a different theta-vacuum. *Everything persists equally*, unweighted possibilities for the match and so equally likely to make a match. The possibilities are symmetrical, all equally possible for a match with input, no favored possibilities, all carrying the same weight.

In recognition the weights on the possibilities become a degree of freedom. The symmetry of equal possibilities is broken by the weights conferred in recognition. The more recognitions of some input the more probable that recognition. Whereas total memory traces bring difference, recognition traces have a different function. Without recognition, time would have its arrow but nothing would persist in time. Recognition breaks the symmetry of possibilities and differentiates out identities over time. *Total memory traces difference and recognition memory traces identity.* When input breaks the symmetry of the vacuum state in the formation of total memory, which gives time its arrow, the dual modes are symmetrical to each other. In the formation of recognition traces, however, the dual mode symmetry is broken, distinguishing identity, the self-same, out of mere difference.

Assignment of the subject

The ~ and non~ quantum modes of dual mode QBD are no tweedledum and tweedledee. One of the modes works beautifully for all dynamics, and we are well accustomed to it for a century now. It is this mode that collapses to classical objects. The other mode is utterly *alter*. Umezawa, out of thermodynamical considerations, introduces an alter mode of universe. Intuitively, since the non~ mode of universe is all of quantum physics up to Umezawa's innovation, we associate the non~ mode with the physical object and assign subjectivity to the strange new alter universe. But this is not Vitiello's take, for reasons habitual to physics.

Now quantum theory is specialized to closed energy systems. A thermodynamical system + its heat bath environment approximate a closed system, enough that quantum theory applies under suitable idealizations. Umezawa's ~universe is conceived as heat bath environment for our non~ universe. The heat bath is an *alter* environment. *Sans* heat bath there is only dynamics; for a quantum field theoretic thermodynamics of dissipative systems, a heat bath is required (Vitiello 1995). The two kinds of heat bath "environment" should not be confounded. One is classical, the surrounding "environment" for a living system, and the other is quantum, the heat bath "environment" for our quantum mode, provided by an alter quantum mode.

If we try assigning subjectivity to the ~mode – put the subject in the *alter* heat bath, indeed – much falls into place. We see now why we can never capture subjectivity, not because it is a different substance or a parallel process or a certain aspect or an emergent property, but because unreachable subjectivity is of an alter universe that defaults our own. We are unable to actually grasp hold of subjectivity because "it" is alter mode. We can bring subjectivity into the discussion obliquely, however, in pointing to the addresses of default. We can "re-mark" subjectivity but never objectify it.

Control of the ~conjugate match and the result

The living brain as open, dissipative, far-from-equilibirum system embedded in the body, specializes in control of the vacuum state. The living brain here has great advantage over silicon, in controlling its between rather than being its creature, like a computer is. The dissipative brain system hoists oscillating, spinning dipoles, whose least energy state is the vacuum state, and controls them. One source of control is the input flow from external world and the body, whose symmetries (invariances) constrain the match in the between. Another source of control is unimode recognition traces. The weights on the re-cognition traces control the possibility of actualization. The more an input invariance has been re-cognized, the greater its weight and the more likely it will make a match and be actualized.

What is controlled, then, is the specific ~conjugate match achieved out of the many possibilities. Weighted possibilities are offered to the match with the input flux. Input offers its current specific possibilities to the vacuum and weighted unimode memory traces offer past recognitions to the vacuum – and so the ~conjugate match is controlled by both present and past. The weighted possibilities to the match with input from the surrounding reality are a ~mode attunement based on traces of re-cognitions.

The mathematical fruit of the ~conjugate match is real numbers, which are associated with presencing observables. World appears in the ~conjugate match between reality and attunement … world is of the between. The *lumen naturale* lights up in the match disclosing world. Presence itself – *Being* – is the between of a ~conjugate match. All Being arises in the belonging-together of attunement and reality.

Review

I have been exploring some ontological implications of dual mode quantum brain dynamics. Our quantum ~attunement of superposed weighted recognition traces meets the flux of input from reality and in the ~conjugate match of the between: world presences, while unmatched possibilities are wasted, withdrawn to the unknowable alter mode. (This loss is the hallmark of a "general economy," as developed by Bataille 1988–1990.) In this formulation the presencing of world – Being – is not a (forgotten) postulate, but is continually generated in the *between-two* by ~conjugate match between reality and attunement. Being is derivative, hoisted in the between of more originary duals, one of which withdraws from us unknowably. In addition to Being, dual mode QBD explains trace. Dual mode and unimode memory traces have different functions: the dual mode provides a total memory of all input and the unimode a memory of re-cognition that breaks the mirror-image symmetry across modes. *The provenance of dual mode QBD is both ontological and epistemological: presence and trace.*

2. Consciousness, subjectivity and the alter universe

Quantum physics and consciousness

The relation of consciousness to quantum physics bears careful scrutiny. By and large consciousness is just assumed as an attribute of the observer, which allows the observer to perceive the meter-readings on which physics is empirically based. The attitude of physics is let somebody else worry about consciousness, while we jot down where the instrument arrow clearly points. This putting aside of consciousness has in fact not impeded the spectacular advance of physics. But once quantum brain theory enters into play, suddenly physics as neurophysics is in the vicinity of consciousness which can no longer be ignored. Suddenly physics by way of neurophysics has a stake in the consciousness it has blithely accepted in quotidian convention.

The dependence of quantum physics on the assumption of consciousness is shown by the Born postulate in QM. Born proposed that the complex numbered coefficients of the Schrödinger wave equation be brutely multiplied by their complex conjugates and the resulting real numbers then be *interpreted* as probabilities. The postulation of multiplication by the complex conjugate and the interpretation of numbers as probabilities works out famously, and so seems entirely justified. The world is surely there after wave function collapse on measurement – the Schrödinger equation just doesn't give us the particulars – and so, it is usually thought, *the probabilities are not ontological but merely epistemic*. Quantum theory is only capable of probabilities of actual presences. For the actual happening, the conscious observer is required to rescue quantum physics from its dull statistical confinement, an observer whose consciousness has been taken comfortably for granted … until quantum theory nudged into brain theory and reopened the question of consciousness.

There have been noteworthy defections from this historical forgetting of consciousness. Early on von Neumann (1955) thought consciousness collapses the wave function. (See also Wigner 1962 and Stapp 1993.) Thus the fate of Schrödinger's famous cat, made to be in a quantum superposition of dead and alive, is decided when the observer makes a conscious observation and collapses the superposition to one or the other alternative. (Otherwise the von Neumann chain of quantum states in the brain under unitary evolution of the Schrödinger wave function would merrily continue as a succession of quantum superpositions which do not presence.) So, von Neumann thought, it must be consciousness that collapses the wave function, brings the chain to an end by abrupt collapse. The measurement that collapses the brain's wave function is by operation of consciousness. This view faded in quantum theory, since the wave function does not need consciousness to collapse. (Except under special conditions the wave function will collapse on its own in picoseconds, by noise interference from outside the quantum system (Zurek 1991).)

Unpacking consciousness

It is time to start unpacking "consciousness." Vitiello (2001) is content with the conventional attitude of the physicist which does not problematize the subjective experience of the world. Subject and object are distinguished in classical ways. Thus Vitiello asks, "How do external inputs and stimuli become our subjective experiences (such as qualia)?" (126).

To take a conventional stance re consciousness is to inherit the mess of problems surrounding it. There remains conscious "subjective experiences" – now formulated in quantum terms – and external world. The classical split between subject/object, immanence/transcendence, remains and so the philosophical load does not shift in this formulation. I instead equate the ~conjugate match to existential world-thrownness. (The "thrown" quality of existence is the ungroundedness of its brute facticity.)

Let's go into the alternative stance in which "consciousness" is not of the between, but "world thrownness" is. For this understanding there is no consciousness *of* an external world, with the perplexing immanent/transcendent divide spanned by Husserlian intentionality (or in Heidegger, a spontaneous *ekstasis*). We simply find ourselves *in the world*, always already amidst pragmata, without prior ground and so "thrown." Our existence is to be such. *The ~conjugate match in the between dual quantum modes is existence as situated world thrownness.* Situatedness is a ~attunement for the match with non~ input, the match in which we find ourselves thrown amidst the world, even including being thrown amidst laboratory apparatus, meters and all, reading them. In this case our situatedness is an attunement to perform experiments, an attunement which fits easily to the laboratory setting.

There is no problem of transcending the immanent to some transcendent external world in this formulation because the distinction is replaced by immanence and the illusion of transcendence (*maya*). It is all "immanent," in a deeper sense, *monadological*: no transcendence is required. (For a well-developed fundamental physics that is monadological, see Nakagomi, this volume and the references therein.) There are weighted interpenetrated possibilities of the ~mode – traces of weighted invariants that comprise a quantum attunement – and there is what belongs to those possibilities in the input flux from physical reality. Input in the non~ mode and ~possibility (recognition traces) belong-together and in that matching state of the between-modes world lights up, "phenomenal world," with us ungrounded amidst it.

Parallel world-thrownness

So we conscious humans are lighted bubbles of world thrownness in disputatious parallel – all the rest is dual mode: our quantum universe and an alter universe that is a ~mode abground. We are bubbles of experience, local quantum monads with dual modes, where the *lumen naturale* switches on in the mode's between when the non~ input flux and the ~attunement, with its shifting weights, belong-together, make a ~conjugate match. In this match we find ourselves thrown, bathed by the *lumen naturale* ... world lights up and shows our journey in the vacuum state succession of belonging-togethers. We are groundless, our subjectivity represented by default in the conventional ground, thrown by the unknowable. The price for this adventure is that our belief in the homey world-in-common becomes radically transformed to the profound isolation of *maya*, and we find ourselves thrown in parallel worlds, lighted bubbles of scattered belonging-togetherness against an unknowable abground.

Loss of the quotidian world-in-common and appreciation of our profound monadic isolation are scary to contemplate. The term *maya* trips lightly from contemporary tongues – where *maya* is thought superficially to be a kind of misinterpretation, an illusory way of thinking to be thrown off so that true reality could show through to whomever "breaks the chains" of *maya*. But *maya*, understood more authentically, hides from an extremal abyss that defaults the knowable, for Heidegger "abground" (*der Abgrund*). For the present discussion abground is alter, a ~mode, that contravenes Heidegger's arrogant dismissal of science, which he would "leave to its mania for its own usefulness" (1999: 108).

~Mode and the subject

Let's equate ~mode to the subject and see how this move turns out. It is immediately clear that a subject who is tilde can never be part of our universe which is under non~ description. Science can never catch hold of the subject, referred to by "I," as a matter of principle, since "I" escapes our universe – the one that experimental science studies – by defaulting it. "I" properly am I~, and so not knowable, only re-markable, having an address that is annihilated.

This ontological dualism of modes is much more economical than the dualisms of Descartes and Leibniz where there are two, interacting and parallel respectively. The new dualism is much effaced, less than two ontologically: a dualism of one and its own defaults, defaults that re-mark the alter. What is a distinctive subjective substance for Descartes, *res cogitans*, becomes here a mode of a thermofield dynamics that applies to open dissipative systems of the living kind, the most advanced form of which is a well-functioning human brain.

"I~"

"I" properly refers to the ~mode attunement. "I am" properly means "I am attuned." "I" am I~. What "I" am is a fluctuating attunement, interpenetrated possibilities with variable weights. I~ can never find itself present in the world because I~ defaults our quantum universe. However, a tight correlate of "I" can be found in the world, viz., the brain which "I" supervenes on. When the brain zones out under anesthesia, "I" go to. In Descartes there was a duality of interacting substances, mental and physical, but now the duality is between participants in a match, unpresent quantum modes that might belong-together. "I" is just such that we know its address but all else lies beyond any objectual conception. "I" is "infinitely near," as Sartre (1957) said, "so we can never encircle it." "I" *am* right here in the dual mode formulation, right here at the address which locates default, but "I" am unknowable, defaulting. Here "I" finally denotes something that science can properly consider. "I" is ontological, in the form of ontological default. "I" am addressible holes in our quantum universe. I am I~.

3. Quotidian and monadological ontologies

Suppose we accept Umezawa's quantum thermofield dynamics and Vitiello's mathematically formulated dual mode quantum brain dynamics. How is this abstract conception to be formulated ontologically? How are the equations to be interpreted with regard to brain and Being? The conventional interpretation is based in quotidian physics, which takes the "Being" in the title of the present book for granted, never problematizing it. (This "natural attitude" is guilty of *Seinsvergessen* – the forgetting of Being (presence) – Heidegger always complained.) The readings on meters can be observed and validated across scientific observers because the meter screen is actually "there," presencing to observers. The good observer is conscious of what is actually there.

The methodology of physics eschews philosophical questioning and so ontological issues are elided. Thus for Vitiello "considering philosophical questions is outside the scope of this book" (125).

> I do not attach any philosophical content to the expression external world". It is only used in a mathematical and physical sense: there are two interacting physical systems, the brain and the external world (whatever it is ...).
> (2001: 135)

He sums up all the "hard problems" into one "very difficult question," "What is the nature of the subjective experience of the world?" (126). But to speak unworriedly of the external world and distinguish it from subjective experience is *already* to take a strong ontological stance, one that is commonsensical. Anyone can see the external world before his or her nose and read its scientific meters. (Let the philosophers be "designated ontological worriers.") The external world is never called into question.

> Of course, I am here assuming the standpoint of the physicist, not questioning the "existence" and "objectivity" of the external world. (141)

Empirical physics is based on conscious observation of meter-readings as an unproblematic part of everyday life, meta to physics, i.e., quotidian metaphysics whose ancient provenance is Socratic. Ontology might be avoided by pure experimental and theoretical physics but not when "subjective experience" enters into the discussion.

Discord is expectable when the extremely uncommonsensical quantum theory is used to develop an extremely revolutionary theory of brain functioning, yet is supposed to match with an extremely conventional ontology. Whereas quotidian ontology seems an unlikely candidate to mesh with dual mode QBD, a monadic conception might be just crazy enough to dance with quantum field theoretic conceptions!

Now in ontology it is crucial to keep the perspective clear. The physicist remains sanguine within the scientific fashion of standing outside the system under study – here outside the two modes. This is a metaphysical stance that allows one to view the dual modes equally and dispassionately. But one of the modes is *ours*; the other mode is *alter*, *tilde*, in principle barred to us. What distinguishes the dual modes is that one of them is ours and the other defaults ours. We can no longer take a meta-stance when it comes to describing our own brain. We must stand with one of the modes, the one that is familiar and successful, indeed, the only quantum mode here-to-fore, which has proved entirely sufficient for the dynamics of quantum systems. This mode can only be the non∼ one, which has been sufficient for all of quantum physics up to dual mode ontology. Since "tilde" here means default, "our" quantum universe must be nontilde, containing defaults that re-mark the unknowable *alter*. (Our mode of quantum universe is riddled with defaults at addresses which re-mark the ∼mode.) So a crucial difference is that Vitiello adopts a perspective meta to the dual modes whereas for me perspective is restricted to the non∼ mode, leaving the ∼mode unknowably alter.

There is a certain tension in Vitiello's (2001, this volume) formulation regarding how such terms as "consciousness," "subjectivity" and "world" are to sort with dual modes and their between. It takes dwelling in terms and a careful reading to resist the enlightened common sense of science and discern the tension. Vitiello continually refers to the first person, subject and self on the one hand and the second person, Double or Sosia on the other hand. There is a "'doubling' of the self" (2001: 141). To sharpen the vocabulary, let's stick with "Self" and "Double." How are Self and Double to be thought together with dual modes and their between?

For Vitiello the informational inputs from the external world are recorded in the non∼ mode which is copied in the ∼mode, the Double. The "existence" of the external world

> is the *prerequisite* for the brain to build up its own 'subjective simulation' of the external world, its *own representation* of the world. (2001: 141)

This representation is non∼ mode and is copied in the ∼mode Double. The ∼mode is "the 'time-reversed copy'" of the non∼ mode (141).

> In some sense, the unavoidable coupling with the external world is "internalized" in the dialectic, permanent relation with the double. (141)

Both external world and ~mode are "environments" for the non~ mode.

So Vitiello sorts Self and Double with non~ and ~ modes respectively. "Consciousness" has "its manifestation as a 'second person' acting as the Double of the subject" (2001: 123). (Note that here consciousness sorts with the Double of the subject, which is inconsistent with the rest of the text and so presumably misspoken.) There is a "dynamical relation between the [non~] system A and its Sosia or Double ~A" (2001: 141, bracket added).

> … the environment is "mathematically represented" as the time-reversed *copy* of the system. However, it is clear that corresponding to different subjects (systems) we will have "different" representations of the environment, each of them being indeed a "copy" of the corresponding subject. (2001: 110)

So the Self, non~ system A, represents the environment, and the Double, ~A, is a copy of A.

> The non-tilde *unveils* its Double and they conjugate in a circular (non-linear) recognition, each being "exposed" to the other's eyes. (this volume, Sec. 5)

Although Vitiello uses such terms as "unveils" – indeed, the book title (2001) is *My double unveiled* – he emphasizes that he doesn't literally mean them. Properly, this is just a vivid manner of speaking. The dual modes are to be understood as physical and mathematical. "… I need to stress that the word 'self' is only used in the limited sense of mathematical nonlinearity" (2001: 140) and doubling has a "limited mathematical meaning" (141).

What happens in Vitiello's between? *Consciousness.* "Consciousness mechanisms," he suggests (Sec. 5), "might be involved in the continual 'trade' (interaction) between the non-tilde and the tilde modes." Consciousness is "grounded" (2001: 141) in the ~/non~ dialectic of the vacuum state. The dual modes are like actors, he explains, forced to act. "The 'one', the subject is the action, the play, their *entre-deux* [between-two]" (Sec. 5). (Note that the subject is no longer a mode.) The subject is the between-two of dual actions. Consciousness "basically resides in such a permanent dialog" (Sec. 5). (Note that the subject and consciousness are not discriminated.) This dialog between the actor-modes is unavoidable, continual, fluctuating and reciprocal. But this is not much of a dialogue, as Vitiello sets it up, with no occasion for reciprocity, because the participants in the dialog are already in full agreement, being identical, since the ~mode copies the non~. What's to say to one's mirror image? The external world is represented in the non~ mode which is copied in the ~mode … which makes consciousness in their ~conjugate match a dull affair.

The monadological ontology has much more flexibility in this regard since the dual modes are no longer given as matching. The alter mode is rich with weighted possibility and our mode is rich with order from physical reality. A match is *achieved* rather than given. "Dialog" has the wrong connotation, two actors dialoguing, who by Hermitean assumption are saying the same thing. There is no dialog for the monadological version, however; instead there is a belonging-together of reality's representatives and certain weighted possibilities.

Now, whether quotitidan or monadological in conception, there is a match in the between, whether the ∼mode is a copy of the non∼ mode or weighted possibilities for matching the non∼ mode. Since a ∼conjugate match is achieved in the between, the state of the match is described by real numbers which are associated with observables. *Then the conscious subject cannot reside in the between's match, since consciousness and subjectivity are not described by real numbers.* The quotidian interpretation of the between turns out untenable.

For the monadological conception, the between is not identified with consciousness or subjectivity but with world thrownness, to which real numbers apply. In the between of the monadic quantum brain our world-thrownness continually lights up; our existence as world-thrown clears in the between. Beyond the monadic bubbles of our living brains with quantum field theoretic degrees of freedom is the unknowable and unthinkable (cf. Plotnitsky this volume). These bubbles are in "post-established harmony" (Nakagomi this volume). The monads sustain the illusion of a world-in-common, *maya*, because they have been similarly tuned by local sociocultural practices, and so when the input is by and large the same across different monads, more or less the same world lights up.

The claim that we are truly monadic beings is surely hard to accept, counter to every moment of our conscious experience of world and its affordances. If we are monadic, then the world is weird, *unheimlich*, sorceric, hoisted by each monad in parallel, hoisted in the ∼conjugate match between quantum attunement and input invariants representative of quantum reality … we are monadic bubbles of disclosure in which we always find ourselves already thrown amidst our world. This position sounds silly to common sense (even more silly than quantum mechanics sounded when first proposed). But as noted above, the quantum revolution extended to brain functioning calls for revolutionary ideas about Being, too, and so the monadological ontology should be taken seriously.

4. Application to the Riemann Hypothesis

Introduction

The ontology of dual mode thermofield dynamics has been explored above. What we end up with finally is the between. This between, in the unique case where the dual modes make a ~conjugate match, is the case of *lumen naturale* which clears a presencing world. Being/presence is not originary but derived from a more primitive, never-presencing, dual mode dynamic of belonging-together. World appears in the particular success of a belonging-together. Different belonging-togethers, different worlds. Being is secondary here, *contra* common sense and empirical science, which takes its measurements unconcernedly.

This way of thinking is certainly unconventional. There is a dual mode upsurge that is primary and a monadic bubble lights up in the dual mode ~conjugate match. So bizarre is this conclusion, so "marginal" (attributed mainly to sorcerers (Castaneda 1973, 1974), who work the margins of socially acceptable reality), and even so terrifying is the monadic conclusion, there is a tendency quickly to dismiss this view as ridiculous.

The productivity of this strange view can be assessed by application to the most outstanding problem in mathematics: proof of the Riemann Hypothesis (RH). Edwards (1974) laments toward the end of his treatise, "Riemann's function,"

> One of the things that makes the Riemann hypothesis so difficult is the fact that there is no plausibility argument, no hint of a reason, however unrigorous, why it should be true. (268)

The present aim is to give some scientifically plausible hints as to why the Riemann hypothesis is true. Of course mathematical proof of RH is a different matter altogether but I shall try to point the way.

Riemann's Hypothesis

It is perplexing to mathematicians that something so basic as the prime numbers appear to be only weakly ordered along the real number line. (An example of "weak" ordering: large primes have a different density distribution along the real number line than small primes.) Riemann's Hypothesis of 1859 is that the prime numbers are indeed most strictly ordered, if you know how to display them, that is, display the primes with Riemann's strong geometrical intuitions, where the order comes out strikingly. Certain zero points that are related to prime numbers are all lined up neatly on a line called the "critical line." Despite longstanding efforts (Derbyshire 2003; du Sautoy 2003), a proof that *all* "nontrivial zeros" of the Riemann function lie on the critical line remains elusive. Surprisingly, there appears to be a deep connection between RH and quantum physics.

To appreciate how dual mode QBD is relevant to RH, a bird's-eye view on RH is helpful. (Ontology suffers up close to the symbols.) The prime numbers are the atoms of arithmetic, divisible only by themselves and one. There is a certain zeta function grounded in powers of primes, due to Euler. Euler's function takes a number as *argument* and converts it to exponents on a series of integer reciprocals. The function gives a *value* for each argument, b, of the zeta function. The equation for zeta of b looks like this:

$$\zeta(b) = 1/1^b + 1/2^b + 1/3^b + \ldots + 1/n^b + \ldots$$

At $b = 1, \zeta(b)$ is just the infinite harmonic series. For a more useful form of $\zeta(b)$ which utilizes the Euler product, all the denominators are prime numbers and addition is replaced by multiplication. That is, the Euler zeta function is a product of prime numbers, each taken to the power specified by the argument.

Riemann's zeta function uses complex numbers as arguments and its values are complex numbers. The argument/value relationship was represented geometrically by Riemann. We imagine an Argand plane whose horizontal axis, X, is the real number line and whose vertical axis in the plane, Y, is imaginary. (See Figure 1.) This is the complex argument plane of the zeta function, a geometric representation of the complex argument put into the Riemann zeta function.

Now there is a topography (not shown) on this complex argument plane, a third axis Z, which is the output of the zeta function, its "value." Every point on the complex plane represents an argument, a complex number which is the exponent on the prime atoms being multiplied. The altitude of Z at each point of the argument plane is the output of the Riemann zeta function; the arguments are points on the complex plane and the values are the topography. There are certain points where the value of this zeta function goes (nontrivially) to "sea level" on the value topography – these are the "nontrivial zeros" of the Riemann zeta function.

The nontrivial zeros all have one feature in common; they all have real part ½. The imaginary component varies. Geometrically, there is a vertical line in the complex plane at real part ½, the "critical line" along which these nontrivial zeros of the Riemann zeta function lie. The Riemann Hypothesis still awaiting proof is simply stated: *All* of the nontrivial zeros lie on the critical line.

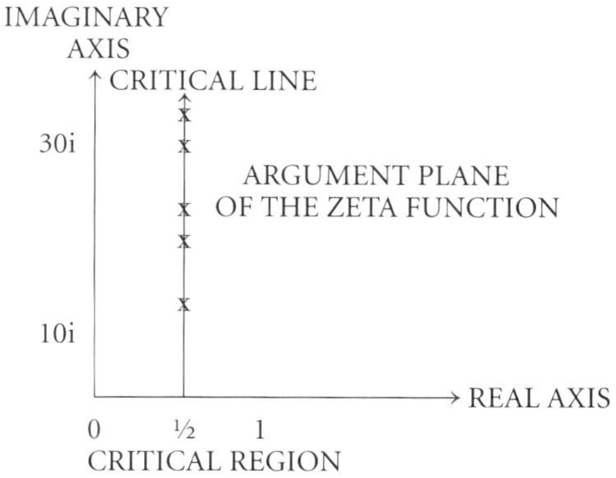

Figure 1. The argument plane of the zeta function is mathematically complex with real and imaginary axes. The first five nontrivial zero points on the zeta value topography are (roughly) located here at the crosses. The rest of the zeta value topography on the argument plane is not shown. These zero points lie on the "critical line" at real ½. The Riemann Hypothesis is that all nontrivial zero points lie on the critical line.

It is of interest to inquire more carefully into the zeta landscape. Strictly speaking, the topography of the zeta landscape does not extend to negative real numbers but stops at zero. Mathematical tricks are used to seamlessly stretch out the negative reals. At an argument of one, the zeta function reduces to the harmonic series and its value diverges to infinity. For arguments greater than 1.0, the zeta function value rapidly decreases and boringly plateaus. So the "critical region" for the zeta function lies between 0 and real 1 and all nontrivial zero points of the Riemann zeta function lie *ex hypothesi* on the "critical line" at real ½.

The points on the argument complex plane can also be conceived as oscillations with amplitude, frequency and phase. Amplitude ("loudness") increases as the location along the real axis increases and frequency ("pitch") increases as the location along the imaginary axis increases. Metaphorically, all nontrivial zeros on the critical line are notes equally loud but differing in pitch. Yet the nontrivial zeros on the critical line are musical, harmonious, belonging-together. Notes on the critical line that are not at sea level are discordant. The "music" of the disordered primes on the real number line becomes the "music" of nontrivial zeros ordered to the critical line.

The argument of the zeta function was for Euler real numbers and for Riemann complex numbers. Now we introduce non~ and ~ mode complex numbers. The argument of the zeta function now has dual modes and the zeta function values also have dual modes. So there are dual pre-landscapes sharing the critical region: a non~ prelandscape and an alter ~prelandscape, too, *pace* Umezawa. Our attention is drawn to points at sea level on the zeta landscape, the nontrivial zeros of the Riemann zeta function, all of which line up along the same imaginary line at real ½, the critical line within the critical region of the zeta topography that lies between 0 and 1 on the real axis.

Despite a reputation for purity, mathematicians use tricks all the time, such as constructing a negative real zeta landscape when the function doesn't actually give one. There is another trick going on, a major simplification that has deep consequences. Recall that the value is a complex number like the argument, but the topography on the argument plane is real. How do you portray two dimensions on one? By the "shadow" of two dimensions, just as our own three-dimensional body throws a two-dimensional shadow on the wall, with the shadow undergoing transformation as we spin. The argument and value planes of the zeta function both are complex planes but the topography is one-dimensional on the argument plane, a one-dimensional "shadow" of two-dimensional values. Information is lost in reduction of dimensions. What is lost in going from the complex value plane to its real line shadow in a mathematically convenient way is phase information. (Attach an arrow at each point of the topography, rotating through 360 degrees. The direction of the arrow represents the phase.)

So the values lie on a complex plane, as do the arguments, but we have only one dimension to represent values, the Z axis of the topography, whose X and Y axes are taken up by the complex-numbered argument. This is easily solved: Just multiply each zeta value by its complex conjugate. This gives a real number, called the "modulus" of the complex number, which occupies the third topographical dimension.

With this sketch of RH, put in a dual mode context, there are two questions that I shall consider. (1) Why are all the nontrivial zeros of the Riemann zeta function lined up on the imaginary line exactly through real ½? (2) What is the distinction between nontrivial zeros that lie on the critical line and the more numerous other points also lying on the critical line but above sea level?

Answer (1). As you increase from zero to real one, the loudness/energy increases. As already noted, in thermofield dynamics the total energy of the dual modes remains constant, so that if one mode gets louder, the other mode gets quieter. The only place the modes are equal in amplitude/energy is at real ½. So the critical line represents imaginary points where the two modes are equally loud, have equal energy, whether or not the points are nontrivial zeros. Most of the points on the critical line are above sea level and some of them are at sea level. What distinguishes the nontrivial zeros from the more numerous non-zeros along the critical line? That is the second question.

Answer (2). The dual modes may have the same energy and the same frequency, but be incoherent, out of phase. The nontrivial zero points are points where the dual modes balance in phase, belong-together, whereas for all nonzero points the dual modes are imbalanced. Prime numbers, too, are "coherent," harmonious. All primes belong-together and all nontrivial zero points, as points of coherence, belong-together too, which is unsurprising, since the nontrivial zero points lined up along the critical line are manufactured by a function of primes.

In discussion of the prehistory of the Euler and Riemann zeta functions, the "sieve of Eratosthenes" (3rd century BC) comes up. It is a mathematical technique for sieving out the prime numbers from the integers. The Riemann zeta function in effect sieves out primes in the form of nontrivial zeros. The nontrivial zeros represent the dual mode ~conjugate belonging-together. All other points, whether off the critical line or nonzero points on it, represent dual mode incoherence.

Nonzero points on the critical line are associated with phase difference between modes. (As noted in Section 1, the dual modes may be locally – read points – imbalanced so long as globally – read across all points between successive nontrivial zeros – there is balance.) Consider the nonzero points on the critical line between two nontrivial zeros on that line. For the first of the nontrivial zeros, phases are balanced; the arrows at that point for the non~ mode points to 12 o'clock and for the ~mode 6 o'clock. As we move to nonzero points moving up the critical line, the non~ mode arrow moves clockwise and the ~mode arrow moves counterclockwise, so that the phases are out of balance. When the second nontrivial zero is reached, the arrows again point to 12 and 6 o'clock respectively. So nonzeros on the critical line mean dual mode phase difference and nontrivial zeros on the critical line mean phase coherence and dual mode balance.

Since the ~conjugate match between modes is associated with presence, the application of dual modes to mathematical thought links primes, nontrivial zeros of the Riemann zeta function and Being. If the present preliminary explanation of the Riemann zeta function proves tenable, then the productivity of the dual mode ontology will be striking, even enough, perhaps, to compensate for the strangeness of monadic *Existenz*.

Afterword

If brain functioning turns out to have significant quantum degrees of freedom which are not involved in quantum computation, then this would be a *bona fide* scientific revolution. The model overthrows the *logos* of metaphysics which still dominates our conception of the kind of beings we really are. Domination by logic is succeeded by the cooperation of belonging-together. Rather than mechanistic action, there is a dynamical welling-up of belonging-together, a spontaneous match in which we always find ourselves already world-thrown, ungrounded amidst the world of pragmata, in the $\sim$conjugate match of the between. Belonging-together underlies prime numbers, nontrivial zeros of the Riemann zeta function, and Being. The extent to which statistical factors enter into the match is variable. If the $\sim$attunement is sharply tuned – one possibility weighted heavily so its probability of actualization in the match approaches 1.0 – not much is left to chance. If the $\sim$attunement is weighted for all invariants equally, then chance has its full opportunity.

The model developed here also gives a fresh account of subjectivity: I am I$^{\sim}$. Subjectivity is not denied, vitiated, or mechanized but is given a new life – not as a substance or an action but as alter. I cannot be observed – the subject is "private" – in that I defaults the domain of public observation, yet takes action. I's action is to participate in constraining the vacuum state match.

If we are such creatures ... spontaneous upsurges! ... autopoietic and autorhoetic (self forming, self-maintaining and self-flowing), variably attuned monads ... monads interiorly bathed in *lumen naturale* ... then brain computation is only a pragmatic capability, not the whole story. Being/presence no longer need be a problematic forgotten by science but quantum brain theoretically understandable.

If we are at heart such amachinative and alogical quantum creatures, we are ultimately "free" of complete control by our universe ... left with some dignity, as poetic and spiritual beings ... we Turing-incompatible I$^{\sim}$s of an alter universe ... empty addresses of default, where not-even-nothing is there ... free of *logos*, *epistemē* and determinism ... for each, our indominable harmonious sovereignty assured.

Note

* I thank Giuseppe Vitiello for pointing out an error in my treatment (Globus 2003) of the Hamiltonian difference between dual modes and misrepresentation of his model on this point, which is corrected here. Communications with him were most helpful in developing this article.

References

Bataille, G. (1988–1990). *The accursed share*, 3 vols. (R. Hurley, Trans.). New York: Zone.

Castaneda, C. (1973). *Journey to Ixtlan*. New York: Simon and Shuster.

Castaneda, C. (1974). *Tales of power*. New York: Simon and Shuster.

Derbyshire, J. (2003). *Prime obsession*. Washington: Joseph Henry Press.

Derrida, J. (1981). *Dissemination* (B. Johnson, Trans.). Chicago: University of Chicago Press.

Edwards, H. M. (1974). *Riemann's function*. Mineola, NY: Dover.

Fröhlich, H. (1968). 'Long-range coherence and energy storage in biological systems'. *Journal of Quantum Chemistry, 2*, 641–649.

Globus, G. (2003). *Quantum closures and disclosures*. Amsterdam and Philadelphia: John Benjamins.

Heidegger, M. (1999). *Contributions to philosophy (from Enowning)* (P. Emad & K. Maly, Trans.). Bloomington: Indiana University Press.

Jibu, M. & Yasue, K. (1995). *Quantum brain dynamics and consciousness*. Amsterdam and Philadelphia: John Benjamins.

Penrose, R. (1994). *Shadows of the Mind*. Oxford: Oxford University Press.

Plotnitsky, A. (2002). *The knowable and the unknowable*. Ann Arbor: University of Michigan press.

Ricciardi, L. M. & Umezawa, H. (1967). Brain and physics of many-body Problems. *Kybernetik, 4*, 44–48.

Sartre, J.-P. (1957). *The transcendence of the ego* (F. Williams & R. Kirkpatrick, Trans.). New York: Farrar, Straus and Giroux.

Sautoy, M. du (2003). *The music of the primes*. New York: HarperCollins.

Stapp, H. (1993) *Mind, matter and quantum mechanics*. New York: Springer Verlag.

Stuart, C. I. J. M., Takahashi, Y., & Umezawa, H. (1978). On the stability and nonlocal properties of memory. *Journal of theoretical biology, 71*, 605–618.

Stuart, C. I. J. M., Takahashi, Y., & Umezawa, H. (1979). Mixed-system brain dynamics: Neural memory as a macroscopic ordered state. *Foundations of physics, 9*, 301–327.

Umezawa, H. (1993). *Advanced Field Theory: Micro, Macro, and Therma Physics*. New York: American Institute of Physics.

Umezawa, H. (1995). Development in concepts in quantum field theory in half century. *Mathematical Japonica, 41*, 109–124.

Vitiello, G. (1995). Dissipation and memory capacity in the quantum brain model. *International journal of modern physics B, 9*, 973–989.

Vitiello, G. (2001). *My Double Unveiled*. Amsterdam and Philadelphia: John Benjamins.

Von Neumann, J. (1955). *Mathematical foundations of quantum theory.* Princeton, NJ: Princeton University Press.

Wigner, E. (1962). Remarks on the mind-body question. In I. J. Good (Ed.), *The scientist speculates.* New York: Heinemann.

Zurek, W. H. (1991). Decoherence and the transition from quantum to classical. *Physics Today, 44*, 36–44.

Chapter 7

Quantum monadology and consciousness

Teruaki Nakagomi
Kochi University, Japan

There are serious conflicts between physics (as described by quantum theory and relativity theory) and consciousness (as our interior experience). In order to solve the conflicts, the author proposes *quantum monadology*, in which fundamental theories of physics are incorporated consistently into monadology. The monadistic structure consists of: (1) monads (individuals) with consciousness and volition, (2) internal worlds in which respective monads live, and (3) communication between monads and correspondences among internal worlds (pre-established and post-established). Although we can find such structure in the human level first, it reaches to the ultimate level of physics, and human-level monadic structure results from this ultimate monadology. This paper describes (1) why monadology is needed, (2) the basic structure of monadology in a clear symbolic formulation, and (3) outlines of quantum monadology and its consequences.

1. The aim of quantum monadology

Monadology is a world model presented by Leibniz (1714), in which matter and mind are integrated carefully and consistently in terms of monads and their preestablished harmony.

Quantum monadology was proposed by the author (Nakagomi 1992), inspired by the idea of Leibniz's monadology, the purpose of which is to solve the apparent conflicts between physics (as described by quantum theory and relativity theory) and consciousness (as our interior experience). Recently, the theory has been elaborated thoroughly (Nakagomi 2003a, b), which consists of a monadistic scheme (NL world) and incorporation of physics into this scheme. The physics adopted there is relativistic quantum field theory in the most general form, which provides a firm basis to investigate the relationship between consciousness and physics.

Before proceeding to quantum monadology, it is worthwhile to review what are the problems between physics and consciousness.

Inside of matter. Consciousness is internal experience. We can experience consciousness through introspection, but cannot observe it externally. When we dissect the brain, we will see only material systems such as neurons, microtubules, proteins, molecules, and so on. What is the experience of consciousness or the direct experiences of colors, sounds, smells, pains, and so on? If these are material phenomena, then matter must have an interior, because we are able to experience these only from the inside of the brain matter. Here note that "the inside" does not mean a spatial concept such as "the inside of the skull". What is the inside of matter? Physics does not have the concept of inside or interior of matter and cannot explain the interior-exterior mystery of consciousness.

Now. Consciousness exists *now* and not at any point of the past or the future. The inside of matter, if it exists, is accompanied by *now* and not by the future nor the past. What is *now*? *Now* is the time point or the duration where consciousness exists. My consciousness and your consciousness seem to share in common the now, which is flowing or passing. The physical picture of time is, however, the fourth component of the four-dimensional space-time continuum. In this picture, the now is a special point moving along the time axis, but physics does not have any special point in the time axis. We cannot understand the now if we start our thinking from the space-time concept in physics.

Volition. Consciousness is not only passive in receiving senses but also active in making indeterministic effects on the material world. The effects occur with now and propagate in the future but not in the past. However, physics does not accept the indeterministic effects due to volition nor asymmetry between the past and the future, because there is the deterministic and time reversible law of change of material states at the bottom level of physics, i.e., the Schrödinger equation or the unitary evolution law of quantum states.

Those three problems come from the side of consciousness, and cause no problem in physics because physics neglects facilely consciousness. The following two are problems in physics itself.

Measurement. The measurement problem of quantum mechanics is usually understood as the conflict between determinism due to the Schrödinger equation and indeterminism caused by measurements. Another important point is in what measures what. In physics everything is matter, and hence matter measures matter. However, physics does not have any principle to divide matter into two parts, what measures and what is measured.

Statistical mechanics. In applying quantum theory to actual phenomena, we use it not in bare form, but wrapped with the filter of statistical mechanics. However, this procedure is an analogical inheritance from classical statistical mechanics and is not derived from first principles. We have not yet understood the ultimate origin of thermal probabilities. This is also concerned to the problem of determinism and indeterminism.

The above five problems are correlated mutually, and must be dealt with all together. Considering the brilliant success of modern physics, the origin of these problems is not in physics itself, but in the world model that underlies physics. Modern physics does not explicitly refer to its world model, but tacitly assumes materialism or Cartesian dualism.

Materialism. In the materialistic world model, everything must be understood as phenomena of matter that has no volition, no consciousness, and no inside. What we call consciousness is also material phenomena, and is to be described by the laws of physics, which does not include any law peculiar to consciousness. However, physics that is interpreted materialistically fails to explain the consciousness as our internal experience, and results in denying its existence as mentioned above.

Cartesian dualism. Matter and mind are assumed to exist separately and to be linked by some mechanism. Consciousness is a phenomenon of mind, being outside the physical world, and is not required to obey the laws of physics. The problems that appear in the materialistic model can be resolved in this model by assuming suitably the laws of mind and the linkage mechanism. The internal experience, consciousness, is in mind and does not belong to the material world. The flowing time, now, is associated with mind, which is linked to a point or a duration in the physical spatial time. Since volitional action can be postulated independently of the physical laws, indeterminism can be introduced into the material world through the linkage between mind and matter.

Though the Cartesian model has these advantages, it includes three crucial defects: (1) Since mind is assumed to be independent of the material world, there is no clue how to investigate the properties of consciousness. (2) Also, there is no clue how to establish the linkage between mind and matter. (3)

Moreover, this model cannot find a reasonable explanation for the fact that the state and function of mind are strongly dependent on those of matter as revealed in the recent development of brain science.

Leibnizian monadology. In this world model, everything is internal phenomena. The world consists of *monads*. Each monad lives in its own internal world. What we call material phenomena occurs in the internal world of a monad, and hence physics provides the laws of the internal world. Consciousness is a part of the internal world and is subject to physics. The consciousness part has both active and passive relations with the other part of the internal world. The passive relation serves as perception, and active one gives the volitional action on the other part. The conscious part is then regarded as the self part.

Although the word "internal" expects "external", there are only internal worlds and no external world in which monads are located. Since monads are not in an external world, there is no physical interaction between internal worlds. However, the internal world of each monad reflects the whole of monads, and there is a correspondence mechanism among the internal worlds, which produce a virtual world that has an appearance of an external objective world. In the original monadology of Leibniz, this correspondence, called *preestablished harmony*, is supposed to be a divine mechanism, or one of the first principles of the theory in scientific terms.

If the law of change of internal worlds is deterministic and the same for all monads, then the harmony will be maintained only if initially the correspondence is established, just as two complete clocks tick always the same time once they are synchronized. However, monads have volitional actions which causes indeterministic changes in their internal worlds, and it is necessary to communicate between monads so as to maintain the correspondences among internal worlds. This communication is non-physical or sub-physical process in the sense that it is not described by the mechanical laws of internal worlds.

The author expects that quantum mechanics can be well interpreted in the Leibnizian model rather than the materialism or the Cartesian dualism. In the ordinary interpretation of quantum mechanics, the two types of state changes, Schrödinger's equation and the reduction of states are not unified consistently. The former describes physical process, while the latter should be considered as relating to the sub-physical communication process between monads. The relativity principle is also adaptable to the Leibnizian model. The Lorentz transformations may be interpreted as the correspondences between internal worlds.

Mathematical formulation. The problems that appear in the former two world models are all expected to be resolved in the Leibnizian model so long as the fundamental theories of modern physics are incorporated into it. There are some philosophers, among which Whitehead (1929) is famous, who take up the Leibnizian world model to interpret physics. However, their monadology remains on the level of a "way of thinking," and they have not been successful in incorporating physics into monadology. The fundamental theories of physics are thoroughly mathematical, and hence the "way of thinking" must be also formulated as a mathematical system if we wish to integrate it with physics. It is only after obtaining the mathematical formulation of monadology that we are able to know whether monadology is consistent with physics or not. The NL world scheme presented in the previous paper (Nakagomi 2003a) is the recent version of author's mathematical formulation of monadology. Incorporating relativistic quantum field theory into the NL world scheme yields quantum monadology (Nakagomi 2003b). The NL world has some differences from the original Leibniz's monadology, among which an important one is in that the correspondences are post-established through an evolutionary process on the basis of a minimum assumption of pre-established correspondences.

Monadistic thinking. There are some observations that help to understand the ideas of monadology.

1. Consider many cars running on a road. The distances between cars vary smaller or larger as they are running, but normally they do not collide with each other except for rare accidents. Observing such phenomena, physicists may consider that there is a repulsive force between cars and they may succeed in explaining average motions of cars. However, the force assumption would not describe the motion of individual cars. We find a similar situation in quantum physics. The motion of a car is determined by the mind of the driver. Decisions of the driver depend on the memory of the past and the prediction of the future and are not affected by those of the other drivers. Also in quantum mechanics, the behavior of a particle is affected by the wave function belonging to the particle and not by the wave functions of the other particles.
2. We have a concept of the nation that we belong to. However, there is no physical entity of the nation, but it exists in our mind. We have a common concept of the nation, then the nation exists. If the members have different concepts of the nation, then the nation will cease to exist.

3. When I speak to you, really I speak to your image in my mind not to your physical body, This mental process causes physical process, and then it causes correspondent mental process in your mind. This mechanism will work well in an ordinary situation in which your image system and mine are correspondent to each other. Otherwise, we could not communicate successfully. Note that the role of "physical" in this case is different from that in the basic monadology.
4. In a computer game through networks, each player battles in a battlefield that his computer produces. The battlefields of different players are correspondent to each other by the communication through networks, and players feel themselves to be playing in a common battlefield. However, it does not exist anywhere.

2. NL world

The monadistic world model is formulated in a mathematical symbolic style called NL world. The formulation presented here is almost an extract from the previous paper (2003a), but includes an important modification in the renewal rule of internal worlds, which also causes small changes in other relating parts. Significance of this modification will be discussed in the last section. For details of mathematics used here, consult the previous paper.

An NL world is specified by the following three sets and five functions

$$W = (V, F, L, \eta, \rho, \omega, \lambda, \beta).$$

Each item is defined as follows: V is a finite set of *monad-images* with a special element v_{self}, *self-image*. F is a set of *internal states*. L is a σ-complete orthomodular lattice of *contents of consciousness*.[1] η is a mapping $\psi \in F \mapsto \eta(\psi) \geq 0$, *appetite*. ρ is a mapping $\psi \in F \mapsto \rho(\cdot|\psi)$ (a probability measure on L or 0 measure), *preferability*. ω is a mapping $\psi \in F \mapsto \omega(\psi)$ (an orthogonal system of elements of L), *list of choice*. λ is a mapping $(r, \psi) \in \mathfrak{S}(V) \times F \mapsto \lambda(r, \psi)$ (an automorphism on L), *interpreter*, where $\mathfrak{S}(V)$ is the symmetric group of V (the group of all one-to-one and onto mappings from V to V). β is a mapping $(\ell, \psi) \in L \times F \mapsto \beta(\ell)\psi \in F$, *state-change operator*. Here the following minimum conditions are assumed

$$\omega(\psi) = \varnothing \Rightarrow \rho(\cdot \mid \psi) = 0 \text{ and } \eta(\psi) = 0, \tag{C01}$$

$$\omega(\psi) \neq \varnothing \Rightarrow \sum_{\ell \in \omega(\psi)} \rho(\ell \mid \psi) = 1, \tag{C02}$$

for any $\psi \in F$.

Rule 1 (Monads and correspondences). With the NL world W, a set M_W of *monads* is associated, whose number of elements is the same as that of V. Each monad m has an entity-image correspondence $c_m : M_W \to V$ (one-to-one and onto) with condition that

$$c_m m = v_{\text{self}}.$$

Between any pair of monads m, m', the image-image correspondence $r_{mm'}$: $V \to V$ is then defined by

$$r_{mm'} = c_m c_{m'}^{-1}.$$

Rule 2 (Current states and renewal cycle). Each monad m has a variable Ψ_m, *current state*, that takes values in F, and follows the renewal cycle of three steps given below:

Step 1. Each monad m is urged once to make a decision, and *one* monad, say m_1, is hit with probability proportional to the appetite $\eta(\Psi_{m_1})$.

Step 2. The hit monad m_1 chooses an item ℓ from $\omega(\Psi_{m_1})$ with probability proportional to the preferability $\rho\big(\ell \bigm| \Psi_{m_1}\big)$.

Step 3. Each monad m interprets the choice ℓ by the monad m_1 as $\ell_m = \lambda(r_{mm_1}, \Psi_m)\ell$, and renews its current state Ψ_m by the substitution formula

$$\Psi_m := \beta\big(\ell_m\big)\Psi_m.$$

Internal worlds and consciousness. Each monad m is supposed to have its own internal world whose current state is represented by Ψ_m. However, the internal world should not be identified with the consciousness of the monad. A monad m makes its decision on the basis of the information given by the appetite $\eta(\Psi_m)$ and the preferability $\rho(\ell \mid \Psi_m)$ distributed over the choice-list $\omega(\Psi_m)$, on which its volition acts. In this sense, these three items should be interpreted as constituting the consciousness of a monad, in particular, the elements of the choice-list are regarded as the contents of consciousness. The set L contains all the possible contents of consciousness, and is supposed to have the structure of a σ-complete orthomodular lattice so as to make a consistent mechanism of decision.

Null monads and active monads. A state ψ is called a *null* state if $\omega(\psi) = \varnothing$ and an *active* state otherwise. A monad in a null state, called a *null monad*, makes no decision of choice, and hence does not participate in the change of the world of Rule 2. However, null monads cannot be eliminated from the NL world for two reasons. First, a null monad is not assured to remain in a null state in the process of state-change within the setting presented so far, though we will give an optional condition to assure it below. Second, null monads have their images in active monads. Null monads appear in an active monad as inert entities or matter.[2]

Chronological description. The value of a current states Ψ_m can be recorded sequentially for each renewal cycle, and be numbered from the start of recording as follows:

$$\psi_m[0], \psi_m[1], \psi_m[2], \ldots, \psi_m[s], \ldots. \qquad (*)$$

The numbering is identified with *time*, and we call $\psi_m[s]$ the state of monad m at time s. Henceforth, we will make temporal statements with temporal terms such as "at some time", "the next time", "before", "after", and so on, when referring to the values of Ψ_m. Note that the time parameter s represents the flow of time and not the time axis of space-time. The flow of time is promoted by volitional actions of monads. The relation between the two time concepts will be discussed in the last section.

Correspondence theorem. The NL world has the basic correspondences given by Rule 1. In order to extend it to higher-level correspondences among internal worlds, we need to put some conditions.

Condition 1. For any $\ell \in L$ and any $\psi \in F$,

$$\rho(\ell \mid \psi) > 0 \iff \omega(\beta(\ell)\psi) \neq \varnothing.$$

If a monad m is in a null state then $\omega(\Psi_m) = \varnothing$, which implies $\rho(\ell \mid \Psi_m) = 0$ for any $\ell \in L$ by (C01), and hence $\omega(\beta(\ell)\Psi_m) = \varnothing$ by this condition. Therefore a null monad remains a null monad in the process of the renewal of current states. On the other hand, an active monad m can change to a null monad in such a situation as $\rho(\ell_m \mid \Psi_m) = 0$ and $\rho(\ell \mid \Psi_{m_1}) > 0$ for some hit monad m_1 $(\neq m)$. The set of active monads can then be preserved or diminished in the process of the renewal.

Condition 2. Each $r \in \mathfrak{S}(V)$ has an operation on F that preserves the group properties

$$r(r'\psi) = (rr')\psi \quad \text{and} \quad I_V\psi = \psi,$$

and allows the following symmetry relations

$$\lambda(rr', r\psi) = \lambda(r, r\psi)\lambda(r', \psi),$$
$$r\beta(\ell)\psi = \beta\big(\lambda(r, r\psi)\ell\big)r\psi,$$
$$\rho(\ell \mid \psi) = \rho\big(\lambda(r, r\psi)\ell \bigm| r\psi\big),$$

for any $r, r' \in \mathfrak{S}(V)$, any $\ell \in L$ and any $\psi \in F$, where I_V is the identity in $\mathfrak{S}(V)$.

Theorem 1. *Let the NL world W satisfy Conditions 1 and 2. If the following relation holds for any pair m and m' of active monads at some time*

$$\Psi_m = r_{mm'}\Psi_{m'}, \tag{**}$$

then, at any time after that time, the set of active monads is preserved and the above relation holds for any pair m and m' of active monads.

Define the subgroup $\mathfrak{S}_0(V) = \{r \in \mathfrak{S}(V) \mid rv_{\text{self}} = v_{\text{self}}\}$. Since any element in V other than the self-image v_{self} has no special meaning, the rules of the NL world W should have the $\mathfrak{S}_0(V)$-symmetry. Condition 2 includes this symmetry for λ, β and ρ. Since Rule 2 uses η and ω, we need to put a further condition.

Condition 3. For $s \in \mathfrak{S}_0(V)$,

$$\lambda(s, \psi) = \lambda(s) \quad \text{(independent of } \psi \in F\text{)},$$
$$\eta(s\psi) = \eta(\psi),$$
$$\omega(s\psi) = \lambda(s)\omega(\psi).$$

Conditions 2 and 3 guarantee that Rule 2 is invariant under the rearrangement of the image-entity correspondences, $\{c_m\}_{m\in M_W} \rightarrow \{r_m c_m\}_{m\in M_W}$ with $r_m \in \mathfrak{S}_0(V)$. Moreover, these conditions make it possible to describe the renewal rule in reference to a fixed active monad m_0 without explicit reference to the other monads in the situation in which (**) is established among all active monads. We can use v-parameterized functions $\eta_v(\Psi_{m_0})$ and $\omega_v(\Psi_{m_0})$ instead of $\eta(\Psi_m)$ and $\omega(\Psi_m)$, where $v = c_{m_0}m$ is the image of monad m appearing in the internal world of m_0, and $\eta_v(\psi)$ and $\omega_v(\psi)$ defined by

$$\eta_v(\psi) = \eta(r\psi) \quad \text{if } rv = v_{\text{self}},$$
$$\omega_v(\psi) = \lambda(r^{-1}, \psi)\omega(r\psi) \quad \text{if } rv = v_{\text{self}}.$$

This is called m_0-focused description, which amounts to the physical description of the world. The replacing m_0 by another corresponds to the transformation of observer's frames. From this point of view, the Lorentz transformations can be interpreted, which was discussed in detail in the previous paper (2003b).

3. Quantum monadology

Quantum monadology is an instance of the NL world. The main ideas are:

internal world	$\leftrightarrow$	quantum state and Lorentz frames associated with monad-images
automatic change	$\leftrightarrow$	unitary evolution law
volition	$\leftrightarrow$	reduction of quantum state
consciousness	$\leftrightarrow$	self-other coupling
correspondence	$\leftrightarrow$	inhomogeneous Lorentz transformation
communication	$\leftrightarrow$	correspondence of reduction

where $\leftrightarrow$ should be read as "be related to" or "be realized by." On the basis of these ideas, the items of the NL world $W = (V, F, L, \eta, \rho, \omega, \lambda, \beta)$ are specified, which will be sketched roughly (see Nakagomi 2003b for details).

The set V of monad-images can be taken as any finite set with an element specified as the self-image v_{self}. The internal world of a monad is described by a quantum state and Lorentz frames associated with the monad-images. Hence the state space F is given by the direct product of a Hilbert space $\mathcal{H}$ and inhomogeneous Lorentz groups associated with monad-images. The σ-complete orthomodular lattice L is defined by the set of projections on $\mathcal{H}$.

The set V is decomposed into two subsets $\{v_{\text{self}}\}$ and $V_{\text{other}} = V - \{v_{\text{self}}\}$. The quantum state of a monad is also decomposed into the tensor product of the self part and the other part (explained below), and the self-other coupling of the quantum state defines the list of choices ω. The choices by monads appear as quantum reduction processes of self-other coupled states. The preferability ρ is related to the reduction probability. The choice-driven part of state-change operator β represents this reduction. The automatic part of β is defined so as to cause translational change of Lorentz frames associated with monad-images. The interpreter λ is specified by frame-frame relation and self-other conversion of monad-images. Finally, the appetite η is given by the entropy of distribution of ρ over ω.

The minimum conditions (C01) and (C02) are fulfilled. From Rule 1, this world has N $(= |V|)$ monads, M_W, and each monad m has an entity-image correspondence $c_m : M_W \to V$ (one-to-one and onto) with condition $c_m m = v_{\mathrm{self}}$, and any pair of monads m and m' has image-image correspondence $r_{mm'} = c_m c_{m'}^{-1}$. From Rule 2, each monad m has its current state variable Ψ_m that takes values in F. The renewal cycle is expressed concretely as follows:

Step 1. One monad m_1 is hit with probability proportional to its entropy $\eta(\Psi_{m_1})$.

Step 2. The hit monad m_1 chooses an item Q from $\omega(\Psi_{m_1})$ with probability proportional to the preferability

$$\rho(Q \mid \Psi_{m_1}) = \frac{\langle \Psi_{m_1}, Q\Psi_{m_1} \rangle}{\langle \Psi_{m_1}, \Psi_{m_1} \rangle}.$$

Step 3. Each monad $m \in M_W$ interprets the choice Q by the monad m as $Q_m = r_{mm_1} Q r_{m_1 m}$, and renews its current state Ψ_m by the substitution formula

$$\Psi_m := \frac{\|\Psi_m\|}{\|Q_m \Psi_m\|} U Q_m \Psi_m.$$

In the above formulation, the Lorentz frames' part of the current states and their renewal law are omitted for the sake of simplicity.

Since $\omega(\psi) = \varnothing$ is equivalent to $\psi = 0$, a null monad m is specified by the condition $\Psi_m = 0$. It is evident that Condition 1 is fulfilled and that a null monad does not convert into an active monad.

Self-other coupling and measurement. In order to make *measurement* meaningful in one system, we need to divide the system into two parts, what measures and what is to be measured. The internal world is divided into the self part and the other part, and hence the Hilbert space $\mathcal{H}$ is decomposed into the form of the tensor product

$$\mathcal{H} = \mathcal{H}_{\mathrm{self}} \otimes \mathcal{H}_{\mathrm{other}}$$

According to the standard theory of quantum measurement (von Neumann 1932), we can calculate the list of items to be measured and weights associated with respective items. In the tensor product expression, a state $\psi \in \mathcal{H}$ can be written as

$$\psi = \sum_k a_k \sum_{i \in A_k} f_i \otimes g_i$$

with $a_1 > a_2 > \cdots > 0$ and suitable orthonormal systems $\{f_i\}_i$ in $\mathcal{H}_{\mathrm{self}}$ and $\{g_i\}_i$ in $\mathcal{H}_{\mathrm{other}}$, where A_k's are disjoint index sets. This expression is unique up to the diagonal expressions within respective blocks of A_k's, and is considered to characterize the coupling between the self part and the other part of the state ψ.

The list of measurement items is then given by the orthogonal set of projections

$$\omega(\psi) = \left\{ \sum_{i \in A_k} |f_i\rangle\langle f_i| \otimes I_{\text{other}} \;\middle|\; k = 1, 2, \ldots \right\},$$

where I_{other} the identity operator on $\mathcal{H}_{\text{other}}$. This set is unique for given ψ. Finally, the *entropy* of the state ψ is defined by

$$\eta(\psi) = -\sum_{Q \in \omega(\psi)} \frac{\langle \psi, Q\psi \rangle}{\langle \psi, \psi \rangle} \log \frac{\langle \psi, Q\psi \rangle}{\langle \psi, \psi \rangle}.$$

4. Discussion

Time. In physics, the time evolution of a quantum state is written as $\psi(t) = e^{-itH}\psi$ with a suitable Hamiltonian H. Each value $\psi(t)$ can be mapped sequentially from $t = -\infty$ to ∞ along the time axis, which seems to be similar to the chronological sequence $(*)$ in the NL world. However, there is an important difference between them. The former is unitary evolution, and hence if one value in the sequence is obtained, then all values from $t = -\infty$ to ∞ in the sequence can be determined. Unitary evolution creates nothing new. Changes are only in appearance.

On the other hand, the sequence $(*)$ is indeterministic. One value in the sequence does not determine the other values. Something new is created at every time in the sequence. Combining the physical evolution law with the monadistic evolution law yields the two-parameter state $\psi[s](t) = e^{-i(t-\tau s)H}\psi[s]$, where s is the index of the sequence $(*)$, while t is the parameter of the function defined on the whole range of time from $t = -\infty$ to ∞ for each value of s. In the renewal cycle of internal worlds, the two-parameter state undergoes the change process

$$e^{-i(t-\tau s)H}\psi[s] \rightarrow e^{-i(t-\tau(s+1))H}\psi[s+1].$$

In this process, the state changes "simultaneously" over the whole range of time t. The time in the sense of the sequence $(*)$ is promoted only when a monad makes decision. It seems possible to consider that the t dependence interpolates between τs and $\tau s + \tau$ in a non-relativistic case, but it causes serious contradiction with the relativity principle. From the point of view of quantum monadology, the reduction of quantum states occurs in the time s, and hence the space-time picture of the reduction is related not to a spatial intersection of space-time but to the whole space-time, by which the reduction of states becomes consistent with the relativity principle. Minkowski space-time does not have "now" and hence there is no past nor future. The time axis in Minkowski space-time is a mathematical entity needed for manipulating the Lorentz group and does not represent the real time in which we live.

Single vs. multiple decision models. The renewal rule of the NL world is modified from that of the previous paper (2003a). In the renewal rule of this paper, only one monad is allowed to make decision at each renewal cycle, whereas, in the previous one, multiple monads can make decision simultaneously. In this point, the theory returns to the first one presented in 1992. By this modification, the law of probability of monads' decision is quite simplified and the compatibility condition of $\omega(\psi)$ between different monads become unnecessary. In spite of these advantages of the single decision model, the more complicated multiple decision model was adopted in the previous paper (2003a), because a satisfactory reason could not be found for the single decision model. Recently, however, the single decision model was inspired by Watanabe's (2002) assertion that there is only one ego in the world which is running through all the individuals existing from the past to the future. This idea, he says, comes from Schrödinger's (1985) thinking about Vedanta. The author has come to consider that the single decision model is not so unnatural. Anyhow, these two models are asymptotically equivalent if the appetite η is sufficiently small, but it is left to future study to decide which model is better.

Human-level monadology. Though monads have consciousness and volition, there is a long way from the basic level of quantum monadology to the mind of the human level. We should not confuse these two levels. However, there is a reason for this confusion, that is, mind has a monadistic structure on the human level by identifying the individual mind with a monad. (Cf. Globus this volume.) Both monadistic structures have good similarity. Apparently Leibniz constructed his monadology by getting a clue from this similarity. The human level monadistic structure should be considered as a higher-order world derived from the basic monadistic world.

It is expected that there is some mechanism that derives the similarity between the higher-order monadistic structures and the basic one. For such a mechanism, an enhancement process of volition and perception of monads is proposed in the previous papers (2003a, b), by which an ensemble of monads with a central monad appears as a higher-order single monad. In particular, the central one is allowed to be a null monad. In this case, it will be a zombie or a robot, but it is difficult to tell whether the central one is null or not. The enhancement process is also used to describe the quantum measurement process with a macroscopic apparatus.

In general, the derivation of the higher-order structures is very difficult, but we can apply the NL world scheme directly to the human level by assuming a suitable approximate mechanism of mind and communication between individuals. The NL world scheme provides only a formal scheme, and we can adopt various kinds of physics, from the most general one to restricted or virtual ones, and apply it to various levels of systems in a hierarchy of monadistic worlds, from the basic physical level to bio-, human, and socio- levels. The evolutionary process of establishing the correspondence among monads makes it possible to discuss the integration process that appear in socio-systems such as political, cultural, and economic systems or in the cyber-world produced by computer networks.

Self-other coupling and consciousness. From the point of view of quantum monadology, consciousness is related to the self-other coupling associated with the self-other division of a quantum state. This coupling is expressed in the form of a linear combination of diagonalized branches of the self-other correlation in the quantum state as shown above. The content of consciousness is assumed to relate to these branches. The preferability is assumed to be proportional to the quantum weights of branches, and the appetite is made to be linked to the entropy of the weights of branches. These assumptions for consciousness seem to be the almost unique possibility that quantum physics allows. In quantum monadology the state space $\mathcal{H}$ has an infinite dimension, and the self part has a possibility to create infinitely fine correlation with the other part, which suggests that consciousness has no finite limitation in its "capacity".

Null monads. Sometimes it is asked why a null monad remains a null monad. In order to obtain Theorem 1, we have imposed Condition 1, which forbids null monads to convert into active monads. However, we may assume waking-up process of null monads as not destroying the correspondence $(**)$ if necessary in the future study.

Physical verification of quantum monadology. The final and most important problem of quantum monadology is how monad-images appear in the world where we live, since quantum monadology is supposed to describe this world on the level of physics. There are no physical interaction between monads, but there are between monad-images. The monad-images introduce a new structure into the physical world, which causes two effects in physics. The first is the decoherence of quantum states. The decoherence has been discussed for a long time in relation to the measurement problem. Recent experimental and theoretical discussions are found, e.g., in Joos et al. (2003), but they lack ontological discussion. The author considers that the ultimate origin of quantum decoherence should be attributed to the monads' volitional actions. The second is the interaction between monad-images that overlaps the uniform interaction of quantum fields, which might be related to the renormalization technique of field theory. We could verify monadology on the physical level by detecting the monad-image structure directly or indirectly through these effects. To this end, we must refine quantum monadology, in particular we need an explicit form of the interaction between monad-images.

Notes

1. An *orthomodular lattice* L is a lattice which has 0, 1, and an orthocomplement operation $a \in L \mapsto a^{\perp} \in L$ in addition to the lattice operations $\vee$, $\wedge$ and $\leqq$ and satisfies the following conditions

$$\begin{aligned} &a^{\perp} \vee a = 1 \quad \text{and} \quad a^{\perp} \wedge a = 0, \\ &(a^{\perp})^{\perp} = a, \\ &a \leqq b \Rightarrow b^{\perp} \leqq a^{\perp}, \\ &a \leqq b \Rightarrow b = a \vee (b \wedge a^{\perp}) \quad \text{(orthomodular law).} \end{aligned}$$

Two elements a and b are *orthogonal* iff $a \leqq b^{\perp}$ (or equivalently $b \leqq a^{\perp}$). Classical logic and quantum logic (defined by the set of all closed subspaces (or projections) of a Hilbert space) are typical examples of orthomodular lattice.

2. According to G. Franck (1994), an idea equivalent to null monads is found also in Leibniz's monadology: "the monads, as souls, cannot simply cease to exist. They persist in the state of death, as they pre-exist in the state of not yet being brought into their world. They make appearance in life by awaking consciousness. They die, but do not disappear, by its extinction. The world each monad lives in is extended only by its own conscious activity. In the state of death this world has no extension. Dead (or transmigrant) souls are 'metaphysical points'."

References

Franck, G. (1994). Physical time and intrinsic temporality. In H. Atmanspacher & G. J. Dalenoort (Eds.), *Inside versus outside. Endo- and exo-concepts of observation and knowledge in physics, philosophy, and Cognitive Science* (pp. 63–83). Berlin: Springer.

Joos, E., Zeh, H. D., Kiefer, C., Giulini, D., Kupsch, J., & Stamatescu, I.-O. (2003). *Decoherence and the appearance of a classical world in quantum theory*. Berlin: Springer.

Leibniz, G. W. (1714). *Monadology*. Japanese trans. by Y. Kōno (Iwanami-Shoten, Tokyo, 1951); *The Monadology and other philosophical writings*, ed. and English trans. by R. Latta (Oxford Univ. Press, 1925); see also e.g. *The Cambridge Dictionary of Philosophy*, R. Audi (Ed.) (Cambridge University Press, New York, 1995) 425–429.

Nakagomi, T. (1992). Quantum monadology: a world model to interpret quantum mechanics and relativity. *Open Syst. Inform. Dyn., 1*, 355–378.

Nakagomi, T. (2003a). Mathematical formulation of Leibnizian world: a theory of individual-whole or interior-exterior reflective systems. *BioSystems, 69*, 15–26.

Nakagomi, T. (2003b). Quantum monadology: a consistent world model for consciousness and physics. *BioSystems, 69*, 27–38.

Schrödinger, E. (1985). *Mein Leben, meine Weltansicht* (Paul Zsolnay Verlag, Wien). Japanese trans. by H. Hashimoto, R. Nakamura, H. Hayakawa, & C. Hashimoto (Kyoritu-Shuppan, Tokyo, 1987; Chikuma-Shobo, Tokyo, 2002).

von Neumann, J. (1932). *Mathematische Grundlagen der Quantenmechanik* (Springer, Berlin), *Mathematical foundations of quantum mechanics* (Princeton University Press, Princeton, 1955).

Watanabe, T. (2002). *The mystery of "My Death": Overcoming the solipsism through "the Psychology of Worldview"* (in Japanese). Kyoto: Nakanishiya.

Whitehead, A. N. (1929). *Process and reality. An essay in cosmology*. New York: Macmillan Publishing Co.

Chapter 8

Quantum connectionism and the emergence of cognition

Eliano Pessa
Università degli Studi di Pavia, Italy

This paper argues that the validity of the usual connectionist claim, according to which the macroscopic cognitive behaviors are nothing but collective effects, emergent from the interactions between brain neurons, can be proved only if we resort to a description of neural brain dynamics based on Quantum Field Theory. This description is denoted as *Quantum Connectionism*. It is proved, through a concrete example, that the direct introduction of quantum-like features within neural dynamics gives rise to the occurrence of long-range correlations, like the ones characterizing the macroscopic cognitive operation.

1. Introduction

One of the main challenges for modern science is to understand how human and animal cognitive behavior emerges from the seemingly chaotic realm of neural excitations. To deal with such a challenge the most promising approach, so far proposed, is to be identified with the *connectionist* one (McClelland & Rumelhart 1986). Namely the latter tries to explain macroscopic cognitive behaviors as collective effects emerging from the cooperative interactions between *microcognitive* units, acting at a microscopic level. It is generally taken for granted that these units, in ultimate analysis, correspond to neural assemblies or even to single neuronal units located in a biological brain. In this way, by resorting to a description of the operation of cognitive systems based on the formalism of *artificial neural networks* (ANN), the connectionist approach throws a bridge between psychology and neurophysiology, between cognition and neural activity patterns, between mind and brain.

Such a bridge, however, is still unreliable. In this regard, this contribution will present a number of arguments supporting the claim that the description of macroscopic cognitive behaviors as collective phenomena works only within the context of a (suitable) quantum theory of the behavior of microscopic constituents of the brain, i.e. a *quantum brain theory*. Moreover, we will introduce the expression *Quantum Connectionism* to denote a reformulation of the connectionist approach based on a quantum theory of neurons, be they cosidered as microcognitive units or as, more or less abstract, counterparts of biological neurons.

This paper will be devoted to the problems to be dealt with when building up quantum connectionist models. Being forced, in this regard, to touch on a number of different, and important, questions, we organized its content as follows. In the second section we will recall some fundamental concepts underlying the connectionist approach and the ANN models, whereas the third section will be devoted to a discussion about the notion of *emergence*. The aim of the fourth section will be to explain why every serious attempt to model emergent phenomena should be based on the principles of quantum theory. Within the fifth section, then, we will discuss possible approaches to a quantum theory of neurons, and we will develop in detail, in this regard, a new proposal. The sixth section will deal with quantum neural networks – networks constituted by interconnected quantum neurons – whereas the conclusions will be the subject of the seventh section.

2. The connectionist approach and ANN models

The main principles underlying the traditional connectionist approach can, in their essence, be reduced to the following two statements:

a. there exist two different levels of description of cognitive processing : the *macroscopic* one, in which cognitive processing is studied on the observational scale of traditional experimental psychology, and the *microscopic* one, in which we focus on the operation of simple detectors of single features of input patterns or of complex cognitive constructs;
b. the cognitive processing features observed at the macroscopic level *emerge*, as a sort of collective effect, from the (cooperative) interactions between the operations of the single detectors working at the microscopic level.

Usually the modeling of the microscopic level is performed by resorting to ANN (for a review see Bishop 1995; Rojas 1996). A generic definition identifies every ANN with a system containing two components:

- *units* (or *neurons*), each of which is an input-output device, characterized, at each time instant, by an *output state* (termed also *activation*), by an *inner state* (the so-called *activation potential*), and by an *input state*;
- *connection lines* between the units; they feed the activation state from the output line of a unit to the input line of another unit; each connection is associated to a *connection weight* (in most cases a real number), which modulates the activation signal crossing it.

This definition entails that an ANN can be viewed as a dynamical system, in which the *state vector*, containing the instantaneous output states of the single network units, changes with time, starting from an initial state, as a consequence both of the form of the activation laws adopted and of the distribution of connection weight values. There is, however, another sense in which an ANN can be considered as a dynamical system: when its parameters (mostly the connection weights) change with time as a consequence of a *learning process*. Namely the most interesting cognitive performance of ANN consists in their ability to *learn only from examples* to approximate whatever form of input-output relationship, or to categorize the input patterns (see, e.g., Bartlett & Anthony 1999).

The fact that ANN are dynamical systems should constitute the basis for substantiating the second fundamental claim of the connectionist approach, that is that cognitive processes are nothing but collective effects emerging from the interactions between the units of an ANN. Namely, if we could show that the dynamical evolution of an ANN is associated to a sort of *phase transition*, like the ones encountered in physical systems, then we could take advantage of the theoretical apparatus, already developed by physicists in this regard, to identify macroscopic cognitive behaviors with collective effects. And, within this context, the attribute 'emergent' would acquire a well defined meaning.

Unfortunately the proof of the existence of a phase transition associated to the dynamical evolution of an ANN is very difficult (if not impossible) to obtain through the actual techniques. In this regard we remind that in recent times a number of authors (Saad 1998; Marinaro & Scarpetta 2000) tried to describe the learning process within ANN by resorting to the methods of Statistical Mechanics. This field of research, known as *on-line learning*, is based on two main hypotheses:

a. the distribution of input pattern features is random;
b. the number of input units tends to infinity.

The interesting results so far obtained concern the individuation of the order parameters and the macroscopic evolution equations. Unfortunately, despite their attractiveness, these results cannot be applied to concrete ANN. Namely the latter violate both the previous assumptions because not only the number of input units is always finite (and generally small), but the input vectors are never derived in a random way from a probability distribution. On the contrary, they are chosen according to precise rules, in turn connected to the very nature of the data domain from which these vectors were extracted.

The above arguments lead to an unavoidable conclusion: the traditional connectionist approach is unable to prove that macroscopic cognitive behavior is a collective effect emerging from the interactions between the microcognitive units. Such an inability is due to the lack of a consistent theory of emergence and, in turn, is the very cause of a number of problems (still unsolved) which plague the connectionist approach (such as the catastrophic interference problem, the binding problem, and the grounding problem). But what form should such a theory have?

3. What is emergence?

We will not trace here a history of the concept of 'emergence' (see, in this regard, Beckermann et al. 1992; Goldstein 1999), by limiting ourselves to sketch the main conclusions of a debate about this concept, lasting for more than a century (the attribute 'emergent' was introduced for the first time by the English philosopher G. H. Lewes in 1875) and involving a number of scientific experts in different fields (for a synthesis cf. Holland 1998).

The first characteristic of emergence on which there is a general agreement is that it isn't an *objective* property of a given system or process, but rather a measure of the *surprise* experienced by an observer when looking at a system, equipped with a suitable model of its behavior, with goals, mental schemata, knowledge, and he/she detects a property not expected, which appears suddenly and cannot be attributed to an observable cause acting within the system.

A second characteristic of emergence is that it is associated to the existence of *different levels* of description of the same system. Namely, all phenomena we usually consider as emergent typically occur on a level (e.g. macroscopic) higher than the one (e.g. microscopic) on which we search for the causes of the behavior of the system under study. Such a circumstance allows for an easier, and often immediate, recognition of the emergent phenomenon, or property, as *emergent*. Of course, a more precise definition of emergence would require a rigorous definition of what we mean by 'level of description'. In this regard, we remark that the easiest way for introducing it should rely on the notion of *timescale*. For instance, very short timescales are associated to microscopic phenomena, whereas very large timescales are associated to macroscopic ones. Thus, we could, roughly, identify each level of description with a particular timescale associated to system's evolution.

A third characteristic of emergence, which helps crucially in its detection, is the *coherence*. Roughly this means that emergence gives rise to integrated wholes, which keep constant their identity in time (on timescales greater than the ones referring to the individual components of the systems exhibiting emergence). A key point is that this identity is rather *stable* against perturbations, so that we can assert that a true emergence must be associated to a mechanism keeping it, in spite of fluctuations and of actions coming from the environment.

In order to introduce a distinction between the *different kinds* of emergence, we will rely on a categorization already proposed by Crutchfield (1994), who speaks of three kinds of emergence:

e.1. *intuitive*, corresponding to the rough identification of emergence with 'novelty'; it reflects the common use of the attribute 'emergent' in most domains of biological and social sciences;

e.2. *pattern formation*, in which a pattern is said to be "emergent" when it occurs as a non-trivial consequence of the model structure adopted, even if it could be forecast in advance on the basis of a sophisticated mathematical analysis of the model itself; this is the case, for example, of some macroscopic models of self-organization making use of bifurcation phenomena;

e.3. *intrinsic emergence*, in which not only the occurrence of a particular behavior cannot be predicted in advance (even if compatible with the model assumptions), but such an occurrence gives rise, in turn, to a deep modification of system's structure, in such a way as to require the formulation of a new model of the system itself; a typical example is the emergence of new computational abilities (see Sipper 1997; Mikhailov & Calenbuhr 2002).

For many years the only kind of emergence taken into consideration by model builders was pattern formation, usually labelled as *Self-Organization*. Within all these models the focus is on the time evolution of suitable macrovariables, ruled by *evolution equations*, containing a number of *parameters*, and associated to given *initial* or *boundary conditions*. It is easy to understand that the most important forecasting one can derive from these models, obtained by resorting to the methods of *dynamical systems theory* (cf. Glendinning 1994), deals with the number and the type of *attractors* of this dynamical evolution.

Within this framework, the most studied phenomenon has been the one of *bifurcation*. This term denotes a change of the number or of the type of attractors occurring, in the most simple cases, when the value of a parameter, the so-called *bifurcation parameter*, crosses a *critical value*. Such a circumstance suggests a close analogy between bifurcation phenomena and *phase transitions* taking place within physical systems. Namely the two different states of affairs, before and after the critical value, can be viewed as analogous to different *phases* of matter, the critical value itself being viewed as analogous to the *critical point* of a phase transition. It has been shown, however (see Nitzan & Ortoleva 1980; Stein 1980; Anderson 1981; Anderson & Stein 1985), that such an analogy breaks down when we take into account the fact that the values of macrovariables undergo unavoidable fluctuations,due both to the limited sensitivity of our measuring instruments, and to the coupling betwen the system and a noisy environment.

From the definitions previously proposed, it is clear, therefore, that the most interesting models of emergent phenomena should exhibit intrinsic emergence. As regards this kind of models, we will subdivide them into two main classes:

m.1. *ideal models of emergence*, characterized by:

- the identification of macroscopic phenomena with the ones corresponding to infinite volume limit (thermodynamical limit),
- the possibility of deriving microscopic dynamical equations from a suitable general maximum (or minimum) principle;

m.2. *non-ideal models of emergence*, characterized by:

- the existence of a finite, predefined and fixed, volume into which the system is contained,
- the derivation of microscopic dynamical equations only from phenomenological arguments.

A very interesting, but still unsolved, question is whether models belonging to the class *m*.2 can be reformulated or not in such a way as to be classified as belonging to the class *m*.1. The interest in such a question stems from the fact that it is very difficult to find criteria letting us decide whether intrinsic emergence exhibited by a non-ideal model results from a particular, fortunate, choice of system's volume (and whence of boundary conditions), and of the form of microdynamical equations, or from general principles underlying the model itself. Without these criteria the intrinsic emergence exhibited by non-ideal models cannot be *controlled*, and this limits crucially their usefulness.

The discussion about the notion of emergence so far presented within this section left unsolved an important question: how to build an ideal model of intrinsic emergence? In the next section we will present some arguments supporting the claim that the answer to this question can be found only within the framework of quantum theory.

4. Why quantum theory?

Before starting our discussion, we must remind that the idea of resorting to quantum theory to describe cognitive processing as an emergent phenomenon has a long history. The latter began in 1967 with the publication of the seminal paper by Ricciardi and Umezawa on the application of methods of Quantum Field Theory (QFT) to study cognitive processes as collective effects in assemblies of brain neurons (Ricciardi & Umezawa 1967). This paper was followed by the building of a number of interesting models of cognitive processes (mainly of memory operation), based on QFT, and proposed mainly by Umezawa and his pupils (Stuart, Takahashi, & Umezawa 1978, 1979; Jibu & Yasue 1995; Vitiello 1995, 2001). Starting from very different premises, in the same period a number of other researchers (cf. Stapp 1993) tried to apply the methods of quantum theory to the study of consciousness. Both lines of research produced, as a consequence, an ever growing interest for the applications of models based on quantum theory to the description of a number of features of cognitive system operation (cf. Pribram 1993; Nanopoulos 1995; Jibu & Yasue 1997).

We will start our analysis by reminding that quantum models can be subdivided into two main categories: the ones based on quantum mechanics (QM), and the ones based on QFT. QM deals (for a textbook see San Fu Tuan 1994) with systems constituted by a finite, and fixed, number of particles, contained within a finite, and fixed, volume. The physical quantities characterizing them, however, cannot be all measured simultaneously with arbitrary precision. A first consequence of such an uncertainty is that a complete characterization of a particle dynamical state with unlimited precision is impossible. One is, then, forced to introduce the concept of *representation* of the state of the system being considered. In rough, nontechnical terms the choice of a representation consists in selecting a subset of the dynamical variables describing the state of the system, such that all variables belonging to the subset can be measured simultaneously with arbitrary precision. In a sense, every representation can offer only a *partial* description of system's dynamics. However, an important theorem, proved by Von Neumann (1955), asserts that in QM all possible representations are reciprocally *equivalent*. This means that they give rise to the same values of probabilities of occurrence of results of all possible measures relative to the physical system under consideration, independently from the particular representation chosen.

A second consequence of uncertainty is the occurrence of a typically quantum phenomenon, named *Bose-Einstein condensation* (BEC). The latter occurs in presence of suitable conditions (e.g. low temperatures) and consists in the fact that all particles belonging to a given system fall simultaneously into the same quantum state. This implies that their behaviors become all correlated, thus giving rise to a macroscopic state, which appears as globally *coherent*. BEC can be considered as the prototype of formation of macroscopic entities emerging, as collective effects, from the laws ruling the behaviors of microscopic particles.

These advantages of QM are, however, counterbalanced by a number of shortcomings. The main ones can be listed as follows:

- the Von Neumann theorem prevents from any application of QM to the description of *structural changes*; namely, as all representations are physically equivalent, it is impossible to have a model based on QM in which a system can exist in two different, nonequivalent, forms; therefore it will be, in principle, impossible to formulate a theory of *phase transitions*, and, *a fortiori*, of emergent phenomena;
- the occurring of macroscopically coherent states, as in BEC, is, in most cases, hindered by the interactions of the system under study with the external environment; namely they act in such a way as to destroy the quantum coherence and, if the *decoherence time* is small enough, the latter becomes unobservable; this circumstance limits the usefulness of QM to particular cases (such as the world of atoms or of molecules).

These drawbacks prevent any application from QM to a description of intrinsic emergence, and suggest that perhaps QFT could be a better framework for modeling it. QFT, contrary to what happens in QM, assumes that the main physical entities are the *fields* and not the particles, the latter being identified with domains in which the field strength is exceedingly high. Such a standpoint has a long tradition, going back to Faraday and Maxwell, and underlies the most powerful architectures ever built in theoretical physics, such as Einstein's General Relativity, and unified gauge theories.

Following such an approach, QFT attempts to treat fields as defined by uncertain quantities, As the fields, in principle, are not restricted to definite volumes, QFT deals typically with infinite volumes. In this way it becomes easier to introduce a sharp distinction between *macroscopic* phenomena (the ones surviving when the volume tends to infinity) and *microscopic* phenomena (which appear as fluctuations when the scale of observation is large enough). Of course, the approach followed by QFT is very difficult to implement in a concrete way, much more than in the case of QM. For this reason, QFT can still be considered an incomplete theory, of which only particular realizations have been so far worked out, at the expense of the introduction of a very complex mathematical machinery.

Despite these difficulties, QFT, first proposed in 1926 by P. A. M. Dirac, obtained in the last fifty years remarkable successes in describing and forecasting phenomena occurring within the domain of particle physics and condensed matter physics (see, in this regard, Itzykson & Zuber 1986; Umezawa 1993). Here we will limit ourselves to mentioning an important feature, which is central for our future discussion about emergence: within QFT, differently from QM, there is the possibility of having different, *nonequivalent*, representations of the same physical system (cf. Haag 1961; Hepp 1972). A consequence of this is that *only* QFT, allowing for the existence of different phases of the system itself, can deal with *phase transitions*, i.e. with global structural changes of the system under study. Such a circumstance entails that the framework of QFT is actually the only one possible if we attempt to model intrinsic emergence:

We will now substantiate the previous arguments through the exhibition of a concrete ideal model of intrinsic emergence based on QFT. In this regard, let us start by stressing that in an ideal model, by definition, we can always introduce a function playing a role analogous to the one of *energy* in physical systems, so that stable and metastable equilibrium states are directly associated to local minima of such a function. The occurring of intrinsic emergence, then, can be identified with a transition, triggered by the change of value of a given parameter, in which a local energy minimum is split into a number of different local energy minima, all *equivalent*, i.e. characterized by the same value of minimum energy (we will refer to these states as *ground states*). The intrinsic emergence is due to the fact that, if the system was, before the transition, in the state corresponding to the old energy minimum, surely the transition will provoke the settling of the system into one of the new energy minima, but we cannot forecast *which* of them will be chosen, on the basis of the model we have, because all minima are equivalent one to another. Such a form of transition is usually called *spontaneous symmetry breaking* (SSB), and appears as the only way, so far known, to introduce intrinsic emergence in systems described by ideal models.

As regards SSB, some remarks are in order. The first one is that, in most cases, the multiplicity of ground states exists only if we go at the infinite volume limit. A second remark is that, both in classical and QFT-based descriptions of SSB, the system will be anyway forced to choose one particular ground state. States corresponding to linear combinations of different ground states are not allowed, even in QFT, because it can be proved that any operator connecting two different ground states vanishes at the infinite volume limit (cf. Huang 1987).

A third remark is that, if we describe SSB within the context of QFT, the occurring of a SSB implies the appearance of collective excitations, which can be viewed as zero-mass particles carrying long-range interactions. They are generally called *Goldstone bosons* (see Umezawa 1993). Such a circumstance endows systems, in which SSB takes place, with a sort of *generalized rigidity*, in the sense that, acting on one side of the system with an external perturbation, we can transmit such a perturbation to a very distant location essentially unaltered. The reason for the appearance of Goldstone bosons is that they act as *order-preserving* messengers, preventing the system from a change of the particular ground state chosen at the moment of SSB transition. Besides, they are a direct manifestation of intrinsic emergence, as none of the forces acting between system's elementary constituents is able to produce generalized rigidity. We thus have that only within a QFT description of SSB we observe the occurring of macroscopic coherent entities which are stable with respect to external perturbations. It is, further, possible to prove (for technical details see Stein 1980; Rumer & Ryvkin 1980) that classical, but ideal, models of SSB, not based on the framework of QFT, cannot give rise to Goldstone bosons and to generalized rigidity. This occurs because the order-preserving messengers in the classical case are absent. As a conclusion, SSB, even if possible in classical cases, is *unstable* with respect to (thermal) perturbations.

The models of SSB based on QFT give a consistent description of the collective effects: they are to be identified with the long-range correlations supported by the Goldstone bosons arising after a system, crossing the critical point of a SSB, chooses a particular ground state between the many ones available. The validity of such a picture is also supported by the fact that the Goldstone bosons associated to most SSB transitions (such as phonons in crystals, or magnons in ferromagnets) have been experimentally detected. Within the next section we will deal with the problem of how to implement the first step needed for the introduction of QFT formalism within the connectionist approach based on ANN: the definition of quantum neurons.

5. Quantum neurons

In order to build connectionist models based on the formalism of QFT two different roads could be followed:

- to show that the description of the dynamics of a given ANN is already equivalent, as it stands, to a suitable QFT-based model;
- to modify from the starting the dynamics of an ANN (viewed as a classical object) in such a way as to endow it with QFT-like features.

The former road is, however, still bristling with considerable technical difficulties. For this reason in this paper we will follow the latter road. In this regard, we will start by remarking that a modification of the ANN dynamics to endow it with QFT-like features can be performed in three different ways:

- by resorting to a microscopic description, of quantum nature, of the activity of a single biological neuron on a molecular level;
- by introducing a quantum formulation of the equations describing neuronal behavior within the traditional ANN (*quantum neuron*);
- by deriving from the microscopic description of ANN dynamics suitable macroscopic neural field equations and quantizing them through the standard procedures of QFT.

The first and the third alternative entail a number of problems which are very difficult to solve. Therefore within this section we will deal only with the second one.

The introduction of a quantum description of the activity of a neuronal unit is, however, a difficult technical problem, notwithstanding the proposals advanced by some authors (cf. Samsonovich 1994; Zak 2000; Gupta & Zia 2001). Namely, within both QM and QFT, the quantization procedures are usually based on the existence of a conserved quantity, that is the total energy of the system under study. On the contrary, within traditional connectionist models the neural units are usually described as *dissipative systems*, in which the energy is no longer conserved. How to find a way out of this problem?

We can adopt, in this regard, three different strategies:

1. to modify the structure itself of Hamiltonian mechanics in such a way as to take into account the dissipative phenomena;
2. to introduce a suitable description of the *environment* (the cause of dissipation) in such a way that the whole system neuron+environment be still conservative;
3. to modify the description of neural dynamics in such a way as to make it conservative.

The strategy 1 is currently pursued by a number of authors (see, for instance, Tarasov 2001). The strategy 2 was implemented in the best way through the so-called *doubling mechanism* (Celeghini, Rasetti, & Vitiello 1992; for an application to cognitive processing we recommend the superb exposition contained in Vitiello 2001). Within this section, instead, we will shortly explore a possible implementation of the third strategy. In this regard, let us start from the usual formulation of the dynamics of a McCulloch-Pitts neuron in terms of a differential equation:

$$ds_i/dt = -s_i + F(\textstyle\sum_j w_{ij}s_j),$$

where s_i denotes the instantaneous output activity of the i-th neuron, w_{ij} are the connection weights, and F denotes a suitable activation function. By supposing that each neuron be characterized by a spontaneous base activity, represented in a fictitious way through the introduction of a suitable *self-connection weight*, denoted by w, we will obtain that a single isolated neuron will obey the dynamical law:

$$ds/dt = -s + F(ws).$$

We can now consider this latter as a *first integral* of the 'true' dynamical equation, which can be easily obtained through a further derivation:

$$d^2s/dt^2 = s - F(ws) - wsF'(ws) + wF'(ws)F(ws).$$

Here the symbol F' denotes the derivative of the function F with respect to its argument. It is then easy to show that the quantity:

$$E = (1/2)(ds/dt)^2 + V, \quad V = -(s^2/2) + sF(ws) - [F^2(ws)/2]$$

is conserved. It is tempting to call this quantity *energy*, but it is deeply different from the physical energy. Namely s doesn't denote a spatial coordinate, but an activation value.

Now we need to take into account that the spatial coordinates, used in physics to specify the position of a particle, are nothing but particular examples of *configurational variables*, that is independent variables, different from the time variable, which allow for a specification of the instantaneous configuration of the system under study. We must, then, recognize that the configurational variable most suited to describe our McCulloch-Pitts neuron is its output activity s and not its spatial location. Within this framework we could say that E is a sort of *configurational total energy*. This lets us introduce a *configurational wavefunction* $\psi(s, t)$. If we identify the *configurational momentum* p with ds/dt, we can now generalize the definition of the quantum operators associated to the configurational coordinate and to the configurational momentum, so that the Schrödinger equation for a single quantum neuron can be written under the form:

$$-(h/8\pi^2)(\partial^2\psi/\partial s^2) + \{-(s^2/2) + sF(ws) - [F^2(ws)/2]\}\psi = i(h/2\pi)(\partial\psi/\partial t).$$

The most complicated issue of the behavior of a network of quantum neurons – a quantum neural network – will be dealt with in the next section.

6. Quantum neural networks

In order to generalize our previous theory, we will allow for an interaction between different quantum neurons. If we suppose that the self-connection weight w be the same for all neurons, then the classical dynamical equation ruling the behavior of a generic neuron will assume the form:

$$d^2s/dt^2 = s - F(ws + I) - wsF'(ws + I) + wF'(ws + I)F(ws + I),$$

where I denotes the contribution coming from other neurons:

$$I = \sum_j w_{ij}s_j.$$

Some mathematical considerations show that, in this case, the total potential energy V of a neural network containing N neurons will be given by:

$$V = \sum_i V_i + V_{int},$$

where:

$$V_i = -(s_i^2)/2 + s_iF(ws_i + \sum_j w_{ij}s_j) - (N/2)F^2(ws_i + \sum_j w_{ij}s_j)$$
$$V_{int} = \sum_{ij} s_is_j[F(ws_i + \sum_k w_{ik}s_k) + F(ws_j + \sum_k w_{jk}s_k)].$$

In order to study the behavior, both of the classical and of the quantum version of this neural network, we should, in the classical case, solve the network dynamical equations, and, in the quantum case, solve the associated Schrödinger equation. Both tasks are impossible to perform. We thus adopted a different strategy: to study, through computer simulations, the behavior of a quantity characterizing the long-range correlations both in the classical and in the quantum case (of course, the long-range correlations in our case are correlations between different activation values). If we should find that the long-range correlations are present in the quantum case and are absent in the classical one, then we shoul have obtained a concrete proof of the usefulness of a direct introduction of quantum features within a classical ANN.

In order to evidence the existence of long-range correlations, we chose the *Pair Correlation Function of Activations* (PCF) $D(s_1, s_2)$, defined in such a way that $D(s_1, s_2)ds_1 ds_2$ is the probability of finding simultaneously a neuron, whose output activation is contained within an interval of amplitude ds_1 and centered on s_1, and a neuron, whose output activation is contained within an interval of amplitude ds_2 and centered on s_2. In the classical case, the Statistical Mechanics (cf. Huang 1987) shows that there is no long-range correlation between neurons characterized by different values of output activation, as their PCF decays very quickly with increasing difference Δs between their activation values.

The numerical simulations confirmed what is expected on the basis of classical Statistical Mechanics. In Figure 1 we reported the plot of PCF vs. Δs obtained by averaging the results of 10 different simulations, each performed on a network of 50 neurons interacting according to the classical form of the dynamical laws introduced before.

As can be seen, the plot evidences, for a classical neural network, an absence of long-range correlations in the space of output activations, as expected. Let us now deal with the quantum version of our neural network. In this case we can obtain the form of the ground state of the system by resorting to numerical simulations based on the Quantum Diffusion Monte Carlo method (QMC) (for a description of QMC see, e.g., Vesely 2001). In Figure 2 we report the plot of PCF vs Δs, evaluated from the numerical data obtained through QMC on the ground state of a neural network constituted by 30 quantum neurons of the kind introduced above.

As we can see from the Figure 2, in the quantum case we have a strong evidence for a long-range correlation of neuron output activities, contrary to what happened in the classical case.

On the basis of these findings, we can assert that the introduction of quantum-like features in a direct way gave rise within neural networks to long-range correlations which were absent in the classical case. Of course, this doesn't mean that we reached our initial goal: so far we worked only within the framework of QM and we need a further step to build a QFT-based model.

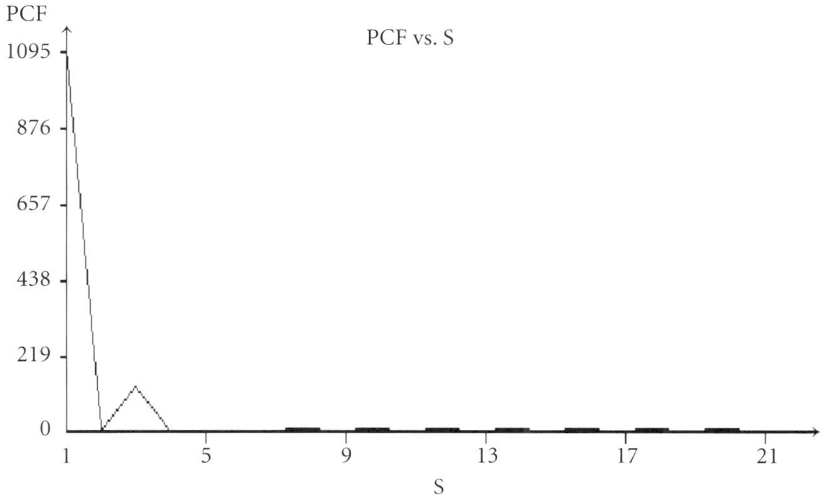

Figure 1. Plot of PCF vs. Δs, obtained from the average on 10 computer simulations, each performed for 500 steps; parameters: $w = 0.1$, amplitude of each time step = 0.001, connection weights chosen randomly between – 0.1 and 0.1; numerical integration performed through the Euler method.

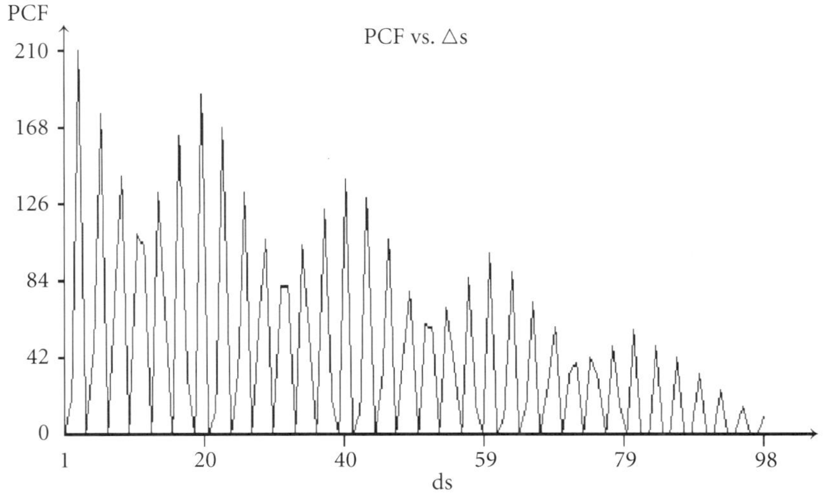

Figure 2. Plot of PCF vs. Δs obtained from the numerical data about the ground state of a network of 30 interconnected quantum neurons. The parameters have the same values as in Figure 1.

Such a step, however, is now easier, as we already have at our disposal a QM-based description of the operation of neuronal units. In this regard we will limit ourselves to mentioning that a possible neural field equation could be given by the Schrödinger equation ruling the quantum behavior of an isolated neuron, obtained at the end of the previous section. If, further, we suppose that the self-connection weights vary with time following a Hebbian law of the form:

$$dw/dt = -w + \chi s^2 + \beta,$$

(here χ, β denote suitable parameters) we will obtain a pair of equations which are nothing but the field equations describing the interacting fields $\psi(s, t)$ and $w(s, t)$. Within the framework of QFT we will interpret the fields ψ and w not as functions but as *field operators*. By using standard methods we could, then, study this first QFT-based connectionist models and try to discover if there is a SSB, if collective phenomena could occur and if there is a place for Goldstone bosons. We will not pursue here such a topic, because within this context it suffices to have shown that it is possible to build a concrete connectionist model based on the framework of QFT.

7. Conclusions

We can now summarize our main findings as follows:

1. the claims of the connectionist approach lack any validity in absence of a rigorous theory of intrinsic emergence; we must state what are the conditions granting for a macroscopic emergence of a given kind and endowed with given macroscopic features; otherwise it will be impossible to have a theory of the cognitive operation, as the one the connectionists pretend to build, which can be compared with the experimental data;
2. the only ideal models of intrinsic emergence existing thus far can be formulated within the framework of QFT; the non-ideal models (such as the traditional neural networks, cellular automata, Artificial Life models, and like) lack any form of *control* on the emergence itself;
3. despite the dissipative nature of neural dynamics, it is nonetheless possible to introduce a QM-based description of the dynamics of a quantum neuron and of a quantum neural network, by using a formalism fully analogous to the one already in use for the quantum description of systems of particles;

4. the direct introduction of quantum features within the dynamics of a neural network gives rise to the occurrence of long-range correlations; such correlations, on the contrary, are absent in the classical counterpart of the same neural dynamics;
5. it is concretely possible to introduce a model of neural dynamics based on the formalism of QFT; this shows that it is possible to introduce a theory of intrinsically emergent behaviors and of collective effects in a neural network based on a QFT-like description of its dynamics. Such a theory is the heart of Quantum Connectionism and allows for a sound grounding of the claims advanced by connectionists.

The final conclusion of this paper can be cast under the form of an answer to the main question from which we started: can we identify the macroscopic cognitive behaviors with collective effects emerging from the interactions between brain neurons? The answer is *yes*, provided, however, we adopt a description of neural brain dynamics based on QFT. Only in this way we can establish a bridge between the brain and the mental operations, and, mostly, individuate what factors and what microscopic neural features *control* the macroscopic features of emergent cognitive behaviors. Thus Quantum Connectionism will help in a substantial way the builders of future Cognitive Science.

References

Anderson, P. W. (1981). Can broken symmetry occur in driven systems? In G. Nicolis, G. Dewel, & P. Turner (Eds.), *Equilibrium and Nonequilibrium Statistical Mechanics* (pp. 289–297). New York: Wiley.

Anderson, P. W. & Stein, D. L. (1985). Broken symmetry, emergent properties, dissipative structures, life. Are they related? In F. E. Yates (Ed.), *Self organizing Systems: The emergence of Order* (pp. 445–457). New York: Plenum Press.

Bartlett, P. L. & Anthony, M. M. (1999). *Neural Network Learning: Theoretical Foundations.* Cambridge, UK: Cambridge University Press.

Beckermann A., Flohr, H., & Kim, J. (Eds.). (1992). *Emergence or reduction?* Berlin: de Gruyter.

Bishop, C. M. (1995). *Neural Networks for Pattern Recognition.* Oxford, UK: Oxford University Press.

Celeghini, E., Rasetti, M., & Vitiello, G. (1992). Quantum dissipation. *Annals of Physics, 215*, 156–170.

Cruchtfield, J. P. (1994). The calculi of emergence: Computation, dynamics and induction. *Physica D, 75*, 11–54.

Glendinning, P. (1994). *Stability, Instability and Chaos: An introduction to the theory of Nonlinear Differential Equations.* Cambridge, UK: Cambridge University Press.

Goldstein, J. (1999). Emergence as a construct: History and issues. *Emergence, 1*, 49–72.

Gupta, S. & Zia, R. K. P. (2001). Quantum neural networks. *Journal of Computer and System Sciences, 63*, 355–383.

Haag, R. (1961). Canonical commutation relations in Quantum Field Theory and Functional Integration. In W. E. Brittin, B. W. Downs, & J. Downs (Eds.), *Lectures in Theoretical Physics*, Vol. 3 (pp. 353–381). New York: Wiley.

Hepp, K. (1972). Quantum theory of measurement and macroscopic observables. *Helvetica Physica Acta, 45*, 237–248.

Holland, J. H. (1998). *Emergence. From Chaos to Order*. Redwood City: Addison-Wesley.

Huang, K. (1987). *Statistical Mechanics*. New York: Wiley.

Itzykson, C. & Zuber, J. B. (1986). *Quantum Field Theory*. Singapore: McGraw-Hill.

Jibu, M. & Yasue, K. (1995). *Quantum Brain Dynamics and Consciousness: An Introduction*. Amsterdam: Benjamins.

Jibu, M. & Yasue, K. (1997). Magic without magic: Meaning of quantum brain dynamics. *Journal of Mind and Behavior, 18*, 205–228.

Marinaro, M. & Scarpetta, S. (2000). On-line learning in RBF neural networks: A stochastic approach. *Neural Networks, 13*, 719–729.

McClelland, J. L. & Rumelhart, D. E. (Eds.). (1986). *Parallel Distributed Processing. Explorations in the Microstructure of Cognition*. Cambridge, MA: MIT Press.

Mikhailov, A. S. & Calenbuhr, V. (2002). *From cells to societies. Models of complex coherent actions*. Berlin-Heidelberg-New York: Springer.

Nanopoulos, D. V. (1995). "Theory of brain function, quantum mechanics and superstrings", preprint ACT-08/95, CERN-TH/95-128, CTP-TAMU 22/95, hep-ph/9505374. Geneva: Cern Theory Division.

Nitzan, A. & Ortoleva, P. (1980). Scaling and Ginzburg criteria for critical bifurcations in nonequilibrium reacting systems. *Physical Review A, 21*, 1735–1755.

Pribram, K. H. (Ed.). (1993). *Rethinking neural networks: Quantum fields and biological data*. Hillsdale, NJ: Erlbaum.

Ricciardi, L. M. & Umezawa, H. (1967). Brain and physics of many-body problems. *Kybernetik, 4*, 44–48.

Rojas, R. (1996). *Neural Networks. A Systematic Introduction*. Berlin-Heidelberg-New York: Springer.

Rumer, Yu. B. & Ryvkin, M. Sh. (1980). *Thermodynamics, statistical physics, and kinetics*. Moscow: Mir.

Saad, D. (Ed.). (1998). *On-line Learning in Neural Networks*. Cambridge, UK: Cambridge University Press.

Samsonovich, A. (1994). Storage capacity of quantum neural networks. In P. Werbos, H. Szu, & B. Widrow (Eds.), *WCNN 94*, Vol. II (pp. 804–807). Hillsdale, NJ: Erlbaum.

San, Fu Tuan (Ed.). (1994). *Modern Quantum Mechanics*. Reading, MA: Addison-Wesley.

Sipper, M. (1997). *Evolution of Parallel Cellular Machines*. Berlin-Heidelberg-New York: Springer.

Stapp, H. P. (1993). *Mind, Matter, and Quantum Mechanics*. Berlin-Heidelberg-New York: Springer.

Stein, D. L. (1980). Dissipative structures, broken symmetry, and the theory of equilibrium phase transitions. *Journal of Chemical Physics, 72*, 2869–2874.

Stuart, C. J. I., Takahashi, Y., & Umezawa, H. (1978). On the stability and non-local properties of memory. *Journal of Theoretical Biology, 71*, 605–618.
Stuart, C. J. I., Takahashi, Y., & Umezawa, H. (1979). Mixed-System brain dynamics: Neural memory as a macroscopic ordered state. *Foundations of Physics, 9*, 301–327.
Tarasov, V. E. (2001). Quantization of non-Hamiltonian and dissipative systems. *Physics Letters A, 288*, 173–182.
Umezawa, H. (1993). *Advanced Field Theory. Micro, Macro, and Thermal Physics*. New York: American Institute of Physics.
Vesely, F. J. (2001). *Computational Physics – An Introduction*. New York: Kluwer.
Vitiello, G. (1995). Dissipation and memory capacity in the quantum brain model. *International Journal of Modern Physics B, 9*, 973–989.
Vitiello, G. (2001). *My double unveiled*. Amsterdam: Benjamins.
Von Neumann, J. (1955). *Mathematical foundations of quantum mechanics*. Princeton, NJ: Princeton University Press.
Zak, M. (2000). Quantum model of emerging grammars. *Chaos, Solitons and Fractals, 11*, 2325–2330.

Chapter 9

The rheomode of language of David Bohm as a way to re-construct the access to physical reality

Maxim I. Stamenov
Georg-August-Universität Göttingen, Germany and
Bulgarian Academy of Sciences

In his most popular book the theoretical physicist David Bohm (1980) proposed *inter alia* an experiment with language. The aim of the new rheomode of language as a radical alternative to natural language is to make the mental representation transparent to itself and, in this way, to attune itself also to the 'totality of all that is', the holomovement. The development of this mode of language is an attempt to make explicit what orders of observation by the cognizing function can reach for the physical reality. It shows the very paradoxical metamorphoses an observation must undergo in re-structuring itself in order to attune to the very dynamics of reality as its double-in-coemergence in self-and-other transparency.

With this orientation, the rheomode goes beyond all available paradigms in the history of science and philosophy that study and model the relationship between mind and reality. This is the case because it does not try to re-present by some symbolic means aspects of the physical reality that are in principle inaccessible to the senses. The rheomode is not an attempt to go outside the 'visible' and/or thinkable and/or knowable. It is, instead, a re-construction aiming in the opposite direction – at making what is capable of becoming 'visible' all the time visible in a seamless way.

1. The rheomode of language – the basic idea

In his most popular book the theoretical physicist David Bohm (1980) proposed *inter alia* an experiment with language, and this experiment is strikingly different from those carried out in linguistics, psycholinguistics, sociolinguistics, communication studies, cognitive psychology and cognitive science, philosophy, logic, etc.[1] He proposed "to *experiment* with changes in the structure of the common language" (Bohm 1980: 27).

From linguistic, psychological and philosophical points of view, the proposal of Bohm for an 'alternative' language is an unprecedented one along the following lines. Nobody before him claimed that in order to comprehend how language contributes to the way thought is constituted it is not enough to follow it; it is necessary actively to interfere with its function in order to discern clearly 'same' and 'different' in its structures. This practice with no alternative in linguistics is to study language in manipulating the acceptability and grammaticality of its units with different sorts of permutations in, deletions of and additions to their structure. To this *status quo* Bohm offers the following alternative:

> [...] one of the best ways of learning how one is conditioned by habit (such as the common usage of language is, to a large extent) is to give careful and sustained attention to one's overall reaction when one 'makes the test' of seeing what takes place when one is doing something significantly different from the automatic and accustomed function. (Bohm 1980: 28)

Bohm starts his argument with the point that subject-verb-object sentence structure is common to the syntax of modern languages and this structure powerfully builds in us the implicit and ever present presupposition that action arises in a separate entity and this action, in the case it is described by a transitive verb, crosses over the space between them (the subject and object) to another separate entity, the object (Bohm 1980: 29). In some ancient languages like Hebrew, however, the verb was given primary, i.e., basic, importance in the grammatical structure of language itself, i.e., not in its description only, as the roots of almost all words in Hebrew were certain verbal forms, while adverbs, adjectives, and nouns were obtained by modifying the verbal form with prefixes, suffixes, etc. In other words, the 'inner form' of these words was directly and explicitly pointing to some action, event, or 'movement' as the 'pedestal' of the sense of the word in question.

The aim of the new mode of language, the *rheomode* (from *rheo*, a Greek verb, meaning "to flow"), is to develop such structures of language "in which movement is to be taken as primary in our thinking and in which this notion will be incorporated into the language structure by allowing the verb rather than the noun to play a primary role" (Bohm 1980: 30). The aim is, *ergo*, to create a mode of language with a new structure that is not prone toward fragmentation, as is the case with all our native ones.

As a cue where to start re-building the bewildering complexity of natural language, the high prominence in contemporary usage of the word 'relevant' is pointed out (cf. Sperber & Wilson 1986, for an analysis of this concept in the context of linguistic pragmatics), but the latter is interpreted in mentalist terms:

> [...] to see the relevance or irrelevance of a statement is primarily an act of perception of a very high order similar to that involved in seeing its truth or falsity. In one sense the question of relevance comes before that of truth, because to ask whether a statement is true or false presupposes that it is relevant (so that to try to assert the truth or falsity of an irrelevant statement is a form of confusion), but in a deeper sense the seeing of relevance or irrelevance is evidently an aspect of the perception of truth in its overall meaning.
> (Bohm 1980:33)

One can further develop this idea by citing another passage from the book under discussion:

> [...] it is not right, for example, to regard the division between relevance and irrelevance as a form of accumulated knowledge of properties belonging to statements (e.g., by saying that certain statements 'possess' relevance while others do not). Rather, in each case, the statement of relevance or irrelevance is communicating a perception taking place at the moment of expression, and in the individual context indicated in that moment. [...] when relevance or irrelevance is communicated, one has to understand that this is not a hard and fast division between opposing categories but, rather, an expression of an ever-changing perception, in which it is possible, for the moment, to see a fit or non-fit between the content lifted into attention and the context to which it refers. (Bohm 1980:34)

In order to make 'relevance' move appropriately we have to "make it fluid again". The fast and strict formal divisions could be made again flexible and fluid by conceptualizing 'relevance' not as a state-bound noun, but to consider it as a movement, an action-bound verb. We are invited to re-build the 'inner form' of the verb 'to re-levate'. It ultimately comes from the root 'to levate', "to lift". The meaning of this verb is defined using as a 'pedestal' the general sense of "to lift" in the following way:

> The spontaneous and unrestricted act of lifting into attention any content whatsoever, which includes the lifting into attention of the question of whether this content fits a broader context or not as well as that of lifting into attention the very function of *calling attention* which is initiated by the verb itself. (Bohm 1980:35)

One must mention at least three characteristics of this most basic 'movement' in the rheomode of language:

a. The *spontaneousness and unrestrictedness of the act of lifting into attention of any content whatsoever*, which means, psychologically, a realization of an intentionally controlled function (attention) that is at the very same time not restricted by the limited intentional potential of any control structure in charge of attention;
b. The ability *to judge the fit* between the spontaneously and unrestrictedly lifted content and the broader context in which it comes up (one 'lifts the lifted by itself' without losing sight of the 'ground');
c. The *self-recursiveness of the act of calling attention* in its ability to call attention not only to the content which is picked up, but to the very function of calling attention, i.e., being attentive to something, as well as being attentive of the function of this being attentive to.

The definition of 'to levate' may look strange; something more, it may look self-contradictory. Can a cognitive act be aware of itself at the very same time as it is aware of its object? That is something different from a recursive loop of several cognitive acts in succession, each being capable of reflecting on the structure of the previous one representing it as its own content. The requirement for a *self-recursive loop of (self-facing) attention* means to pay attention to the function of attention at the very moment it is activated. Bohm himself comments that the aspects (a)–(c) of 'to levate' are not nouns, and, *ergo*, they can be 'fluidized' and 'merged' into a cognitive act in which you are aware of the object as well as of the subject of consciousness in one and the same time with direct immediacy! This seems logically impossible. But this is impossible if and only if the states are construed as noun-like cognitively impenetrable entities that are taken under the scope of the predicate as its subject and direct (immediate) object.

2. The way of the rheomode of language

Next I will represent the way Bohm developed the idea of the rheomode of language following him step by step. The only difference with the original is that here only the first verb paradigm will be given in full and in a more grammar-like way, compared with the original. For all other verb paradigms only the form of the basic verb will be provided as the derivation in all cases is along the lines of the first paradigm.

Acting in accordance with the logic envisaged above, we can now introduce the full paradigm of the verb 'to levate':

To levate = the spontaneous and unrestricted act of lifting into attention any content whatsoever, which includes the lifting into attention of the question of whether this content fits a broader context or not as well as that of lifting into attention the very function of calling attention which is initiated by the verb itself;

To re-levate = lifts certain content into attention again, for a particular context, as indicated by thought and language. The prefix *re-* signifies a new occasion of 'to levate', as well as similarity between the two occurrences – of levating some content and re-levating it. But the re-levated content should not be considered to be simply identical, because it implies time, another occasion which cannot only be similar to the first one, but is also different;

To re-levate is re-levant = to enact a perceptual act proving whether the content lifted again fits the observed content. When this perception reveals a fit we are entitled to say that 'to re-levate is re-levant' [please note that property ascription cannot stand alone but is derivative of the possibility to form a proposition with a verb to which it is ascribed];

To re-levate is irre-levant = when the act of perception discloses actual non-fit of the content lifted again to the context in question we say that 'to re-levate is irre-levant';

Re-levation = a continuing state of lifting again and again a given content into attention;

Irre-levation = to continue with a state of re-levation where it is irrelevant to do so or doing that from the very start. In other words, inappropriate could be either the first act of re-levating and next this state is continued because of the inability and/or of the desirability of remaining in the state of irre-levation, or re-levation can turn into irre-levation because of a change of the internal and/or external context of use of the mental content while, being inattentive to the actual change, we continue to lift through the application of our attention a habitual pattern to match the stimuli;

Levation = a sort of generalized and unrestricted totality of acts of lifting into attention. (Bohm 1980: 35)

This way of using the structure of language gives us the possibility "to discuss what is commonly meant by 'relevance' in a way free of fragmentation, for we are no longer being led, by the form of the language, to consider something called relevance as if it were a separate and fixed quality" (Bohm 1980: 35–36). The meaning construction here is grounded in some action or event in the mental continuum. This is the referent of the root verb form. Finding out the meaning of the derivative forms of rheomode presupposes tracing their sense to the basic internal action implied plus the means for their 'solidification' and becoming static objects or qualities in the mental continuum. All this is enacted 'here-and-now' on each occasion of the use of the verb and all its derivatives.

The way of use of the new language as the most appropriate cognitive means for 'entering the realm of consciousness and reality' is given in the following passage:

> Even more important, we are not establishing a division between what the verb 'to levate' means and the actual function that takes place when we use this verb. That is to say, 'to levate' is not only to attend to the thought of lifting an unrestricted content into attention but it is also to engage in the very act of lifting such an unrestricted content into attention. The thought is thus not a mere abstraction, with no concrete perception to which it can refer. Rather, something is actually going on which fits the meaning of the word, and one can, at the very moment of using the word, perceive the fit between this meaning and what is going on. So the content of thought and its actual function are seen and felt as one, and then one understands what it can mean for fragmentation to cease, at its very origin. (Bohm 1980: 36)

Bohm gives only seven more word-formation paradigms, all of the latter referring to mental activities in the perceptual cum cognitive mode of functioning. It is not clear exactly how many have to be included in the mini-max set of them; in order to decide on this question we need, according to Bohm, a 'flash of very penetrating insight' yet to be enacted.

For the formation of the second paradigm is taken not without good motivation the Latin verb *videre* meaning "to see". The root verbal form in the rheomode for "seeing" will be 'to vidate':

> *To vidate* = calls attention to a spontaneous and unrestricted act of perception of any sort whatsoever, including perception of whether what is seen fits or does not fit 'what is', as well as perception even of the very attention-calling function of the word itself. (Bohm 1980: 36)

One can point out on this occasion that in the rheomode perception and attention are not two different functions but two complementary aspects of one and the same 'stream of consciousness':

> In an act of vidation, it is necessary to levate a content into attention, and in an act of levation, it is necessary to vidate this content. So the two movements of

> levation and vidation merge and interpenetrate. Each of these words merely emphasizes (i.e., re-levates) a certain aspect of movement in general. It will become evident that this will be true of all verbal roots in the rheomode. They all imply each other, and pass into each other. (Bohm 1980: 37)

The verb next to consider is 'to divide'. It is analysed as a combination of the verb *videre* "to see" and the prefix *dis* meaning "asunder". 'To divide' could be re-etymologized as meaning "to see as separate":

> *To di-vidate* = calls attention to the spontaneous act of seeing things as separate, in any form whatsoever, including the act of seeing whether or not the perception fits 'what is', and even that of seeing how the attention-calling function of this word has a form of inherent division in it. The *form* of the word makes clear that it is different from 'to vidate' from which it is derived. 'To di-vidate' implies not only a *content* (or meaning, or operation) of division; the very word-formation structure of this word corresponds to the function for which the notion of division is seen to provide a description capable to fit the actual structure of inner action. (Bohm 1980: 37)

Of specific methodological purport becomes the relation between 'to vidate' and 'to di-vidate'. If we consider them different, we foster fragmentation, opening the Pandora's box leading 'from the primordial oneness (or non-differentiation)' to 'duality (= splitting-into-two in potentially endless recursive loops)'. Bohm points out in this respect, that the difference between them holds true only in some limited context and should not be taken as an actual break between the meanings and functions of these two words: "Ultimately, wholeness is primary, in the sense that these meanings and functions pass into each other to merge and interpenetrate" (Bohm 1980: 37).

The meaning and function of 'to di-vidate' is for the sake of developing more and more articulated and detailed descriptions of the whole. Surprisingly, Bohm (1980: 38) declares that the opposite movement, the movement from division to one-ness of perception is through the action of *ordering*. This seems from the first sight to run evidently wrong, as ordering is considered an analytic cognitive operation, through which we can only multiply objects available in the mental representation. Bohm here speaks however not about ordering of objects (internal or external), or operations with them, but about *ordering aiming at one-ness of perception*, and the 'direct (intelligent) perception' in the rheomode is something very special. (Cf. below for further discussion re. the notion of 'intelligence'.)

Attention-movement itself is conceptualized as a very high, practically open-end hierarchy of potential orders (far exceeding *simple sequential* ones): "This movement [of attention; M.S.] has to have an order that fits the order in that which is to be observed, or else we will miss seeing what is to be seen" (Bohm 1980: 39).

The next word-formation paradigm is introduced so as to better understand the nature of ordering in language and thought:

> *To ordinate* = calls attention to a spontaneous and unrestricted act of ordering of any sort whatsoever, including the ordering involved in seeing whether any particular order fits or does not fit some observed content, and even the ordering which arises in the attention-calling function itself. It does not primarily mean 'to think about an order' but, rather, to engage in the very act of ordering attention, while attention is given also to one's thoughts about order. (Bohm 1980: 39)

We can now see that *ordination* evidently implies *levation*, *vidation* and *di-vidation*, and vice versa. But in the relations between them a hierarchy seems to be implied, which Bohm did not discuss explicitly. For example, it is with 'to ordinate' that we face for the first time the problem of sequence and time and its way of representation in the rheomode; the previous three verbs represent spatial structure that can be visualized without stretching the boundaries of the immediate present.

With the re-patterning of the way of linguistic meaning and thinking comes a revision of the nature of 'truth' and 'fact' in them. In ordinary language truth is taken as a noun, i.e., as a static representation which stands for something that can be grasped once and forever or to be approached, as the goals of science (at least in the contemporary mainstream paradigm) by approximation step by step (to an unfathomable future when, supposedly, all the facts collected to the date miraculously will self-arrange through some *objective* procedure into a representation of the world as a whole as it really is).

In the context of linguistic semantics and analytical philosophy of language, even worse, truth is considered to be a *property* of statements. But "truth and falsity *in content* of a statement is apprehended by observing whether or not this content fits a broader context which is indicated either in the statement itself or by some action or gesture (such as pointing) that goes together with the statement" (Bohm 1980: 42). We face still greater problem when we come to statements formulating world views, i.e., which have to express the nature of 'the totality of all that is'. With them we lose the possibility (how fundamental this is one can start to appreciate from occasions like this) to have a clearly definable context (= background or ground in the context of cognitive linguistics; cf. Talmy 1988; Langacker 1987) as a 'horizon' for the referent represented in consciousness.

What can we do under such circumstances? Bohm proposed the appropriateness of truth in function, i.e., "the possibility of free movement and change in our general notions of reality as a whole, so as to allow a continual fitting to new experience, going beyond the limits of fitting of older notions of this kind" (Bohm 1980: 42). The role of the "executive in charge" realizing it is ascribed to the 'spontaneous and unrestricted' attention-calling function. These characteristics of the latter guarantee it will work in a way making sure that there will be a resultant truth in function.

In the rheomode, the semantically driven derivation of 'truth' is built as follows:

> *To verrate* (from Latin *verus* "true") = calls attention to a spontaneous and unrestricted act of seeing truth in any form whatsoever, including the act of seeing whether this perception fits or does not fit that which is perceived actually to happen in the apprehension of truth, as well as seeing truth in the attention-calling function of the word itself; i.e., 'to verrate' is in the act of perceiving truth, as well as in attending to what truth means. (Bohm 1980: 42)

Next to the fore comes the question what is a 'fact'. It is pointed out that its root meaning (diachronically) is "that which has been made". In the rheomode the 'factuality' is established in the following way:

> *To factate* = enacts spontaneous and unrestricted attention to consciously directed human activity in making or doing any sort of thing whatsoever, including the action of the attention-calling function of the word itself. (Bohm 1980: 43–44)

A basic constitutive characteristic of any fact is its capacity to remain *constantly* valid. Turning to the attribution of the adjective 'constant', Bohm once again uses as a point of departure the history of its meaning for the sake of constructing afterwards a paradigm in the rheomode:

> *To con-statate* = to give spontaneous and unrestricted attention to how any sort of action or movement whatsoever is established in a relatively constant form that stands together relatively stable, including the action of establishing a body of facts that stands together in this way, and even the action of this very word in helping to establish the fact about the function of language itself. (Bohm 1980: 45)

Bohm systematically makes use in his book of linguistic etymologies of the basic concepts with which he works. The root meaning of 'order' is to be found in "giving attention to similar differences and different similarities" (Bohm 1980: 115–116). 'Measure' in ancient times meant "limit" or "boundary" (Bohm 1980: 118), i.e., 'to measure' means to delimit the scope of appropriateness of certain order. The appropriation of measure and order is done in a 'flash of very penetrating insight' which is 'poetic' in its nature. 'Poetry' is traced to the Greek *poiein* meaning "to make", or "to create", *ergo*, in its most creative aspects, science takes on the quality of poetic communication of creative perception of new order (Bohm 1980: 114).

Let's in this context turn now to the notion of 'structure', which is a superordinate concept defined by its 'measure' and 'order'. The essential meaning of the Latin root *struere* indicates "to build", "to grow", "to evolve". The suffix forming in Latin the corresponding noun is *ura* which originally meant "the action of doing something", i.e., the action of 'structuring' is not a finished product, ultimate result, but an open ended development. Following the example from Latin, here the following verb is introduced in the rheomode:

> *To structate* = to create and dissolve structures. (Bohm 1980: 120)

Appropriate also in this context is Bohm's etymology-coining of 'analysis' which comes from the Greek root *lysis* meaning "to break up or dissolve". This is rather surprising, as 'to break up' (i.e., 'to analyse', 'to differentiate') is considered today to be a cognitive operation incommensurable with that of 'to dissolve'.

3. A structural and functional characterization of the first tetrad of rheoverbs

Let's now try to investigate the problem why the first four verbs could be differentiated as a possible mini-max set at the deepest phenomenologically discernible level of thought formation. (I will have no space here to consider the other four verbs.) The specificity of each of them as an aspect of the 'movement' of consciousness is repeated here once again for the sake of their direct juxtaposition:

> *To levate* = an act of *lifting* into attention;
> *To vidate* = an act of *perception* (e.g. in the modality of vision or hearing);
> *To di-vidate* = an act of seeing things as *separate*;

To ordinate = an act of *ordering* (putting together again in a sequence what was previously separated for analytical purposes).

The four verbs in question seem to implement complementary movements along two dimensions: of attention vs. perception of a mental content and dividing vs. merging certain mental contents into an evolving order.

The definition of each of these four verbs is constituted of the following structurally isomorphic parts (if we compare their definitions, as given above):

a. 'calls attention to';
b. 'a spontaneous and unrestricted act of X of any sort whatsoever';
c. 'including the X involved in seeing whether any particular X fits or does not fit some observed content';
d. 'and even the X which arises in the attention-calling function itself'.

Where the variable X can have the following values:

i. lifting into attention;
ii. perception of any sort whatsoever;
iii. seeing things as separate;
iv. ordering of any sort whatsoever.

The definition (a)–(d) is fourfold. Something more, it constitutes a cycle, or is rounding itself to a wholeness (circle) because (d) returns us to the execution of (a). During any circle one is supposed simultaneously to attune one's own inner movement along the two dimensions identified above (attending-seeing and dividing-ordering) in a spontaneous and unrestricted way to the movement of the observed internal 'object' via filling in of an appropriate value for X. The co-movements of the subject (attention) and object (mental content) constitute and represent the whole (the totality of what is). It is impossible also to enact one of the verbs in separation from the others, as they 'flow' into each other; the content of each of the verbs mirrors the content of the other ones; each of them represents effectively the whole cycle.

Let me point out in passing that the status of the movements in the first tetrad already displays some deep paradoxes in the way of performance of a cognitive system aiming at its own re-construction in order to manage to apprehend reality in a veridical way. It is enough, for the sake of illustration, to consider the status of 'the spontaneous and unrestricted **lifting** into one's attention of any mental content whatsoever'. This is a 'lifting' quite unlike any other act of the volitionally controlled attention; it is a lifting that is identical with 'self-lifting of the content itself'. This is the status of the movement of 'lifting' in the rheomode – it signifies 'a lifting that is a self-lifting'. This may, indeed, become possible when the reality itself starts to respond to the attempts to be lifted from the perspective of the lifting at the very time it is supposed to be enacted. The reality and observer coincide in their spontaneous dynamics in a mirror-like completely mutually transparent way.

Bohm evidently gestated the idea of an alternative to natural language as a mirroring of the expressions in a 'language' and reality for many years. This becomes evident from his correspondence with the American painter Charles Biederman (cf., e.g., his letter from December 22, 1961 in Pylkkänen 1999: 135), i.e., the rheomode was by all means not conceived as an *ad hoc* funny illustration of a profound physical principle in order to amuse the reader and make her/him more easily grasp something on a completely different plane. Much more was at stake. From the very beginning the problems Bohm faced were: How language, thought and reality can be made to coincide in mirroring each other's dynamics? Can we represent reality if reality is 'all that is', i.e., a totality that includes things, thoughts and words? Under what conditions can a part (word, assertion, sentence, rule, equation, law) represent the whole?

It is in formulating and trying to find answers to questions like these that he came with the idea of the rheomode of language as a reorganization of the way of performance of the mind in fitting to the way the totality-reality is. This line of thought eventually resulted in a *tour de force* without precedent in the history of thought – an attempt to formulate under what conditions an expression of a language and/or a mental representation can stand for 'all that is', the totality of existence.

4. The 'direct referents' of the first tetrad of rheoverbs or what makes the rheomode of language so difficult to appreciate

What are the 'referents' of the first, the founding tetrad of rheoverbs? There are several psychological referents Bohm (1980) uses that definitely require further elucidation. These are first and foremost attention, perception, intelligence, and thought. It is this aspect of his proposal that makes the rheomode so difficult to comprehend for linguist, theoretical physicist and philosopher alike. On the other hand, a psychologist with a sophisticated professional knowledge of the subject is not well oriented to the nature of natural language and the possibilities to observe the way it functions introspectively. We must, of course, add that for a linguist, cognitive psychologist and the vast majority of philosophers the specifically physical aspect of the proposal is no less a *terra incognita*.

The mutual impenetrability of the psychological, philosophical, linguistic and physical aspects of this proposal makes its appreciation especially difficult.

4.1 Attention

Attention in the rheomode serves a double function. It acts as the 'executive subject', as the subject in charge of awareness formation, as well as, recursively, as its object as if in the very same point-instant of consciousness. From the logical point of view, this evidently involves a paradox. To pay attention to the function of attention at the time one is attentive to an object (an other) requires the enaction of the function performing the double action of the type *I see seeing myself*. For the implementation of such a paradoxical super-system we actually need two mutually juxtaposed systems which 'mirror' each other in the instant there occurs some change in any one of them due to the input from the 'object behind the object-mirror' (the undescribable flow) and the 'subject-mirror' (attention function).

Is this actually possible to achieve? Apparently 'yes' under the conditions of blocking the cognitive functioning (the conceptual thinking), of developing and maintaining a highly concentrated mode of an exclusively perceptual functioning and of rheo-moving within it in a seamless way.

4.2 'Intelligent' perception

From the psychological point of view, another troublesome concept of Bohm is that of 'intelligent perception'. Apparently, the attention function operates on/with perception-like mental representations. Normally, when we think, we do not operate with percepts directly, but with their 'vaguer' analogues in spatial cognition, i.e., cognitive schemata. The 'intelligent' perception is a mode of spatial thinking that is closer to, even identical with, perception from the point of view of the richness of the corresponding mental representation, as well as the spontaneity of its formation. It seems to center upon the flow of conscious visual experience on a lower level of mental processing but not on the relatively high-level abstract mental schemata we tend to use when we fulfil the injunction: "And now imagine a typical bird!". (For further discussion of some of the features of this 'direct' mode of perception cf. Stamenov 1996.)

4.3 Blocking natural-language-specific recursion

A third specific feature of the rheomode of language was already mentioned above in passing. It consisted in the blocking of the self-alienating recursion of the mental functioning where the result of the previous processing is taken as a nominal-like (ready) product for further processing, a difference epitomized in the natural language in the relationship between the transitive verb and its complement, e.g., between "eat" and "something eatable" – banana, toast, bread. It is this difference between nominalized pre-given vs. the actual 'movement' of constructing a mental representation that is supposed to be abolished in the rheomode. Here also come the many possibilities to embed recursively a sentence within a noun phrase and a sentence within a sentence as the language-specific way of embedding of part/whole relationships, the possibilities to extract from and/or to control some aspects of clause structure from outside of the clause, as well as 'barriers' (cf. Chomsky 1986) to extraction and control, etc.

4.4 Thinking vs. intelligence

Another special feature of the proposal of Bohm (1980: 51–52) from psychological point of view is based upon the differentiation he makes between conventional thinking (with natural language) and the creative thinking he calls 'intelligence'. He points out that regular thinking we are accustomed to is a mechanical (automatic) process with *no inherent reason in it* why the thoughts that arise should *be relevant* to the occasion to which they are actually associated. Language-specific representation, in other words, is associated with a specific mental model of the world, while the relationship of the mapping between language-specific thinking pattern and the model of the world (focus + context) remains indirect even at its best. This dissociation, definitely, served a distinct function in the phylogenesis of thinking and language in humans, but it has not only positive but also negative effects on the way of cognitive performance. Bohm makes us aware of the potential drawbacks of the automatically enacted (conventional) patterns of thinking as given in language structure. One must still keep in mind that 'bad' and 'good' are qualifications dependent on the function an object is supposed to perform. The natural language looks 'bad' from the point of view of a seeker of the rheomode of language, but a seeker of the rheomode of language would be an easy prey for a tiger or a couple of wolves if left in the natural habitat (where the natural language evolved for hundreds of thousands of years).

The perception whether or not some particular thought is relevant and fitting to the current context requires the operation of another type of energy, which is not automatic (unconsciously enacted) in its nature. This 'energy' Bohm names, rather loosely, intelligence. The most specific characteristic features of it could be summarized as follows:

a. Intelligence operates via the concrete medium of 'direct' perception, i.e., it operates exclusively using as a medium the qualia of conscious experience;
b. Intelligence is also immediate in its way of performance on-line; it is an action happening here-and-now in the perceptual present all the time;
c. Intelligence is not only concrete and immediate, it is also experienced directly as illuminating the perceptual present of consciousness, of the experiential world as it unfolds here-and-now (the *Lichtung* of Globus 2003:62). The operation of intelligence as a creative (spontaneous in its action) energy is experienced as a 'flash of understanding', in which we see the relevance or irrelevance of our thinking to the problem we are trying to solve.

The flash of understanding Bohm is talking about has the character of directly perceiving some mental content, but is not a result of a chain of thoughts, i.e., it is not a matter of presuppositions, implications, entailments, inferences, deductive or inductive reasoning, etc.

In a sense, one can interpret 'intelligence' as an aspect of consciousness that has an 'illuminating' effect upon certain set of mental contents. If this is the case, there is nothing curious in claiming the perception-like character of intelligence, as it operates on modality-specific qualia of conscious experience.[2]

It is also appropriate to point out that intelligence is not a process like thought, though more subtle in nature (Bohm 1980:52), i.e., they are qualitatively different in the way of their implementation. The difference, apparently, boils down to the following: while thought forms and connects with its mental operations successive mental spaces, intelligence is functioning within a single mental space that embeds within itself the successive foci of the intelligent (or direct) perception. The singleness and self-illumination of this space makes it transparent to itself at any time.[3]

5. Conclusion: The challenge of the rheomode of language

The aim of the rheomode of language is to make the mental representation transparent to itself and, in this way, to attune itself also to the 'totality of all that is', the holomovement. In order to serve its purpose it must fulfil at least three extraordinary requirements:

i. it must consist of performative (just do it) expressions only (cf. Searle 1969 for the concept of performative speech acts);
ii. all expressions must be self-transparent in their form and meaning (even when they come out in a set of recursively embedded orders that would be the potential functional equivalents of phonetic, phonological, morphological, lexical, syntactic, semantic and pragmatic levels of the formation of structure in natural language); and
iii. every structure must provide the possibility for a part faithfully to represent the whole, both in the relation between a whole expression and its components, as well between an expression and the reality part of which it is.

The development of the rheomode of language is an attempt aiming at making explicit what orders of observation by the cognizing function can reach for the physical reality. It shows the very paradoxical metamorphoses an observation must undergo in re-structuring itself in order to attune to the very dynamics of reality as its double-in-coemergence in self-and-other transparency.

The rheomode goes beyond all available paradigms in the history of science and philosophy that study and model the relationship between mind and reality. This is the case because it does not try to re-present by some symbolic means the purported transcendental reality that is in principle inaccessible. The rheomode is *not* an attempt to go outside the 'visible' and/or thinkable and/or knowable (cf. Plotnitsky 2002 for a discussion of this idea). It is, instead, a re-construction aiming in the opposite direction – at making what is capable of becoming 'visible' all the time visible in a seamless way.

But we should still answer the most critical question in our context: **Why we need at all a** language **in order to manage to re-construct the way of access to physical reality? Why just 'seeing it in a different way' is not enough?**

I think the answer should be along the following lines. 'Seeing' is, indeed, not enough, because one must **move** within the 'immediate present' in order to 'stay tuned' to the holomovement. This means, psychologically, that one is supposed to use the resources of the very short lived echoic and/or iconic memory (or both of them in a coordinated way) and to try to code and maintain simultaneously in the so formed unified mental space as many 'movements' as possible. In order to keep them in these two immediate memories with quite fast rhythms of updating and renewal, one must rehearse the material to be kept there. The only mechanism for rehearsal of a material distributed in a time series is, as a matter of fact, the one related to speech articulation (if you try to remember a seven-digit phone number, the only way to keep it in your mind and not forget it without writing it down is to repeat it all the time you need it in order to dial successfully). We do not possess a comparable mechanism for rehearsal of visual material. This seems to be the true key to the necessity to have a 'language' in order to reconstruct the way of access to physical reality.[4] The rheomode provides the subject with the possibility to form and rehearse (participationally enact) a circle of rheo-utterances in a single unified 'mental space' for a time that can be prolonged outside of the limits of the spontaneous rhythm of the iconic and/or echoic memory. One needs a spoken language in order to integrate meaningfully the time variable into the self-transparent movement. Attune the rehearsed to the holomovement and, lo, in an instant the co-articulation of finite and infinite starts to unfold.

Notes

1. The earliest version of this article under the title "Introspection and the Rheomode of Language of David Bohm" was prepared as a paper for the International Congress commemorating the 100th anniversary from the publication of "The Principles of Psychology" by William James, August 13–17, 1990, Amsterdam, The Netherlands.

2. This aspect of Bohm's thinking was at least partially influenced by his long-term dialogue with Jiddu Krishnamurti (cf. e.g. Krishnamurti & Bohm 1985, 1999). The two books in question are transcripts of discussions between them during 1975–1980, exactly at the time when Bohm was writing *The Wholeness and the Implicate Order*.

3. Let me point out in passing that the self-illumination of consciousness by intelligence (the *Lichtung*) is not enough in the quantum brain dynamics account of consciousness, because we miss the 'other' of Vitiello (2001). The 'other' comes through the 'tain of the mirror' (Globus 2003: 163–165, after Jacques Derrida) while the bubble-mirror-from-within is bathing in the immeasurable dynamics-light of the holomovement from without. The 'other' appears through the tain via the attunement (the collective mode of Jibu & Yasue 1995) of intelligence to the immeasurable.

4. Bohm himself never made a claim like this but after critical analysis I do not see any other psychologically viable alternative.

References

Bohm, David (1980). *Wholeness and the implicate order*. London: Routledge and Kegan Paul.

Chomsky, Noam (1986). *Barriers*. Cambridge, MA: MIT Press.

Globus, Gordon G. (2003). *Quantum closures and disclosures: Thinking-together postphenomenology and quantum brain dynamics*. Amsterdam & Philadelphia: John Benjamins.

Jibu, Mari & Kunio Yasue (1995). *Quantum brain dynamics and consciousness.* Amsterdam & Philadelphia: John Benjamins.
Krishnamurti, Jiddu & David Bohm (1985). *The ending of time.* San Francisco: Harper & Row.
Krishnamurti, Jiddu & David Bohm (1999). *The limits of thought: Discussions.* R. McCoy (Ed.). London: Routledge.
Langacker, Ronald (1987). Nouns and verbs. *Language, 63,* 53–94.
Plotnitsky, Arkady (2002). *The knowable and the unknowable.* Ann Arbor, MI: University of Michigan Press.
Pylkkänen, Paavo (Ed.). (1999). *Bohm-Biederman correspondence.* Vol. 1: *Creativity and science.* London: Routledge.
Searle, John R. (1969). *Speech acts.* London: Cambridge University Press.
Sperber, Dan & Deirdre Wilson (1986). *Relevance: Communication and cognition.* Cambridge, MA: Harvard University Press.
Stamenov, Maxim I. (1996). The fractal-like roots of mind: A tutorial in direct access. In Earl Mac Cormac & Maxim Stamenov (Eds.), *Fractals of Brain, Fractals of Mind: In search of a symmetry bond* (pp. 273–322). Amsterdam & Philadelphia: John Benjamins.
Talmy, Leonard (1988). The relation of grammar to cognition. In Brygida Rudzka-Ostyn (Ed.), *Topics in Cognitive Linguistics* (pp. 165–205). Amsterdam: John Benjamins.
Vitiello, Giuseppe (2001). *My double unveiled: The dissipative quantum model of the brain.* Amsterdam & Philadelphia: John Benjamins.

Chapter 10

Can quantum analogies help us to understand the process of thought?

Paavo Pylkkänen
University of Skövde, Sweden / University of Helsinki, Finland

A number of researchers today make an appeal to quantum physics when trying to develop a satisfactory account of the mind, an appeal still felt to be controversial by many. Often these "quantum approaches" try to explain some well-known features of conscious experience (or mental processes more generally), thus using quantum physics to enrich the explanatory framework or *explanans* used in consciousness studies and cognitive science. This paper considers the less studied question of whether quantum physical intuitions could help us to draw attention to new or neglected aspects of the mind in introspection, and in this way change our view about what needs explanation in the first place. Although *prima facie* implausible, it is suggested that this could happen, for example, if there were *analogies* between quantum processes and mental processes (e.g. the process of thinking). The naïve idea is that such analogies would help us to see mental processes and conscious experience in a new way. It has indeed been proposed long ago that such analogies exist, and this paper first focuses at some length on Bohm's (1951) formulation of them. It then briefly considers these analogies in relation to Smolensky's (1988) analogies between cognitive science and physics, and Pylkkö's (1998) aconceptual view of the mind. Finally, Bohm's early analogies will be briefly considered in relation to the later analogies between quantum processes and the mind he went on to propose in his later work.[1]

1. Introduction

There are by now many approaches that seek to explain conscious experience, and mental processes more generally, in terms of a conceptual framework involving ideas from quantum and relativity physics (see e.g. Van Loocke (Ed.) 1999 and Vitiello 2001, and the references therein). In this paper I will explore another, less studied possibility, namely that quantum physics might not only help us to explain commonly acknowledged features of the mind, but also to draw attention to neglected but important aspects of our inner experience. In other words, I am interested in the question of whether the consideration of quantum physics can change our view of what there is to be explained about the mind in the first place (the *explanandum*), over and above the more obvious role it might play as affecting the *explanans* or the conceptual framework we are using when trying to explain some well-known features of conscious experience and mental processes (cf. Van Gulick 1995).

Insofar as it is a quantum physical framework as the *explanans* that helps us to see mind/consciousness as the *explanandum* in a new way, we could say that "*explanandum* is *explanans*-laden". Robert Van Gulick (1995) does not consider this possibility in his interesting paper "What would count as explaining consciousness?", although he makes otherwise a very useful attempt to "divide and conquer" the problem of consciousness. But if we want to acknowledge at the outset any possible bias in our attempts to explain conscious experience and mental processes, then we ought to consider not only how our *explanans* deals with the *explanandum*, but also the more difficult issue of whether the *explanans* we always *already* have affects the way we perceive and define the *explanandum* in the first place.

In consciousness studies, introspection plays a central role as a method. But when studying conscious experience in introspection, how much is what we "see" affected by the scientific theory, and more generally, the world-view or paradigm we happen to hold? To what extent is introspection theory- or paradigm-laden?

To unpack the idea of "paradigm-ladeness of introspection", the first point to note is that in the philosophy of science it is a household fact that "observation is theory-laden", that what we observe in general and also in our scientific experiments is affected by the theory we are using, and other assumptions, and our (more or less unconsciously held) *Weltanschaung* or paradigm (see e.g. Hanson 1958; Kuhn 1962/1970; and Feyerabend 1975/1993; see also Suppe (Ed.) 1977). A nice summary of the Feyerabend-Kuhn view of theory-ladeness is given by Suppe (1977:689):

> [Shapere] finds that the Feyerabend-Kuhn view makes the following chain of inferences:
>
> i. Observation, if it is to be relevant, must be interpreted.
> ii. That in terms of which interpretation is made is always theory.
> iii. The theory that interprets is the theory to be tested.
> iv. The theory to be tested is "the whole of science" (or a branch thereof).

v. This whole forms a unity ("paradigm" or "high-level background theory").
vi. This unified whole not only serves as a basis of interpretation, but also determines ("defines") what counts as an observation, problem, method, solution, and so forth.

For the purposes of this paper the question is whether it makes a difference to consciousness studies (and introspection in particular) whether "the whole of science" includes quantum physics. In other words, can the radically different "whole of science" we get as a result of quantum physics affect the way we interpret our observations in introspection? Can it affect it so that we can obtain in some ways a more complete and accurate explanation of conscious experience and mental processes?

Note that the Feyerabend-Kuhn approach raises the issues of *relativism* and *circularity* in a powerful way. If the theory we are allegedly testing is also the one we use to interpret the results, and the one which defines what counts as observation, problem, method and solution, how objective, neutral and impartial can such testing be judged to be? There is a risk of a deep circularity that arises if we take seriously the idea that observation – including introspection – is theory- and paradigm-laden. In this paper, however, I will not try tackle the difficulties raised by this circularity and relativism. For the sake of the argument, I will assume that there is a sense in which a given theory or paradigm can give us a more complete and more accurate description and explanation of a given phenomenon or domain. And thus, I assume that it is at least in principle possible that, for example, a scientific world-view that takes into account the results of new quantum and relativity physics could help us to describe and explain the mind in a better way. Let us now proceed to explore in more detail how this might be possible.

It seems fairly obvious that people who are familiar with quantum physics develop a whole new set of intuitions or a new "paradigm" about, for example, what it can mean for a phenomenon to be physical, or about general principles that prevail in phenomena. The possibility to be explored here then is that when studying conscious experience in introspection these people might well make use of these new intuitions and see new aspects of the mind, or give importance to aspects that others in a sense see but tend to neglect.

If introspection is theory- or paradigm-laden, then people who hold the classical, mechanistic worldview as the only relevant truth in this context might well tend to see and emphasize only the classical and mechanistic aspects of conscious experience and mental processes, while people equipped with classical *as well as* quantum intuitions might see broader aspects. What are such "classical and mechanistic" aspects? The sorts of thing I have in mind are the dominance of separable objects in experience, causal relations between them, the idea that (more or less) Euclidian/Newtonian space-time is the only arena where conscious experience takes place. In the domain of thought and language, some cognitive scientists emphasize that thinking is essentially mechanical symbol manipulation. Presumably they find such characterization introspectively accurate.

I do not claim that when we ordinarily and pre-theoretically introspect we would automatically find only such mechanistic features. The idea is more that much of contemporary cognitive science and philosophy of mind and cognitive neuroscience etc. involves strongly mechanistic assumptions ("paradigm") that help to draw attention to the mechanistic aspects of the mind, while making it more difficult to see other aspects. Someone once said: "when you have a hammer in your hand, everything in the world looks like a nail". Analogously, to an introspectionist equipped with the mechanistic conceptual tools of modern cognitive science and philosophy of mind, conscious experience and mental processes may well look more mechanistic than to someone with a different, less mechanistic paradigm. This does not, of course, mean that the mind has no mechanistic aspects. But it should make us more open to the possibility that the mind has also other kinds of aspects that might be better seen with a different theory.

The above, if correct, suggests an interesting way in which "quantum approaches" to consciousness and mental processes can be relevant. For they might draw attention to important aspects of mind that tend to be neglected (or simply "not seen") by the more mechanistic prevalent approaches to the mind. The above suggestion may, of course, sound *prima facie* very implausible. How on earth could quantum physics which deals with atomic phenomena help us to more accurately introspect conscious experience and mental processes, which appear to be completely different phenomena at a different, higher, perhaps neurobiological level of organization? One way in which this could happen is if there were *analogies* between quantum phenomena and mental phenomena.[2] Suppose, for the sake of the argument, that quantum phenomena and some features of our inner experiences resembled each other in some important respects. For example, let us assume that the way conscious thought typically "proceeds", or changes its state from moment to moment, would resemble the way a quantum system typically moves. Then suppose that someone familiar with quantum processes would begin to systematically consider the nature of inner experience and thought processes. The seeing of the resemblances between the two domains might well constitute a non-trivial, exciting discovery.

Now, it so happens that many physicists have in fact proposed that there are strong resemblances between quantum processes and our inner experiences and thought processes. Such analogies were proposed to exist early on by the "founding fathers" of quantum theory, for example by Niels Bohr (1934). In contemporary research similar analogies still play an important role. Consider, for example, Globus' (2003) interesting suggestion that Vitiello's dual mode quantum brain dynamics resembles Heidegger's dynamical *der Ereignis* whose modes are Being and time. In this paper we will focus upon a fairly detailed early discussion of analogies between quantum processes and thought presented by David Bohm (1951). Interestingly in view of the questions we started off with, Bohm's analogies seem to draw attention to certain aspects of the mind that, although at least potentially fundamental and important when noticed and considered, nevertheless tend to be neglected in many contemporary academic studies of the mind.

In this paper my aim is thus, via considering Bohm's analogies, to explore whether "quantum intuitions" can help us to understand conscious experience and mental processes in a new and better way – better in the sense that quantum intuitions would draw attention to and help to explain certain important but neglected characteristics of the mind. In order to realize that aim I will first describe and discuss at some length Bohm's analogies between quantum processes and thought processes; I will then consider these in relation to Smolensky's (1988) analogies between physics and cognitive science; I next interpret Bohm's analogies in terms of Pylkkö's (1998) aconceptual view of the mind; and in concluding reflections I briefly consider some of Bohm's later interpretations of quantum theory and the way he used them to develop new analogies to understand the mind.

2. Analogies between quantum processes and thought processes

As mentioned above, already the "founding fathers" of quantum theory, in particular Niels Bohr, drew attention to the possible relevance of quantum physics to our understanding of the mind (see e.g. Bohr 1934). A particularly clear early statement about certain close resemblances between quantum processes and thought processes, influenced by Bohr's ideas, can also be found in the physicist-philosopher David Bohm's (1951) text-book *Quantum theory*. That book, written when Bohm was still an advocate of the so called "orthodox" or

"Copenhagen" interpretation of quantum theory, puts a strong emphasis on making clear the *physical meaning* of quantum theory, as opposed to focusing on the mathematical formalism, which tends to be the case in many textbooks. When writing this 646-page book Bohm thus certainly had the more general physical and philosophical significance of quantum physics strongly in his mind (see especially Chapter 8, "An attempt to build a physical picture of the quantum nature of matter", pp. 144–172). This put him into a good position to consider "wide ranges of experience in which occur phenomena possessing striking resemblances to quantum phenomena". Bohm's basic claim in this regard was that there is a *close analogy between quantum processes and our inner experiences and thought processes.* After discussing such analogies he also provided some speculations of the underlying reasons for the existence of the analogies (pp. 168–172). Let us now consider Bohm's discussion in some detail. I have included fairly long quotations from Bohm and added explanatory comments in order to make the paper more accessible to those without a strong background in quantum physics. Also, I have let Bohm speak on the physics issues whenever this has seemed reasonable. A closer examination of Bohm's (1951) analogies is also useful from the point of view of understanding the historical roots of the idea that quantum physics might play an important role in the study of the mind.

2.1 An uncertainty principle for the process of thought

Bohm starts off by considering the uncertainty principle of quantum theory and certain aspects of our thought processes:

> If a person tries to observe what he is thinking about at the very moment that he is reflecting on a particular subject, it is generally agreed that he introduces unpredictable and uncontrollable changes in the way his thoughts proceed thereafter. Why this happens is not definitely known at present ... If we compare (1) the instantaneous state of a thought with the position of a particle and (2) the general direction of change of that thought with the particle's momentum, we have a strong analogy. (1951: 169)

In classical, Newtonian physics one can, in principle, measure momentum and position of a particle accurately at the same time – a special case of that is when we look at a stationary object where momentum equals zero and the position is where we see and measure it to be located. However, at the quantum level of accuracy it is not possible to measure position and momentum accurately simultaneously, not beyond the limits set by the uncertainty principle. Bohm implies that in this respect thought is more quantum-like than classical-like. Given that the uncertainty principle is one of the most characteristic features of quantum physics (and underlies Bohr's principle of "complementarity", see e.g. Plotnitsky 1994 and 2002), it is of course at least *prima facie* interesting if a kind of uncertainty principle (and thus complementarity) also applies to aspects of our thought process.

Bohm continues:

> ... however, ... a person can always describe approximately what he is thinking about without introducing significant disturbances in his train of thought. But as he tries to make the description precise, he discovers that either the subject of his thoughts or their trend or sometimes both become very different from what they were before he tried to observe them. Thus, the actions involved in making any single aspect of the thought process definite appear to introduce unpredictable and uncontrollable changes in other equally significant aspects. (Bohm 1951: 169)

This, again, is reminiscent of quantum physics. It is possible to make "unsharp measurements" where one obtains an approximate idea of the position of a particle, without making the momentum completely unknown. But should one want to measure the position accurately, the momentum becomes undefined. Thus, even if we were able to make "unsharp measurements" of both the direction and content of our thought process at a given instant, this would still be analogous to measurements in quantum physics.

2.2 Holistic features of thought and quantum processes

Bohm further develops the above analogy by suggesting that the "significance of thought processes" appears to have *indivisibility* of a sort:

> ... if a person attempts to apply to his thinking more and more precisely defined elements, he eventually reaches a stage where further analysis cannot even be given a meaning. Part of the significance of each element of a thought process appears, therefore, to originate in its indivisible and incompletely controllable connections with other elements. (1951: 169)

In a footnote, he adds:

> Similarly, part of the connotation of a word depends on the words it is associated with, and in a way that is not, in practice, completely predictable or controllable (especially in speech). In fact, the analysis of language, *as actually used*, into distinct elements with precisely defined relations between them is probably impossible.

Bohm is here concerned with the nature of meaning. We may customarily think that elements of our thought and language, such as sentences or words, carry their meanings autonomously. Just as classical physics assumed that the physical world consists of some basic elements (particles and fields), whose "intrinsic nature" is not affected by the relationships they enter, so we might assume that thought and language can be analyzed to some basic elements which have determinate and well-defined meanings, independently of the relations that such elements have to other such basic elements or the surrounding context. The idea of "elementary propositions" and "names" in Wittgenstein's *Tractatus* can perhaps be seen as an attempt to conceive language in such a way. Wittgenstein writes (4.221 in *Tractatus*, quoted in Jones 1975:204): "It is obvious that the analysis of propositions must bring us to elementary propositions which consist of names in immediate combination". Jones further describes Wittgensteins's view as follows: "...unless a sentence can be analyzed into a series of simple symbols ("primitive names"), each of which refers to a simple object that can be "elucidated" by primitive propositions, the sentence is meaningless" (1975:204).

As is well known, the later Wittgenstein gives up such an atomistic view of meaning and emphasizes, for example, that when in search of a meaning of a term we ought to consider how the term is *used*. It is also obvious that Bohm did not think that the structure of thought and language is atomistic. Instead, he emphasized the holistic nature of meaning. Elements of our thought process and language do not have their meanings completely autonomously, but instead the meanings originate in the connections with other elements. And, Bohm suggests, these connections are both indivisible and incompletely controllable. For him this implies that it is not possible to analyze language beyond a certain stage and expect to find elements with well-defined significance.

Quine has also emphasized the holistic nature of language. He holds that we cannot define concepts and words individually, for language is a holistic system. Quine has, following Duhem, famously analyzed the implications of this holism for the empirical verification of propositions:

> ...our statements about the external world face the tribunal of sense experience not individually but only as a corporate body (1961/1951:41)

> The idea of defining a symbol in use was ... an advance over the impossible term-by-term empiricism of Locke and Hume. The statement, rather than the term, came with Bentham to be recognized as the unit accountable to an empiricist critique. But what I am now urging is that even in taking the statement

> as unit we have drawn our grid too finely. The unit of empirical significance is the whole of science. (1961/1951:42)

> Any statement can be held true come what may, if we make drastic enough adjustments elsewhere in the system. (1961/1951:43)

These brief quotations provide some broader context for Bohm's ideas about indivisibly of meaning. Interestingly, Bohm's remarks were published at the very same year, 1951, when Quine published his famous article "Two dogmas of empiricism", from which the above quotations are taken. Of course, Bohm is not specifically concerned with propositional thought or truth conditions, but more with meaning as we encounter it with actual thought processes and actual language use. Bohm's way of looking at meaning and thought as psychological phenomena seems more similar to, for example, *Gestalt* psychology, (see e.g. Sundqvist 2003: 177–181) than to analytical philosophy of language which latter is more concerned with the logical and semantic properties of language. It is an interesting question whether the holism of thought and language when seen as an actual psychological phenomenon is connected with the holism of the semantic properties of propositions (which some see as mind-independent). This question, however, will not be pursued here. Instead, let us go on to consider in what way Bohm thinks that there is an analogy between the holistic features of thought/language and quantum processes:

> Similarly, some of the characteristic properties of a quantum system (for example, wave or particle nature) depend on indivisible and incompletely controllable quantum connections with surrounding objects. Thus, thought processes and quantum systems are analogous in that they cannot be analyzed too much in terms of distinct elements, because the "intrinsic" nature of each element is not a property existing separately from and independently of other elements but is, instead, a property that arises partially from its relation with other elements. In both cases, an analysis into distinct elements is correct only if it is so approximate that no significant alteration of the various indivisibly connected parts would result from it. (1951:169)

To get a better idea of the quantum physical side of the analogy, it is useful to consider another description of the indivisible quantum connections that Bohm gives. At the quantum level of accuracy, he says,

> [t]he quanta connecting object and environment constitute irreducible links that belong, at all times, as much to one part as to the other. Since the behaviour of each part depends as much on these quanta as on its "own" properties, it is clear that no part of the system can be thought of as separate.

> If, in a classical experiment, we discovered the presence of irreducible "links" between objects, we should then postulate a third object, the link, and thus re-establish the old type of description [analysis into parts], this time in terms of *three* parts to the system. In quantum theory, however, these quanta do not constitute separate objects, but are only a way of talking about indivisible transitions of the objects already in existence. (1951: 166)

Consider, for example a situation in which a hydrogen atom in the ground state absorbs a quantum of energy from an electromagnetic field: "[d]uring the process of transition, both systems are coupled because they are exchanging an indivisible quantum of energy belonging as much to the electron as to the electromagnetic field" (1951: 166–167). More generally, because strictly speaking all physical "parts" that interact with each other are connected by such indivisible quanta to other "parts", quantum theory implies a fundamentally holistic view of the physical universe. Bohm concludes that

> [t]he entire universe must, on a very accurate level, be regarded as a single indivisible unit in which separate parts appear as idealizations permissible only on a classical level of accuracy of description. This means that the view of the world as being analogous to a huge machine, the predominant view from the sixteenth to nineteenth centuries, is now shown to be only approximately correct. The underlying structure of matter, however, is not mechanical. (1951: 167)

Bohm thus suggests that thought processes and quantum systems have in common a certain *ontological holism*, which means that they cannot be analyzed too much in terms of distinct elements. This is so, he suggests, because an analysis beyond a certain point changes the "intrinsic" nature of the element in question. Elements can have certain characteristic properties (e.g. individual words have meaning; an individual electron exhibits either a wave or a particle nature), but they have such properties partly in virtue of the relations they have with other elements. Change those relations, and you may profoundly change the characteristic properties. For example, an electron that just exhibited a wave-like property may suddenly exhibit a particle-like property, if it is made to interact with an apparatus that measures its position. Analogously, in the spirit of Quine's "meaning holism", a statement that previously seemed false may suddenly seem true if we make drastic enough adjustments elsewhere in the theoretical system it belongs to.

Bohm implies that the context-dependence of properties is no anomaly in a quantum universe. On the contrary, the context-dependence of properties seems to be a very fundamental feature of our physical universe. Although the context-dependence we find with e.g. meanings of words or statements is not necessarily the same type of context-dependence that we with the properties of quantum systems, I would say that the similarities between thought/language and quantum systems in this respect are at least *prima facie* interesting and worth further exploration (see also Maxim Stamenov's discussion of language and Bohm's "rheomode" this volume).

2.3 The classical limit of quantum theory and the logical aspect of thought processes

Bohm next points out yet another analogy, namely that there is also a similarity between the thought process and the classical limit of the quantum theory. Before discussing this analogy, let us briefly consider the physics side of the issue.

When we say that the quantum theory has a classical limit, we acknowledge that although at the fundamental quantum level movement seems discontinuous, there is a domain of physical phenomena where Newton's laws of motion that are continuous and deterministic provide an approximately correct description. Similarly, although quantum theory emphasizes the indivisible unity of the world, it seems that in our everyday experience we encounter a world that can, for all practical purposes, be analyzed into distinct elements. But how to reconcile the classical and quantum "worlds" – after all it seems that the world in which we live has *both* aspects. This question is connected with the *correspondence principle* that Bohr developed, described by Bohm as follows:

> [t]his principle states that the laws of quantum physics must be so chosen that in the classical limit, where many quanta are involved, the quantum laws lead to the classical equations as an average. The problem of satisfying the correspondence principle is by no means trivial. In fact, the requirement of satisfying the correspondence principle, combined with indivisibility, the wave-particle duality, and incomplete determinism … define[s] the quantum theory in an almost unique manner. (1951:31)

The discontinuous, indeterminate quantum level and the continuous, determinate classical level are reconciled by noting that

> …first, the discontinuities are too small to be seen on a classical level and, second, that so many quantum processes take place in any classical process that the deviation of the actual results from the statistical average is negligible. (1951:142)

Yet another quotation illustrates how to reconcile the unpredictability of an individual quantum with the causal laws on a macroscopic scale involving many such quanta:

> As for the appearance of apparently exact causal laws on a macroscopic scale, when only the probability of each elementary quantum transfer is determined, we merely note that, where many quanta are involved, the probability becomes *almost* a certainty (but not quite). This is very similar to the exact prediction, by insurance statistics, of the *mean* lifetime of a person within a large group, even though an *exact* prediction of the lifetime of a single individual in the group is not possible. (1951: 30)

Having now a better idea of what is meant by the classical limit of the quantum theory, let us move on to explore what Bohm means when he says that there is "...a similarity between the thought process and the classical limit". His basic idea is that "[t]he logical process corresponds to the most general type of thought process as the classical limit corresponds to the most general quantum process" (1951: 169–170). This implies that the general structure of thought is analogous to the general structure of physical reality. In physical reality, as seen via the quantum theory, there is the level of general quantum processes with characteristic properties (indivisibility, wave-particle duality, uncontrollability, unpredictability etc.). Also, there is the classical limit where analysis into distinct elements is possible, as well as the mathematical description of the movement and interaction of these elements in terms of the causal laws of classical physics. Bohm suggests that the *relation* between the logical process to the most general type of thought process is analogous to the *relation* between the classical limit and the most general quantum process. How does this analogy work?

> In the logical process, we deal with classifications. These classifications are conceived as being completely separate but related by the rules of logic, which may be regarded as the analogue of the causal laws of classical physics. In any thought process, the component ideas are not separate but flow steadily and indivisibly. An attempt to analyze them into separate parts destroys or changes their meanings. Yet there are certain types of concepts, among which are those involving the classification of objects, in which we can, without producing any essential changes, neglect the indivisible and incompletely controllable connection with other ideas. Instead, the connection can be regarded as causal and following the rules of logic. (1951: 170)

Bohm implies that there is a general type of thought process in which wholeness prevails. The component ideas are not separately existing elements with well-defined meanings. These ideas do not necessarily transform according to the rules of logic, but instead they "flow steadily and indivisibly". This general thought is a process, but not necessarily a process having an order and necessity characteristic of logical thought.

However, just as the physical world has a classically describable domain, so the process of thought includes the domain of logical thought process. In this "classical limit of thought" the indivisibility and uncontrollability between ideas that typically prevail in the general thought process have such a small effect that they can be *neglected* for all practical purposes. This makes it possible for relatively autonomous elements, for example concepts that classify objects, to arise. And it also makes possible for such elements to have causal relationships with each other, for example those causal relationships required for an actual thinking process to proceed according to the rules of logic. In this way the "classical limit of thought", or the emergence of separate concepts and causal connections between them, makes logical thinking process possible.

Of course, the classical limit is fundamentally important for both the physical world *as we know it* and for the very thought process that tries to have knowledge about the physical world. Bohm describes the role of the classical limit of both thought and quantum theory as follows:

> Logically definable concepts play the same fundamental role in abstract and precise thinking as do separable objects and phenomena in our customary description of the world. Without the development of logical thinking, we would have no clear way to express the results of our thinking, and no way to check its validity. Thus, just as life as we know it would be impossible if quantum theory did not have its present classical limit, thought as we know it would be impossible unless we could express its results in logical terms.
> (1951: 170)

It is important to note that Bohm does not deny the importance of the "classical limit of thought", any more than he would deny the importance of the classical limit of quantum theory. On the contrary, he emphasizes that logical thinking is fundamental for the enterprise of science, and for thought in general. But his approach implies that it would be a mistake to assume that logical thinking is the most general essence of the thought process, just as it would be a mistake to assume that classical physics reflects the essential nature of the physical world. He writes:

> Yet, the basic thinking process probably cannot be described as logical. For instance, many people have noted that a new idea often comes suddenly, after a long and unsuccessful search and without any apparent direct cause. We

> suggest that if the intermediate indivisible nonlogical steps occurring in an actual thought process are ignored, and if we restrict ourselves to a logical terminology, then the production of new ideas presents a strong analogy to a quantum jump. In a similar way, the actual concept of a quantum jump seems necessary in our procedure of describing a quantum system that is actually an indivisible whole in terms of words and concepts implying that it can be analyzed into distinct parts. (1951:170)

Bohm thus implies that the basic thinking process is non-logical. Logical thinking then emerges out of such a process in certain conditions, analogously to the way causal physical processes emerge out of the general quantum process at the classical limit. Interestingly, he suggests above that the production of new ideas involves the more quantum-like thought process that essentially involves indivisible, non-logical steps. Galileo Galilei famously said that the path of discovery is different from the path of proof. Bohm gives a new expression to this old idea when he implies above that the discovery of new ideas may require a quantum-like, general thinking process, while their justification has to take place in the "classical limit of thought" and make use of the logical thinking process.

Notice also that Bohm's above quote implies that there is yet another feature which quantum processes and the general thought process have in common: it is difficult to talk about both of them when using the logical thinking process which employs well-defined concepts. Bohm emphasizes that the notion of "quantum jump" is an example of a notion that we have to use in quantum physics when we try to talk about something indivisible in terms of words and concepts implying that it can be analyzed into distinct parts.

The new proposal that comes out of Bohm's analogies is the idea that our thought process has a "quantum-like" aspect, and even more strongly, that the *basic*, most *general* type of thinking process is quantum-like. This basic thinking process is characteristic of quantum-like complementarity, in the sense that making one aspect of the process definite inevitably changes other equally significant aspects. It is also characteristic of quantum-like wholeness, in the sense that the characteristic properties (e.g. meaning) of elements of thought depend on indivisible connections with other elements. Further, the suggestion is that the way the general thought process changes from moment to moment is also quantum-like, for it involves indivisible non-logical steps. The component ideas in such a process are not separate but flow steadily and indivisibly. Finally, the basic thinking process seems to have a "classical limit", namely thinking in terms of well-defined concepts, including the logical thinking process.

Suppose, for the sake of the argument, that Bohm is correct when suggesting that the most general type of thinking process is quantum-like. We can then ask *why* this is so. One possibility is that it is a mere co-incidence. But it seems also natural to ask whether the quantum-like features of the basic thinking process could be an indication that the physical aspect of the basic thinking process literally involves quantum processes. When considering this question, Bohm refers to Bohr's (1934) suggestion that "...thought involves such small amounts of energy that quantum-theoretical limitations play an essential role in determining its character" (1951: 170). Bohm further writes:

> There is no question that observations show the presence of an enormous amount of mechanism in the brain, and that much of this mechanism must probably be regarded as operating on a classically describable level. In fact, nerve connections found thus far suggest combinations of telephone exchanges and calculating machines of a complexity that has probably never been dreamed of before. (1951: 170–171)

Bohm thus acknowledges that a great deal of neural mechanisms is classically describable. But unlike contemporary cognitive neuroscientists who tend to assume that *all* neural mechanisms relevant to understanding cognition and conscious are classically describable, Bohm, following Bohr, was looking for a role for quantum processes in neural functioning:

> In addition to such classically describable mechanism that seems to act like a general system of communications, Bohr's suggestion involves the idea that certain key points controlling this mechanism (which are, in turn, affected by the actions of this mechanism) are so sensitive and delicately balanced that they must be described in an essentially quantum-mechanical way. (We might, for example, imagine that such key points exist at certain types of nerve junctions.) It cannot be stated too strongly that we are now on exceedingly speculative grounds. (1951: 171)

There are by now a number of different and much more detailed suggestions about where such "quantum sites" could be located in the brain (see e.g. Loocke (Ed.) 2001; Penrose 1994; Hameroff & Penrose 1996). What is important in Bohm's suggestion, however, is the general scheme. We are to envision two different kind of levels of physical activity in the brain, one of them classically describable, while the other one needs to described in a quantum theoretical way. This leads naturally to the assumption that the physical correlate of the logical thinking process is at the classically describable level of the brain, while the basic thinking process is at the quantum-theoretically describable level. Bohm also implies above that there is a two-way traffic between these two levels. A typical state of mind then includes both levels. It is also interesting to speculate that different types of states of mind could correspond to physiological states that differ with respect to the relative contribution made by "classical" and "quantum" neural processes. In some altered states of consciousness, for example, the classically describable neural processes might make a relatively small contribution to the content of experience, and as a result the holistic features typical of the quantum-theoretically describable neural processes would dominate.

Although Bohm emphasizes that these ideas are speculative, he does not think that they are therefore not worth pursuing further:

> Bohr's hypothesis is not, however, in disagreement with anything that is now known. And the remarkable point-by-point analogy between the thought processes and quantum processes would suggest that a hypothesis relating these two may well turn out to be fruitful. If such a hypothesis could ever be verified, it would explain in a natural way a great many features of our thinking. (1951: 171)

This illustrates what is at stake here. The idea is that our thinking process has many features that are difficult to explain. Of course, in the end we use the thinking process to explain the thinking process, so there is a circularity to begin with. But the sort of thinking process that we commonly use in science is the logical thinking process. Thus we are using the logical thinking process when trying to describe the most general thinking process. The logical thinking process works best when we use it to describe the "classical limit" – whether the classical physical domain, or the logical aspect of the thinking process. It is more difficult to describe quantum processes with the help of the logical thinking process – just as it is difficult to describe the general thinking process with it. But we have already a fair amount of experience of dealing with quantum processes, both mathematically and conceptually. Now, if the general thinking process is analogous to quantum processes, we could make use of our experience with the quantum domain when trying to explain the general thinking process. For example, we saw above how Bohm characterized the production of new ideas as being analogous to a quantum jump.

Bohm also considers the alternative that the general thinking process does not literally involve quantum processes:

> Even if this hypothesis should be wrong, and even if we could describe the brain's functions in terms of classical theory alone, the analogy between thought and quantum processes would still have important consequences: we would have what amounts to a classical system that provides a good analogy to

> quantum theory. At the least, this would be very instructive. It might, for example, give us a means for describing effects like those of the quantum theory in terms of hidden variables. (It would not, however, prove that such hidden variables exist). (1951: 171)

Suppose that the general thought process is classically describable, and suppose that it is closely analogous to quantum processes. This opens up the possibility that quantum processes might, after all be classically describable; or at least it might be possible to describe them more fully than what standard quantum theory allows.

Another reason why Bohm thinks the analogy could be helpful, even in the absence of experimental data, is that it can give us a better feeling for quantum theory:

> For instance, suppose that we ask for a detailed description of how an electron is moving in a hydrogen atom when it is in a definite energy level. We can say that this is analogous to asking for a detailed description of what we are thinking about while we are reflecting on some definite subject. As soon as we begin to give this detailed description, we are no longer thinking about the subject in question, but are instead thinking about giving a detailed description. In a similar way, when the electron is moving with a definable trajectory, it simply can no longer be an electron that has a definite energy. (1951: 171)

Here one is using the analogy between quantum processes and thought process as a tool that helps to understand quantum theory. The kind of example Bohm gives above (about asking someone to describe what they think) is, in principle, easily understandable to all of us. If such familiar features of the thought process are analogous to quantum processes, this makes it easier for us to understand quantum processes. Given that quantum theory is notoriously difficult to understand, this underlines the usefulness of the analogy.

The upshot is that not only might quantum physics help us to understand the mind (as was suggested at the outset of this paper), but we might also use our understanding of thought processes as a tool to understand quantum processes! Thus, if quantum processes are relevantly analogous to thought processes, our understanding and familiarity of quantum processes can help to understand thought processes and vice versa. We are intimately familiar with our thought processes, although our acquaintance of their more subtle features can be highly tacit. Insofar as our thought processes have quantum-like aspects and we are familiar with these aspects, this can make some aspects of the *prima facie* strange quantum processes easier to understand. At the same time we encounter in quantum processes certain important principles (e.g. complementarity and wholeness) in very elementary and prototypical form. This makes it easier to abstract such principles and conceptualize them theoretically. Such theoretical principles can then be useful when trying to understand the quantum-like features of thought processes which are typically much more complex than elementary quantum processes. So although we may all be in some tacit way familiar with the quantum-like features of our thought processes, quantum theory can help us to recognize them and to describe them explicitly and theoretically.

Let us next consider the possibility that it is the case that the thought process *literally* involves quantum processes. This would open up yet another possible way for us to understand quantum processes:

> If it should be true that the thought processes depend critically on quantum-mechanical elements in the brain, then we could say that the thought processes provide the same kind of direct experience of the effects of quantum theory that muscular forces provide for classical theory. Thus, for example, the pre-Galilean concepts of force, obtained from immediate experience with muscular forces, were correct, in general. … We suggest that, similarly, the behavior of our thought process may perhaps reflect in an indirect way some of the quantum-mechanical aspects of the matter of which we are composed. (1951: 171–172)

This is an extremely interesting possibility. Remember that quantum effects are often thought to lie in a domain that is not at all accessible to us in ordinary experience. As a consequence, it is assumed that we should not be surprised that it is difficult for us to understand the quantum domain – after all, we have no experience of it prior to the scientific experiments that probe the domain. Bohm's above suggestion turns this familiar scheme upside down. For it might be the case that all of us are, after all, directly familiar with some quantum-mechanical aspects of matter, in virtue of being familiar with an important part of ourselves, namely the behaviour of our thought processes! Quantum effects, which were supposed to lie in some mysterious domain that only physicists have access to, may lie much closer to home than we thought. If we are, psychologically, partly quantum-theoretical beings, then by being aware of ourselves we might, in principle, be aware of quantum effects. In philosophy, we sometimes speak about "maker's knowledge", implying that someone who has made or constructed something has a special kind of knowledge about it which others may lack. In a similar vein we might speak about "be-er's knowledge" – the knowledge someone has in virtue of *being* a certain kind of system (so I do not here mean to imply that a barley drink has knowledge!).[3] Thought is a part of our being, and if thought processes reflect in an indirect way some of the quantum-mechanical aspects of the matter of which we are composed, we might, in principle have or be able to obtain "be-er's knowledge" of quantum aspects of matter.

Something like the idea of "be-er's knowledge" figures prominently in philosophical discussions about conscious experience. For example, Thomas Nagel (1974) famously argued that there is something it is like to be a bat; and that we cannot find out what it is like just be studying the bat's brain and behaviour. Only the "be-er" has direct experience of what it is like to be that system. In the case of the bat, it cannot communicate what it is like to be a bat in terms of concepts and logical thought to us. For Nagel the example of the bat underlines the subjective nature of consciousness. But even in the case of humans, it is not at all clear that we can communicate in any exhaustive sense in terms of concepts and logical arguments *what it is like to be* a human being. The problem is not merely that a description of the objective, physiological correlates of conscious experience does not necessarily capture what it is like to be that system. There may be a part of our being that simply is so holistic, unpredictable etc. that it is difficult to capture in terms of conceptual and logical thought.

We can interpret Bohm's above suggestion as saying that part of what it is like to be a thinking human being is to have direct experience of the effects of quantum theory. Of course, this is not to say that we all thereby have conceptually organized knowledge about the effects of quantum theory. Bohm argues that the general thinking process is holistic and uncontrollable by its nature, and thus difficult to describe in terms of our usual scientific language that is organized in terms of separate concepts and logical arguments. Insofar as we are all familiar with the quantum-like general thinking process, our familiarity or understanding may be "pre-conceptual" and "pre-logical" (cf. Pylkkö 1998). We can, of course, try to improve this "pre-conceptual" familiarity and try to develop new concepts and principles to capture the holistic and uncontrollable aspects of our thought processes. We will return to the issue of pre-conceptual experience later when considering the relation of Bohm's suggestion to post-phenomenology.

In summary, we have seen above that Bohm's analogies emphasize that the process of thought can be easily disturbed by introspective observation; that there is a limit in the extent to which significance of elements of thoughts can be analyzed; and that besides this general incontrollable, unpredictable and indivisible character, the mind also has a domain of separable concepts which can be connected causally, e.g. following the rules of logic. Bohm further suggested that the analogies would get a natural explanation if it turned out that the neural processes that realize thought processes in the brain would involve quantum processes.

I hope that the above lengthy presentation and discussion of Bohm's analogies has also given the reader an example of what can be meant by the idea that was raised in the introduction of this paper, namely that a "paradigm" can draw attention to new aspects of the mind in introspection. The fairly new idea contained in Bohm's analogies is the suggestion that our general thinking process is quantum-like, whether or not this is a mere co-incidence or the result of underlying quantum-physical correlates of thought. Although the above discussion is admittedly sketchy and speculative, one should realize that the scientific and philosophical implications of this line of thought are potentially very significant. There is a possibility of a revolution in our understanding of the mind that might parallel the significance of the quantum revolution in physics.

Let us now move on to compare Bohm's description with some other descriptions of inner experience and thought in contemporary philosophy of mind and cognitive science. I think it is interesting to do such brief comparisons in order to further evaluate the suggestion we made at the outset of this paper, namely that quantum intuitions might help us to see new or neglected features of the mind in introspection. To fully explore this suggestion would require a much more thorough study than is provided here, and the following ought to be thus taken as a very preliminary first attempt, which hopefully motivates a further consideration of this issue in other contexts.[4]

3. Cognitive science and quantum analogies

Traditional cognitive science was for a long time dominated by the so-called symbolic paradigm in which cognition was assumed to be mechanical symbol manipulation according to a set of rules or a "program". This resembles the domain of separable concepts in Bohm's description. Within cognitive science the symbolic paradigm was subjected to heavy criticisms, and as is well known, connectionist modelling was offered as an alternative way to describe cognition.[5] Yet the advocates of the symbolic paradigm, most notably Fodor and Pylyshyn (1988), argued that connectionism lacks certain important features of the symbolic paradigm. As an interesting attempt to reconcile the tension between connectionism and the symbolic paradigm, Smolensky (1988) proposed that the relation between them is analogous to the relation between quantum theory and classical mechanics. The idea is that a theory typically works in its proper domain, and that often a more general theory that applies in a fairly broad domain can give rise to another theory that applies in a narrower domain as a limiting case. According to Smolensky's proposal we ought to view connectionism as the more general theory, describing a subsymbolic level of the mind, while the symbolic paradigm can be seen as a special, limiting case, describing those aspects of the mind where rule following and symbol manipulation seem to take place.

Smolensky's analogy is in some ways similar to Bohm's above analogy. Both recognise that the mind has a "classical limit", as it were, a domain we can describe in terms of separable symbols, related in definite ways. And both agree that another, more general description is required to give a fuller description of the mind, and that the relation between the two descriptions is analogous to the relation between quantum theory and classical physics.

However, there are also differences between the analogies provided by Bohm and Smolensky. A particularly important difference is that Bohm's analogy suggests that cognition is more radically holistic, unpredictable and uncontrollable than what Smolensky's approach implies. The reason for this is that Bohm proposes that there is a close analogy between quantum processes and processes of thought more directly, whereas Smolensky makes a more methodological analogy between the relation of quantum and classical physics on the one hand, and of connectionism and the symbolic paradigm on the other.

Smolensky is proposing that important, general features of cognition can be captured by connectionist networks. Connectionist networks have some holistic properties, and the mathematical formalism of connectionism has similarities to the mathematical formalism of the quantum theory (see Perus 1995). However, it has been emphasized that traditional connectionist models do not go beyond the symbolic paradigm when it comes to mechanical computability. Pauli Pylkkö, for example, suggests that "... most probably, all existing artificial neural networks and artificial models of chaotic systems are Turing-computable and, therefore, *mechanical*, in the obvious sense of the word" (1998: 94). Thus the suggestion that cognition can be described in terms of connectionist models is not as different from the ideas of the symbolic paradigm as the proponents of connectionism, including Smolensky, seem to assume. In contrast, Bohm's claim that our inner experiences and thought processes are closely analogous to quantum processes constitutes a much more radical suggestion than connectionism. For it is implied that mental processes have radically uncontrollable, indeterministic and semantically holistic features, which cannot be adequately modelled by either the symbolic or the connectionist paradigms, in so far as these are embedded in the framework of classical physics which implies controllability, predictability and separability. Bohm's use of quantum analogies thus led him already in 1951 to propose an outline of a much more radical view of the mind than what cognitive science, inspired by computational models of the symbolic and connectionist type came up with during the 1980s.

Now, the fact that Bohm's view of the mind is more radically holistic than that of either symbolic or connectionist cognitive science does not, of course, mean that Bohm's view is correct. But how do we find out which view is correct? Perhaps we ought to listen what the introspectionists and the phenomenologists have to say about the mind, then combine this with relevant computational models and empirical research in cognitive neuroscience and then make our judgement. But this brings us back to our starting point. If introspection, including the introspection used in phenomenology, is theory- or paradigm- or intuition-laden, then there is a clear danger that introspection or phenomenology will, in a self-serving way, produce evidence to the view of the mind that the practitioner of introspection *already* had – consciously or unconsciously – before the introspection. For example, Pylkkö (1998: 80) writes:

> ...classical phenomenology (say, as it is developed in Husserl's *Ideen* 1913/1976) is not free of the intellectualist bias of the scientific-technological attitude because classical phenomenology clearly sides with the conscious subject and its allegedly autonomous rationality and quite openly acknowledges the rational subject's right to dominate the rest of the mind.

In particular, introspectionists and phenomenologists not familiar with quantum physics are unlikely to suggest that the kind of unpredictability, uncontrollability and indivisibility they may encounter in introspection has the radical, non-classical character of quantum processes, simply because they may lack the conceptual tools to recognize such features, and to evaluate their difference from more classical-type features. Of course, this does not prove that mind has quantum-like aspects. Perhaps an introspectionist equipped with a "quantum paradigm", such as Bohm, will be likewise biased in their introspection so that they attribute quantum-like behaviour to aspects of mind that do not really call for it. But as I said in the beginning of the paper, I am assuming that it is at least in principle possible that, say, a quantum paradigm will provide a more adequate characterization of the mind than a mechanistic paradigm. In other words I am assuming that human inner experience and mental processes have certain features that are not completely determined by the "paradigm", and which the different paradigms manage to deal with in varying degree of success. What the paradigm does is then to focus our attention in a particular way – so that we "see" certain aspects of what is "there" (in some sense "given" in experience). So I am suggesting that it is better when doing phenomenology to have a "classical" *plus* "quantum" paradigm, instead of just a "classical" one. That way one can "see" more of the conscious experience and thought processes – especially their unpredictable and holistic features, the sorts of features that are difficult to conceptualize. And the idea is that a yet richer and more inclusive paradigm can reveal yet further features. I do not claim that my above assumption is unproblematic, but I will not attempt to argue for it here.

One might also note here that the connection between physics and the nature of human experience is an interesting theme in the history of philosophy. Most notably, Kant assumed that human experience necessarily has to exhibit certain features of Newtonian physics, such as spatiality, temporality and causality (see Kant 1787/1991; Strawson 1966). With his analogies Bohm is drawing attention to the non-Newtonian features of human experience, and in this sense broadening the Kantian notion of what kinds of human experience are conceivable and possible. Kant was not wrong in claiming that human experience has Newtonian features, but he was perhaps wrong in his estimation of what are the limits within which human experience can vary. Today, armed with the resources of post-Newtonian physics we are in a position to see new analogies between physics and human experience, and consequently encouraged to articulate our view of the limits within which human experience can vary in a new, broader way.

4. Post-phenomenology and quantum analogies

Bohm is, of course, not alone in suggesting that human inner experience, and especially conscious thought, has unpredictable, uncontrollable, indivisible and non-logical features. In particular, the philosophical movement called "post-phenomenology" emphasizes such features. Pylkkö (1998: 78) associates post-phenomenology to the work of the later Heidegger, Merleau-Ponty, Patocka and Bataille. The connection between post-phenomenological ideas and quantum physics has been emphasized by e.g. Plotnitsky (1994, 2002), Pylkkö (1998) and by Globus (1995, 2003), who latter also emphasizes the role of quantum brain dynamics developed by e.g. Jibu and Yasue, as well as Vitiello, which is a major focus of this volume. I have found Pylkkö's views particularly helpful when trying to make sense of the relevance of Bohm's (1951) analogies to cognitive science and the philosophy of mind.

Pylkkö developed a radical philosophical view that differs in some important ways from Bohm's philosophy of nature. In particular, Pylkkö advocates an *antirealist* view, in which one does not assume physical reality to exist in a well-defined way independently of human experience. What is primary for him is *aconceptual experience* in which there is no sharp division between concepts and objects. How do then concepts and objects emerge? Pylkkö's idea is that they arise simultaneously as aconceptual experience divides itself into two aspects, concepts and the corresponding objects. However, he does not want to give too strong an ontological status for either the objects or the concepts. In contrast, Bohm had a tendency to defend a realist viewpoint in his natural philosophy, although his realism got increasingly modified, weakened and problematized as he kept on developing the epistemic implications of his holistic and processual view of nature (see e.g. Bohm 1980: Ch. 3). Here is a more detailed characterization of aconceptual experience given by Pylkkö himself:

> Let us take immediate and unstructured primitive experience as our philosophical starting point. This unarticulated and prelogical experience which we call *aconceptual* is what mind and language primarily *is*. It is not yet organized by concepts. Because we associate subjectivity strongly with the conceptual organization of experience we say also that, in aconceptual experience, there is no such hierarchy and perspective which characterize the subject's presence. The experience is, so to speak, holistically everywhere, without center, or it has a center which is not yet fully organized. (1998: 13)

When viewed from the perspective of Pylkkö's post-phenomenology, Bohm's (1951) analogies clearly draw attention to some important aspects of the "aconceptual mind". For as we saw above, according to Bohm the general thinking process is non-logical, uncontrollable, unpredictable, and its semantic elements are indivisible in a sort of way that makes it difficult to analyze it in conceptual terms. It seems to me that this fits fairly well with the view of the mind a post-phenomenologist like Pylkkö advocates, as long as one bears in mind the difference between Pylkkö and Bohm, say, on the issue of realism. No doubt Pylkkö's view of the mind differs also in other important respects from that of Bohm, but I think it is fair to say that the use of quantum analogies helped Bohm to capture some important and neglected features of the mind, which are today described by the post-phenomenologist in a more sophisticated way.

5. Concluding reflections

The question underlying this paper was that if introspection is paradigm-laden and if quantum physics gives rise to a new paradigm, might then someone armed with this new paradigm see new features in introspection, and thus have new things to tell us about conscious experience and the mind more generally?

As an example, I considered some analogies which Bohm already 1951 proposed to hold between quantum processes and inner experiences, especially thought processes. These analogies draw attention to certain quantum-like features of the most general type of thought process, such as uncontrollability, unpredictability, semantic indivisibility, inseparability, non-logicality and non-conceptuality, while also doing justice to the more "classical" features of the thought process such as semantic separability, conceptuality, logicality and causality. The analogies acknowledge that these classical features are indispensable in many ways, but they also underline the fundamentality of the more quantum-like aspect, for example, for the production of new ideas.

We have seen that Bohm's analogies are in some important respects similar to Smolensky's analogy that tries to reconcile symbolic and connectionist paradigms in cognitive science, but we have also emphasized that Bohm's view of the mind, as a result of being guided by his quantum intuitions, is more radically holistic than that of Smolensky's hybrid cognitive science. We also pointed out that it is difficult to judge which view is correct. If introspection is indeed paradigm-laden, whose paradigm are we going to use when introspecting in order to decide which paradigm is correct? The possible paradigm-ladeness of introspection gives rise to a circularity that ought to make us careful in our judgements regarding the nature of human experience. However, my proposal is that it is worth further considering the idea that human inner experience, and the general type of thought process in particular, has some quantum-like features. This opens up the possibility of a less mechanical and in my view more accurate description of human experience than what cognitive science can currently offer. We also considered the view of mind implicit in Bohm's analogies in relation to Pylkkö's post-phenomenological view of the mind as aconceptual experience, and saw a fairly good fit between some aspects of these views.

What lessons can we draw from our brief study? I think the above discussion provides tentative evidence that quantum physical intuitions can in fact help to introspect human experience in a new, productive way. Introspection seems to be theory-laden, but this need not be seen merely as an epistemic limitation. It is clear that such theory-ladeness requires us to become much more cautious when making statements about the "nature of the mind". But it seems to be at least a reasonable possibility that new theories can help us to see the phenomena we are exploring in a new light. And a theory originally developed to deal with a particular domain may prove useful in a *prima facie* very different domain. I write *prima facie*, because there are currently a number of different hypotheses proposing that in one relevant way or the other, the physical correlates of mental processes literally involve quantum processes, which, if correct, would make it less of a puzzle if quantum physics turned out to be relevant to describing the mind. In other words, the two *prima facie* very different domains may turn out to be partly the *same* domain, as Bohm indeed speculated already in 1951.

I recognize that it sounds somewhat far-fetched to many researchers that the study of physics could help us to understand the mind. But note that such a possibility is implicit in the whole programme of philosophical atomism that underlies many of the successes of modern science. The traditional idea has been that physics studies the simple constituents of nature and the regularities in their behaviour. This understanding of the behaviour at the micro-level may then help to understand features of more complex, higher-level systems. It is important to note, however, that with quantum physics the whole scheme of philosophical atomism is challenged, and one is forced to consider some radically holistic basic principles. It is also those principles that Bohm is referring to when making the analogies between quantum processes and the mind. But surely, if the mind has such holistic features it has had them all along, so does it really take quantum physics to notice and theorize about them?

The mechanistic view of the mind that dominates contemporary cognitive science and philosophy of mind – which is basically still just a hypothesis about the mind – has got part of its legitimacy and plausibility from the successes of the mechanistic view in physics and biology. In a similar vein, a radically holistic basic physics raises the possibility of a holistic biology and a holistic psychology (cf. Gierer 2002). Bohm's analogies suggest that just as the physical world has two aspects – the general holistic "quantum world", and as a special case, the mechanistic "classical world", so the human mind has two analogous aspects, the holistic general thinking process, and the more mechanical, e.g. logical thinking process. Quantum physics can play an important role for psychology in suggesting a simple *prototype* of how a general holistic level and a special case of a mechanistic level can be reconciled. Thus, although the holistic aspects of the mind can no doubt be discovered without quantum physics (and indeed have been), I suggest that the quantum analogies can enrich both our introspective experience of the mind and the theories we construct in psychology.

Of course, Bohm's 1951 analogies are only a limited illustration of the way quantum physics can be useful when trying to understand the mind. For one thing, Bohm himself went on to interpret quantum physics in different ways, and not surprisingly tried to invent new analogies between these new interpretations of quantum physics and the mind. For example, Bohm and Hiley's "ontological interpretation" of quantum theory suggests that electrons are guided by a new type of field containing "active information" (Bohm & Hiley 1993). Bohm further suggested that the way such information acts is analogous to the way information acts in subjective human experience (see Bohm 1990; Pylkkänen 1992; Hiley & Pylkkänen 2001; Hiley 2004). His idea was that such active information could help us to understand what the mental and the physical sides of reality are and how they can affect each other, thus using the quantum theory to tackle both the general mind-body problem and the more specific problem of mental causation. In the context of his "implicate order" framework, Bohm (1980) likewise discussed the relation of mind and matter in a new way. The idea here is that mind and matter are analogous to non-locally connected quantum systems. They ought to be seen as *correlated projections* from a common multi-dimensional ground, rather than as separate substances in causal interaction. This is a radically new version of "neutral monism", which again makes use of the resources of quantum theory.

It is important to note that Bohm's analogies differ from each other in important ways, although they have in common that they all originate from quantum physics (cf. Guarini 2003). This means that it is not a trivial task to construct a unified view of mind and matter on the basis of his analogies between quantum physics and the mind. For example, the implicate order scheme underlines the discontinuity of movement and suggests that the basic mathematical algorithm needed to describe movement is an algebra rather than the differential calculus (Bohm 1980: Ch. 6). If we apply the implicate order scheme to describing cognition and conscious experience, then we give up the idea of cognition as a dynamically describable phenomenon as fundamental (insofar as dynamical modelling considers the differential calculus as fundamental). In contrast, Bohm and Hiley's ontological interpretation of quantum theory (which allows for the hypothesis of quantum particles moving continuously along trajectories) seems to fit much better with the spirit of dynamical systems theory.

My suggestion it that the different Bohmian schemes, when applied to cognition and consciousness, can be seen as different tools which each can provide a useful way of looking at some aspect of the mind (cf. Murphy 1998). The implicate order scheme, I suggest, ought to be seen as the more general and fundamental tool, but this need not exclude the use of the ontological interpretation scheme, as long as one remembers that the latter provides a more limited view. Whether or not these tools, and other similar tools developed by other researchers, help us to construct a more satisfactory theory of mind is currently an open question. But I hope that this brief study, which has focused on some of Bohm's very early ideas, has illustrated some ways in which the consideration of quantum physics when studying the mind can be fruitful and open up radically new possibilities.

Perhaps some ideas discussed in this paper can also be useful when evaluating the relevance of quantum brain dynamics to humanities. For example, it seems clear that Vitiello's (2001) focus on the "double" structure of the mind has been inspired by his consideration of dissipative quantum field theory. This, I think, constitutes yet another example of how quantum analogies can guide us in our search for a new and richer view of the mind. Notice also that Vitiello's approach builds upon a more sophisticated scheme than quantum mechanics, namely quantum field theory. This, when applied to biological systems, opens up the possibility of developing a more empirically accurate description of the physical aspect of the brain. At the same time, quantum field theory is often felt to be more difficult to understand than quantum mechanics, and as a consequence, it can be more difficult to understand the analogies one might draw between quantum field theory and mental processes. Hopefully our brief consideration of some more simple analogies between thought processes and quantum processes can be helpful when trying to understand the fascinating synthesis of quantum field theory, biology, neuroscience and (continental) philosophy of mind that is emerging from the work of Vitiello, Globus and others.

Notes

1. A part of this paper was written in October 2003 while I was a visiting scholar at the Department of Philosophy, Stanford University, based in the Metaphysics Research Lab at the Center for the Study of Language and Information. I would like to thank the director of the lab, Dr. Edward N. Zalta, for providing me a stimulating environment in which to work. I also thank the various people who have commented on this paper. In particular, my colleague at the University of Skövde, Dr. Stefan Berglund made some critical comments that prompted me to considerably develop an earlier draft. Of course, he is not to blame for any mistakes or unclarities that may remain! I am also grateful to Professor Gordon Globus and Mr. Per Hansell for their comments.

2. For an interesting paper discussing analogies between modern physics and cognitive psychology, see Shanon (1991). For a defense of the role of analogy in scientific reasoning, see Campbell (1957) and Hesse (1966, 1974); see also Pickering (1984) and Cushing (1990), all quoted in Guarini (2003).

3. The physicist John Bell came up with the term "beer" in this sense when in search for an ontological counterpart to the term "observable" in quantum theory. Bell famously opted for the term "beable".

4. In a recent paper on introspection Anthony Marcel makes some points similar to this paper, and develops them in more detail. Consider, for example, the following: "*Attention can influence its object*. Attending to one's experience, introspecting, changes the content, nature and form of the experience. It is also widely accepted that the content, nature and form of the experience that constitutes the content of awareness depends on the way that we attend" (2003: 179). In this recent article Marcel further provides an interesting discussion of the way (a) attention can be directed to components or to a whole; (b) how one's stance toward the object of attention can vary between immersion and detachment; (c) how attention can create its object; (d) how awareness distorts its object; and (e) how our theories can mask our experience. I think the way Marcel's ideas stand out as radical ideas is an indication that the sorts of features of the mind that Bohm's analogies raise are not that commonly noticed and acknowledged in contemporary psychology. For example, Marcel (2003: 179) writes: "John Lambie and I (2002) have recently emphasized what we call the mode of attention, the manner in which one attends at any time – an aspect of attention stressed by William James (1890) but largely ignored by most current psychology".

5. In recent years dynamical modelling of cognition has become important and can thus be seen as a third approach alongside symbolicism and connectionism. For an interesting recent discussion of these approaches, see Eliasmith (2003).

References

Bohm, D. (1951). *Quantum Theory*. Englewood Cliffs, NJ: Prentice-Hall. Reprinted by Dover science books.

Bohm, D. (1980). *Wholeness and the Implicate Order*. London: Routledge.

Bohm, D. (1990). A new theory of the relation of mind and matter. *Philosophical Psychology*, *3*(2), 271–286.

Bohm, D. & Hiley, B. J. (1993). *The Undivided Universe: an Ontological Interpretation of Quantum Theory*. London: Routledge.

Bohr, N. (1934). *Atomic Theory and the Description of Nature*. London: Cambridge University Press.

Campbell, N. (1957). *Foundations of Science: The Philosophy of Theory and Experiment*. New York: Dover Publications.

Cushing, J. T. (1990). *Theory Construction and Selection in Modern Physics: The S Matrix*. Cambridge and New York: Cambridge University Press.

Eliasmith, C. (2003). Moving beyond metaphors: Understanding the mind for what it is. *The Journal of Philosophy*, Vol. C, *10*, 493–520.

Feyerabend, P. K. (1993). *Against Method: Outline of an Anarchist Theory of Knowledge* (3rd ed.; 1st ed. 1975). London: Verso.
Fodor, J. & Pylyshyn, Z. (1988). Connectionism and cognitive architecture: A critical analysis. In S. Pinker & J. Mehler (Eds.), *Connections and Symbols* (pp. 3–71). Cambridge, MA: MIT press.
Gierer, A. (2002). Holistic Biology – Back on stage? Comments on post-genomics in historical perspective. *Philosophia naturalis*, Band 39, *1*, 25–44.
Globus, G. (1995). *The Postmodern Brain*. Amsterdam and Philadelphia: John Benjamins.
Globus (2003). *Quantum closures and disclosures*. Amsterdam and Philadelphia: John Benjamins.
Guarini, M. (2003). Bohm's metaphors, causality and the quantum potential. *Erkenntnis*, *59*, 77–95.
Hameroff, S. R. & Penrose, R. (1996). Orchestrated reduction of quantum coherence in brain microtubules: A model for consciousness. In S. Hameroff et al. (Eds.), *Toward a Science of Consciousness – The First Tucson Discussion and Debates*. Cambridge, MA: MIT Press.
Hanson, N. R. (1958). *Patterns of Discovery*. Cambridge: Cambridge University Press.
Hesse, M. B. (1966). *Models and Analogies in Science*. Indiana: University of Notre Dame Press.
Hesse, M. B. (1974). *The Structure of Scientific Inference*. Berkeley and Los Angeles: University of California Press.
Hiley, B. J. (2004). Information, quantum theory and the brain. This volume.
Hiley, B. J. & Pylkkänen, P. (2001). Naturalizing the mind in a quantum framework. In P. Pylkkänen & T. Vadén (Eds.).
Husserl, E. (1913/1976). *Ideas. General Introduction to Pure Phenomenology* (W. R. Royce Gibson, Trans.). London: George Allen and Unwin.
James, W. (1890). *The Principles of Psychology* (1981 edition). Cambridge, MA: Harvard University Press.
Jones, W. T. (1975). *A History of Western Philosophy. The Twentieth Century to Wittgenstein and Sartre* (2nd edition, revised). New York: Harcourt Brace Jovanovich, Inc.
Kant, I. (1787/1991). *Critique of Pure Reason* (2nd edition, originally published in German in 1787). London: J. M. Dent & Sons Ltd.
Kuhn (1970). *The Structure of Scientific Revolutions* (2nd ed.; 1st ed. 1962). Chicago, IL: University of Chicago Press.
Lambie, J. A. & Marcel, A. J. (2002). Consciousness and emotion experience: A theoretical framework. *Psychological Review, 109*, 219–259.
Marcel, A. J. (2003). Introspective report: Trust, self-knowledge and science. *Journal of Consciousness Studies, 10*(9–10), 167–186. Special issue: Trusting the Subject?, Part 1, A. Jack & A. Roepstorff (Eds.).
Murphy, T. S. (1998). Quantum ontology. In E. Kaufman & K. J. Heller (Eds.), *Deleuze and Guattari – New Mappings in Politics, Philosophy and Culture*. Minneapolis: University of Minnesota Press.
Nagel, T. (1974). What Is It Like to Be a Bat? *Philosophical Review, 83*(4), 435–450.
Penrose, R. (1994). *Shadows of the Mind*. Oxford: Oxford University Press.

Perus, M. (1995). Analogies of the Quantum and Neural Processing – Consequences for Cognitive Science. In P. Pylkkänen & P. Pylkkö (Eds.), *New Directions in Cognitive Science* (pp. 115–123). Helsinki: Finnish Artificial Intelligence Society.

Pickering, A. (1984). *Constructing Quarks: A Sociological History of Particle Physics.* Edinburgh: Edinburgh University Press.

Plotnitsky, A. (1994). *Complementarity. Anti-Epistemology after Bohr and Derrida.* Durham and London: Duke University Press.

Plotnitsky, A. (2002). *The Knowable and the Unknowable. Modern Science, Nonclassical Thought, and the "Two Cultures".* Ann Arbor: The University of Michigan Press.

Pylkkänen, P. (1992). *Mind, Matter and Active Information. The Relevance of David Bohm's Interpretation of Quantum Theory to Cognitive Science.* Reports from the Department of Philosophy, University of Helsinki.

Pylkkänen, P. & Vadén, T. (Eds.). (2001). *Dimensions of Conscious Experience.* Amsterdam and Philadelphia: John Benjamins.

Pylkkö, P. (1998). *The Aconceptual Mind: Heideggerian Themes in Holistic Naturalism.* Amsterdam and Philadelphia: John Benjamins.

Quine, W. V. O. (1961/1951). *From a Logical Point of View* (2nd edition, revised). New York: Harper & Row.

Shanon, Benny (1991). Cognitive Psychology and Modern Physics: Some Analogies. *European Journal of Cognitive Psychology, 3*(2), 201–234.

Smolensky, P. (1988). On the proper treatment of connectionism. *Behavioral and Brain Sciences, 11*, 1–74.

Strawson, P. F. (1966). *The Bounds of Sense. An Essay on Kant's Critique of Pure Reason.* London: Routledge.

Sundqvist, F. (2003). *Perceptual Dynamics: Theoretical foundations and philosophical implications of Gestalt Psychology.* Acta Philosophica Gothoburgensia 16. Göteborg: Acta Universitatis Gothoburgensis.

Van Gulick, R. (1995). What would count as explaining consciousness? In T. Metzinger (Ed.), *Conscious Experience.* Schönigh: Imprint Academic.

Van Loocke, P. (Ed.). (2001). *The Physical Nature of Consciousness.* Amsterdam and Philadelphia: John Benjamins.

Vitiello, G. (2001). *My Double Unveiled. The Dissipative Quantum Model of Brain.* Amsterdam and Philadelphia: John Benjamins.

Chapter 11

Information, quantum theory and the brain

B. J. Hiley
Birkbeck College, London, UK

As is well documented the lack of any notion of meaning in Shannon information limits its use in understanding how the brain processes meaning. We suggest that the notion of active information introduced by Bohm and Hiley in their analysis of quantum phenomena is sufficiently general to be applied even to classical signalling processes as suggested by Gabor. We propose a new way of signal processing that directly confronts the question of the transfer of meaning and show how this opens up new possibilities of information processing in the brain.

1. Introduction

One of the words that one finds in constant use, not only in neuroscience, but also in many other branches of science today is 'information'. It is used in so many different contexts, information technology, biological information processing, quantum information and so on, but it is not at all obvious that the term has the same meaning in each of these contexts. Indeed even within the debate as to how the brain functions, I find the word 'information' is being used in a number of rather different ways which makes a difficult topic even more difficult.

Is it information for us, like a list of instructions that we follow should we feel the inclination to do so? Is it merely information capacity, or as Shannon (1948) puts it 'information without meaning'? Indeed is information simply another aspect of entropy as Brillouin (1956) has proposed? Is it some form of objective information that is active in physical processes themselves as suggested by Bohm and Hiley (1987, 1993)? In the latter case there is no free choice as to whether the information is to be followed or not. The process itself is 'driven' by the information. Or is it something more fundamental acting in the Universe out of which everything ultimately emerges as Wheeler has proposed (1990)? What I would like to do in this paper is to address some of these questions and to make some suggestions as to how we can understand the functioning of information in brain processes as a carrier of a form of objective meaning.

The work in this paper has its origins in a sequence of papers where we exploited the notion of active information in an attempt to understand the mind/matter relationship (see Hiley 1995a, 1995b, 2001; Pylkkänen 1985; Hiley & Pylkkänen 1997, 2001). Indeed I had already made a preliminary attempt to bring out the different meanings associated with the word 'information' elsewhere (Hiley 1997). There I concentrated on how the resolution of the problem created by the Maxwell demon seemed to provide a very strong case for introducing the notion of information into physical processes (for comprehensive review see Leff & Rex 1990). As is well known it has long been argued that the second law of thermodynamics can only be saved in this example by taking into consideration the information gained by the demon.

However there have always been dissenting voices such as those of Denbigh and Denbigh (1985) who have found an alternative solution from within thermodynamics itself without the need to consider any notion of information. This together with the recent work of Maroney (2001) leaves me unconvinced by the attempts to justify the use of information in resolving the problem presented by the Maxwell demon. Maroney has shown that the dilemma can also be completely solved from within quantum mechanics itself without appealing to any notion of information. If information is to play an objective role in the evolution of physical processes then it must come from other considerations.

2. The notion of information

As we have indicated in the introduction Shannon's information capacity is not rich enough to include an element of meaning. Furthermore as the colloquial use of the term makes clear, information is usually understood to have 'meaning for us' and we can be expected to act on that meaning. This implies that meaning can be unfolded in a given context only by an intelligent being, thus suggesting a strong subjective element. A key question that will concern us here is whether information can have a purely objective significance that is devoid of any subjective element. An answer to this question is of fundamental significance in clarifying the relation between mind and matter.

Already as far back as the sixties Donald MacKay (1968), who first introduced me to the subject of information theory, was drawing attention to the fundamental problem of the relation between 'information' as used by Shannon and the 'meaning' contained in this information. Let me try to summarise his arguments and for this I will begin by first considering 'information for agents' and later show how the discussion can be generalised to a wider context.

A speaker wishing to communicate some meaningful information to a colleague will first activate mechanical vibrations in the air by using his/her larynx. These vibrations will be transported mechanically through the air until they reach the listener's ear where they will be transformed into mechanical vibrations of the eardrum. These will then be electrochemically transmitted to the auditory cortex where the original information will 'somehow' be converted into meaning once again. The problem then is how meaning can be transformed into meaningless airwaves, which are transmitted and received mechanically and then converted back into meaning once again.

Why do we insist on regarding meaningful and mechanical processes as mutually exclusive? Surely meaning lies in the relationship between the speaker and listener. It is a relationship that is not merely mechanical. Increasing the volume of the sound will not generally alter the significance of the message. There is no physical force exerted on the listener by the speaker. Of course there are mechanical aspects to the process. Without the mechanical vibration of the larynx there would be no sound and without the eardrum there would only be silence. Without mechanical energy there would be no message and nothing would be transmitted. Yet in addition to mechanical aspects of the energy there is the meaning and the problem is to understand how this meaning is communicated.

Notice that there are two aspects in communicating meaning that need to be emphasised. (1) Meaning seems to establish an inseparable link between speaker and listener since the context in which the meaning makes sense must be shared by both parties. (2) An 'agent' exists at either end of the process, the agent in this case being human. If we are not to regard 'mind' as something special in the way a dualist would, we must face the problem of where mechanism ends and meaning begins if indeed a sharp separation does exist. It seems that if we search for the appearance of meaning within the physical processes themselves we must inevitably find some form of proto-meaning active in less complex physical processes. In other words it is unlikely that meaning will suddenly be relevant only at some critical level of complexity.

Once one begins to talk about apparently 'inseparable links' between agents one is immediately reminded of Bohr's notion of the indivisibility of the quantum of action that is central to his understanding of quantum phenomena. In quantum theory this implies that we are unable to make a sharp separation between the system under observation and the means of observation. Could it be that this indivisibility is related to the indivisibility that arises in our discussion of meaning? Bohr refused to go that far and felt it necessary to talk about 'detached observers', but was he right?

While investigating how this indivisibility arises in the quantum formalism Bohm was able to show that the Schrödinger equation throws up what appears to be an extra term in the equation of motion, traditionally called the 'quantum potential'. I feel this is the key to the problem. Indeed Bohm and Hiley (1993) have argued that it behaves like an 'information potential'. We have shown in some detail how this idea works in quantum processes by systematically analysing all the standard 'puzzles' and paradoxes of quantum processes presented in the elementary texts on quantum mechanics. What emerges from this investigation is some remarkable similarities with the discussion in the previous paragraph on the transfer of information with meaning. I want to propose that these simple physical processes provide more than a mere metaphor for what is going on in the transfer of information with meaning.

These ideas have received relatively little attention in the literature mainly because of the strong opposition to the mechanical aspects of the Bohm approach by those who use the term "Bohmian mechanics". But even given the limitations of the arguments that sustain this mechanical view, I fail to understand why there has been such a strong reaction against the approach. The predicted results of the formalism are exactly those of the standard theory and no experimental test can possibly show it to be wrong without also contradicting conventional quantum mechanics. All attempts to show differences are based on a misunderstanding of how the approach works. After all the approach uses the quantum formalism without adding anything new to the mathematics so it is impossible to produce new results simply because of the way the interpretation is abstracted from the quantum formalism. Even those who realise how the approach works can only come up with the statement that they don't believe nature behaves like that!

But my discussion here does not depend on whether the Bohm approach is the 'correct' way to think about quantum processes. All I want to emphasise is that the approach uses a mathematical structure that has features that are common to both classical physics and quantum physics and it is these features that I want to bring out in this paper. Such connections are not that new as they have already been noticed by Gabor (1946), Schempp (1993) and more recently by Binz, Pods and Schempp (2003).

3. Common mathematical symmetries

I want to start the main discussion of this paper by returning to consider in more detail a signal, keeping the example of auditory communication in mind. Here we need to analyse both the time and frequency aspects of the signal. However we immediately experience a problem since analysing a signal at a sharp instance of time requires an infinity of frequencies while analysing a sharp frequency requires an infinitely long wave train. All of this is well known in classical signal analysis and lies behind all the problems associated with bandwidth transmission. I recall all this here simply to remark that the reason for these problems arises from the nature of the acoustical wave equation, which give wave solutions of the form

$$\Psi(r,t) = \sum_{n=1}^{N} a_n e^{-2\pi i \nu_n t}.$$

Thus in order to transmit a signal there is always a trade off between the duration of the pulse, Δt, and the frequency band, $\Delta \nu$, required to form the pulse. The trade off is summarised by the expression

$$\Delta t \Delta \nu \approx 1$$

Consequently in the background of signal theory there is always an ambiguity relation that has strong similarities with the uncertainty relationship of quantum mechanics. Of course there are differences but these differences are in the way we interpret the relationship.

For Gabor (1946) this relationship was interpreted by first dividing the $\nu - t$ space into cells of unit area. The number of these cells then gave a measure of the number of independent data that an instrument can extract from a signal. In other words this approach can define the amount of information a signal carries which ties in directly with Shannon's expression for the information capacity.

I will not develop this argument further here because I want to draw attention to another connection with Gabor's work. In 1968 I was discussing with David Bohm the possibility of developing a purely topological theory of quantum mechanics based on a cellular structure of phase space (Hiley 1971). In my case the cells being considered were in $x - p$ space rather than $\nu - t$ space. My reason for doing this was simply because the $x - p$ uncertainty relationship is considered as much simpler than the $E - t$ uncertainty because of the difficulty that there is no time operator in quantum theory (see Wigner 1972).

Although my rather sketchy and preliminary discussion did eventually lead to some interesting results both in electrodynamics (Bohm, Hiley, & Stuart 1970) and eventually in the Dirac theory of the electron (Bohm & Hiley 1983), it was not until 1981 that Bohm and I related this cell structure to the Wigner-Moyal approach to quantum mechanics (see Bohm & Hiley 1981).

Before going into the details of this connection I think it is important to be clear why there is this similarity between signal analysis and the quantum formalism. It does not mean that the physics is the same. It means that the two are using different aspects of the *same mathematical structure*. For example the Schrödinger equation can be regarded simply as a differential equation, which can be treated like any classical differential equation. It is only when we try to interpret individual terms in the equation that we find observables are turned into operators, a step that is unnecessary in signal analysis. Quantum mechanics associates eigenvalues of the operators correspond to the values found on measurement, whereas in signal analysis t and v are treated as classical commuting variables.

Now let me continue with the cell structure of phase space and show how it leads to the Wigner-Moyal (Moyal 1949; Wigner 1932) distribution, which also plays a key role in signal analysis (Schempp 1993). We will begin by talking about quantum mechanics of thermal systems that must be described by mixed states. There cannot be described by a single wave function because they consist of the incoherent mixture of pure states. They are described by a density matrix, which is written in the form $\rho(x,t) = \sum_n c_n \psi_n^*(x,t)\psi_n(x,t)$. As we will show below we can find a transformation that enables us to introduce a phase space distribution function $F(x,p,t)$ which will enable us to calculate the correct quantum probabilities while x and p remain commuting variables. The key point to note here is x and p are *not* operators. The enables us to include the t–v variables in the same mathematics. The message I am trying to get across here is that there is another way to calculate quantum probabilities without having to resort to non-commuting dynamical variables.

Indeed the Bohm approach also uses commuting x and p and again produces the correct quantum probabilities. Although x and p used in the Bohm approach commute they are not the same x and p appearing in the Wigner-Moyal distribution even though the are given the same symbols! Not surprisingly there is a close connection between the two sets of symbols that has not been brought out fully before. I have recently discussed the relation between the two approaches in Hiley (2003).

The reason why both these approaches are related and can be used in quantum situations is because underlying both classical mechanics and quantum mechanics is the same symmetry group, namely, the *symplectic group*. Indeed this same symplectic group also underlies both classical wave theory and the Schrödinger equation. For example in classical mechanics the infinitesimal canonical transformations that keep Hamilton's equation of motion invariant are simply elements of the symplectic group (de Gosson 2001). Furthermore and even perhaps more surprisingly the classic optics ray formula for a lens system of focal length, f

$$\frac{1}{u} + \frac{1}{v} = \frac{1}{f},$$

can be derived directly from the symplectic group (see Guillemin & Sternberg 1984).

The wave properties in both cases arise by going to the covering group of the symplectic group, namely, the metaplectic group. This group double covers the symplectic group in the same way that the spin group, SU(2), double covers the rotation group SO(3). In fact we can make an even stronger connection in the case of quantum systems with Hamiltonians of up to quadratic powers in x and p. In this case we can show that the Schrödinger equation is exact, being a one parameter sub-group of the metaplectic group. This means that the Schrödinger equation can be derived *rigorously* from classical mechanics (see de Gosson 2001 for details).

The mathematics that lies behind all this work will probably be unfamiliar to readers of this publication so let me try to summarise the implications of these formal results in simple terms. We have a mathematical structure that is basic to both classical mechanics and quantum mechanics. At this level there is no basic difference between the dynamical equations of classical and quantum mechanics. The difference arises once one asserts there is a minimum value for this action and equates this value to Planck's constant h. Thus as long as we do not ask questions that are sensitive to the finite nature of this quantum of action[1] the similarity between classical and quantum mechanics is remarkable.

What this means is that certain results in this area may look as if they are quantum in origin but in fact have nothing to do with quantum mechanics *per se* but actually arise from the group structure that is common to both forms of mechanics. For example the Fourier transformation is common to both classical and quantum situations. Indeed the Fourier transformation is at the heart of Gabor's discussion of information transfer. Thus any results that emerge from an analysis of either the Wigner-Moyal approach or the Bohm approach may not necessarily be to do with quantum phenomena *per se*, and for that reason I would like to call the emerging dynamics that I will discuss below '*information dynamics*'.

4. The Wigner-Moyal approach

Let us consider a function, $\psi(x,t)$, that contains information embedded in it. This function may have nothing to do with quantum mechanics. To abstract the information we must form the correlation function $\rho(x_1, x_2, t) = \psi^*(x_1, t)\psi(x_2, t)$. Here we begin to see already the need for doubling that Vitiello (2001) discusses.

Let us now write

$$\psi(x,t) = \frac{1}{\sqrt{2\pi}} \int \varphi(p,t) e^{ipx}\, dp$$

so that

$$\rho(x_1, x_2, t) = \frac{1}{2\pi} \int \int \varphi^*(p_1, t)\varphi(p_2, t) e^{-i(p_1 x_1 - p_2 x_2)}\, dp_1 dp_2$$

Let us now regard the two points in phase space, (x_1, p_1) and (x_2, p_2) as defining a square with sides $\eta = x_2 - x_1$ and $\pi = p_2 - p_1$, and centre $X = (x_2 + x_1)/2$ and $P = (p_2 + p_1)/2$ so that

$$\rho(X, \eta, t) = \frac{1}{2\pi} \int \int \varphi^*(P - \pi/2, t)\varphi(P + \pi/2, t) e^{i(\pi X - \eta P)}\, dP d\pi$$

Now we can write

$$F(X, P, t) = \frac{1}{2\pi} \int \varphi^*(P - \pi/2, t)\varphi(P + \pi/2, t) e^{i\pi X}\, d\pi$$

or by Fourier transformations we can write this as

$$F(X, P, t) = \frac{1}{2\pi} \int \psi^*(X - \eta/2, t)\psi(X + \eta/2, t) e^{-iP\eta}\, d\eta \tag{1}$$

This is known as the Wigner function, which has been used for information processing in the brain by Schempp (1993) and Arecchi (2003).

The correlation function (density matrix) then becomes

$$\rho(X, \eta, t) = \int F(X, P, t) e^{i\eta P}\, dP$$

so that expectation values of any variable $A(X, P, t)$ can be written as

$$\langle A(X, P, t)\rangle = \int F(X, P, t) A(X, P, t) dX dP \tag{2}$$

Notice that no operators appear in these equations in spite of the fact that when applied to quantum systems the expectation value for the operator $\hat{A}(\hat{x}, \hat{p}, t)$ agrees with the quantum expectation value provided we write

$$\hat{A}(\hat{x}, \hat{p}, t) = \left(\tfrac{1}{2\pi}\right)^2 \int\int\int\int d\eta d\pi dX dP A(X, P, t) \exp\left[i\eta\left(\hat{p} - P\right) + i\pi\left(\hat{x} - X\right)\right]$$

The expression (2) suggests that $F(X, P, t)$ could be treated as a probability density in phase space for quantum systems. Indeed this was the original reason for pursuing this approach. Although this quantity is real the problem is that it is not necessarily positive. Indeed in those regions where quantum interference is occurring this distribution becomes negative. Thus it is strictly not a probability distribution, although it is often referred to as a quasi-probability distribution. These worries tended to detract from the validity of the Wigner-Moyal method as an exact way to treat quantum mechanical problems. However it should be noted that it exactly reproduces the expectation values of operators in quantum mechanics.

5. Gabor information and the radar ambiguity function

In this article we are not specifically interested in quantum mechanics *per se*. We are interested in the nature of information, information transfer and how that information can initiate physical processes in the brain. To this end we should now show how the above discussion links up with Gabor's approach, which we outlined in Section 3.

Having removed the operator status of $x - p$ in Section 4, it is not difficult to adapt the arguments so that they apply to $t - v$ space (see Binz, Pods, & Schempp 2003 for many details). Rather than repeating these details here, I find it is more illuminating to consider an example originally discussed by Schempp (1986) where he showed that these ideas could be explained in terms of radar waves which are then used to abstract information from the returning signal.

Suppose a carrier wave is emitted towards, say an aircraft. Assume this takes the form $f(t)e^{ivt}$. The returning signal will take the form $f(t + \tau)e^{i(v+y)t}$. Here τ measures the delay time of the returning signal from which the distance the aircraft can be calculated using $d = c\tau/2$. From the change of frequency y, the speed of the aircraft can be found using the Doppler shift formula

$$\frac{v}{v_0} = c\frac{v - v_0}{v} = c\frac{y}{v}$$

The auto-correlation function formed by superimposing the returning signal with the original signal is

$$H'(f, \tau, y) = \int f^*(t')e^{-ivt'} f(t' + \tau)e^{i(v+y)t'}\, dt'$$

Changing variables, we find up to a constant $\exp[-iy\tau/2]$

$$H(f, \tau, y) = \int f^*(t - \tau/2)f(t + \tau/2)e^{iyt}\, dt \tag{3}$$

Thus we see the form of this expression is identical to that of the Wigner distribution except the variables are now time and frequency. It is this expression that underlies Gabor's (1946) signal theory and Schempp's (1993) approach to quantum brain dynamics.

6. The Bohm approach to quantum mechanics and information dynamics

Now I want to move on to consider how the information in such a wave can be converted into an explicit physical process so as to abstract directly the *meaning* of the information. The obvious way to decode the signal is to follow the procedure used in radar engineering. But this simply displays the information so that a *human* operator can interpret the meaning of the information in the signal.

What we are interested in is how the meaning of the information carried by an audio signal is understood in the listener's brain. Thus the key question is whether there is some other way of transferring the meaning without the need for an homunculus ultimately to interpret the meaning. Here I will describe a very simple and crude model whose purpose is simply to indicate the principles involved. For this we must turn to the Bohm interpretation that was outlined in Bohm and Hiley (1993). There are two features of this approach that I want to bring out.

1. I want to exploit the suggestion that the quantum potential that arises in this approach should be considered as an information potential.
2. I want to exploit the fact that the Bohm approach is already implicitly contained in the Wigner-Moyal approach (Hiley 2003).

Let me begin by reminding the reader how the Bohm picture emerges. We start by writing the wave function in the form $\psi(x, t) = R(x, t)e^{iS(x,t)}$ $(\hbar = 1)$ so that the real part of the Schrödinger equation becomes

$$\frac{\partial S}{\partial t} + \frac{(\nabla S)^2}{2m} + Q + V = 0 \tag{4}$$

where the information (quantum) potential takes the form $Q = -\frac{1}{2m}\frac{\nabla^2 R}{R}$.

In Bohm and Hiley (1987, 1993) we discussed in detail why we were led to consider this potential as an information potential. For example we found that the potential was totally different from any classical potential that we know. It has no external source in the sense that the electric field has its source in a distribution of charges. Thus it does not emerge from an interaction Hamiltonian as does a classical force. In other words it does not act mechanically on the system. In this sense it cannot be thought to act like an efficient cause. It is more like a formative cause that shapes the development of the process. Indeed as we explored its properties in many different physical systems it reminded me very much of the morphogenetic field generated in biological systems as discussed by Thom (1975). The information field is shaped by the environment in a way that is very similar to the way the development of a plant is shaped by its environment. Thus we can think of the information as active from within giving shape to the whole process and this shape depends on the environment in key ways. In other words the meaning in the wave is expressed through the form that develops.

That this potential carries information about the environment is clearly brought out in the two-slit experiment. Here we see that the mathematical expression for the potential contains information for the nature of the slits, their size, how far they are apart and on the momentum of the particle. In other words it contains information about the whole experimental arrangement just as Bohr insisted (Bohr 1961). Alter the experimental arrangement by, say, closing one of the slits and the quantum potential changes so that the interference fringes change to those produced by a single slit. Thus we have a way of feeding information into the system to actively change the behaviour of the system.

I want to exploit these ideas to account for information processing in the brain. I want to propose that the information of an incoming signal acts as an information potential changing directly the behaviour of the appropriate part of the brain, say the dendritic fields. Thus the changes are not brought about by decoding the signal mechanically as one would in a radio set. They are brought about by a subtler means, namely, through the generation of an appropriate informational potential carrying the meaning directly into the brain. This part of the brain will then function in ways that directly react to the meaning of the information.

Let me spell this out more clearly by first showing how the information potential works in a simple system. Of course this system is far too simple to have any direct relevance to the brain but is intended simply to illustrate the principles involved. Suppose there is a particle that responds to different meanings in different ways. We take the momentum of this particle to be $p_B = \nabla S$ and use equation (3). Here the momentum is clearly influenced by the information potential Q. If Q is modified by some incoming information then the subsequent behaviour of the particle is changed. In this case we say the information is active in the process. If the system does not change the behaviour of the particle we say the information is passive but we leave open the possibility that at some later time this information may become active again. We also leave open the possibility that the information will never become active such as information communicated in a language that will never be known to the listener.

What is missing from this discussion so far is how we can relate this process to the way Gabor deals with information and information capacity. To bring out this relation we need to know how p_B is related to the P appearing in the Wigner distribution so that the information can be abstracted from the correlation function $F(X, P, t)$. Recall the P in the Wigner distribution is the mean momentum of one of the cells in the phase space, the number of cells enabling us to count the number of degrees of freedom carried by the signal.

Let us define $\bar{P}$ through

$$\rho(X,t)\bar{P} = \int PF(X,P,t)dP \tag{5}$$

Here $\rho(X) = \int F(X,P)dP = |\psi(X)|^2$. If we write the wave function in polar form $\psi(X,t) = R(X,t)e^{iS(X,t)}$ in (4) we find $\bar{P} = \nabla S = p_B$, the Bohm momentum. Thus the information is affecting the mean momentum of responding particles and the response is clearly related to the information capacity of the signal but now it responds directly to the active information contained in the incoming signal.

We can take the connection even further since Moyal shows that this momentum $\bar{P} = p_B$ is transported through the system by equation (4). In other words the momentum $\bar{P}$ is transported through the system by the Schrödinger equation confirming that it must be the same as the momentum used by Bohm. Indeed this also ties in with the fact that the trajectories calculated from p_B are just the probability current streamlines.

In his original proposal Bohm simply made a further assumption that each individual particle follows one of these streamlines. This assumption gives rise to a simple interpretation, which gives a consistent explanation of quantum phenomena.

What is important here is that the information that is encoded in $F(X, P, t)$ is translated into the behaviour of the particle of the physical system that has to interpret the message. In the proposals I am making here, this is done directly through equation (4). This does not mean that that the process is quantum in origin. It is merely exploiting a form of mathematics that it shares with quantum mechanics.

7. A more detailed discussion of information and meaning

Finally we must try to understand how the movement of a particle or a group of particles generated by the information potential can create meaning directly. For this discussion I will rely on the arguments presented in Bohm (1994). There he proposes that meaning is in the process itself. To bring this notion out we have to go to a deeper level and think of mind and matter as being two aspects of one undivided process, as two poles of one indivisible process. He wanted to think of this as a generalisation of Bohr's indivisibility of the quantum of action in ways that we bring out below.

To discuss this indivisibility he proposed that all processes have two sides. The manifest, relatively stable side, which can initially be thought of as the material side, and a subtle side, which is more mind-like. Here the notion of the manifest is being used to describe the more robust, stable aspect of the process. The notion of 'subtle' is being used in the sense of its Latin origins, sub-texere, meaning 'woven from underneath' or 'finely woven'. Thus it can give form to the manifest. Both sides are always present even at the mechanical level where the laws of classical physics suffice. Remember that quantum aspects of processes are always present because without the quantum of action matter would not be stable.

In the quantum world the manifest side is the classical material world of measuring instruments in which the subtle quantum processes are displayed. Physicists are happy to talk about the manifest world but would much rather remain silent about the subtle side, at times implicitly denying its existence. Hence they are quite content to talk simply about the result of measurement, leaving the physical processes responsible for these effects as not describable, justifying their position by appealing either to operationalism or appealing to some dubious quasi-philosophical notion of reality, or rather lack of it. I feel this approach, while being 'safe', leaves out the essential core of what physicists should be concerned with.

When we turn to consider the question of mind we find the manifest side involves a complex of electrochemical processes occurring in neurons, between neurons and in other process in the brain. The more subtle side concerns the *significance* of these manifest processes. For example if we are walking down a dark lane at night and we hear an unrecognisable sound behind us we can immediate feel the rush of adrenaline and other chemical discharges throughout our tensing body. We then see that there is the fox scurrying off in the distance and realise what was responsible for the original sound. This new information immediately effects the body returning it to a normal relaxed state. The noise is an input of information whose meaning is not clear and could signify danger so the appropriate chemical responses are activated. When new information arrives the meaning of which is clear it activates a new set of chemical responses. Thus information is active in producing physical responses. Thus the material and mental sides are indivisibly linked by active information.

Thus in processes like these there is a constant interplay between the more manifest side of a process and the more subtle side. As we penetrate into the more subtle side we find that it too can be analysed into a more manifest side comprising those ideas that are more stable and more robust, while in the background we become aware of more subtle nuances, which can themselves be further analysed. There is no end to these processes as we penetrate to deeper and deeper levels of understanding. We are never able to make a sharp separation between mind and matter.

Thus ultimately there is no 'mind' that can be separated from 'matter' and no 'matter' that can be separated from 'mind'. There are merely many degrees of enfoldment. As a given level under consideration changes, a particular content of what is somatic (manifest) and what is significant (subtle) changes. Thus what is somatic at one level may have significance at another level which is then carried into the next more subtle level of soma which in turn has a deeper level of significance and so on. We can call this process soma-significance.

There is also a process where significance can change the soma as in the example of the fox given above. Here the realisation that there is no danger immediately changes the soma towards a relaxed state. Thus there is also the counter movement that establishes a signa-somatic relationship.

But all of this is not merely confined to the activity of humans. This enfolding process permeates right down to the basic levels of what is generally called 'inanimate Nature'. There is no material bottom level in Nature from which everything else follows. The search for the ultimate constituents of Nature will not enable us to reconstruct the Universe as it appears to us. Both Bohr (1961) and Heisenberg (1976) recognised the limitation imposed by quantum mechanics. They did not set out to show how quantum theory could be used to explain the classical world. Rather they realised that what was unambiguous in the classical world, such as position and momentum, must become ambiguous in the quantum domain. Here 'ambiguity' signifies an incomplete definition of meaning. This explanation must not be regarded as 'cop-out', but stems from the fact that all meaning must have a certain degree of ambiguity. What Bohr brings out is that ambiguity of meaning that we normally associate with mind also has a crucial role to play in our understanding of the behaviour of fundamental quantum processes.

The use of Bohr's ideas in this discussion may seem inappropriate particularly in view of the common perception that Bohm's approach to quantum mechanics was an attempt to criticise the 'Copenhagen' interpretation. It is true that an approach called 'Bohmian mechanics' has been advocated as a counter to the 'Copenhagen Hegemony' (see Cushing 1994). However it should be noted that this 'mechanics' that Cushing embraced only shares some of the mathematical structure used by Bohm and Hiley (1993) but none of the ideas that Bohm held to be the important are retained in this approach. Bohm certainly did not share Cushing's view expressed in his book. Indeed in the article I am extensively using here we find Bohm writing,

> This can best be brought out in terms of Niels Bohr's treatment of the subject (sic quantum theory), which is the most consistent interpretation that has been given so far, and one that has been accepted by most physicists (though it must be noted that few physicists have studied Bohr deeply enough to appreciate fully the revolutionary implications of his extremely subtly expressed arguments).

For me the approach to quantum theory that uses the quantum potential actually clarifies some of the more ambiguous statements made by Bohr (1961). This is particularly so in his notion of the 'wholeness of quantum phenomena', which he considers necessary to understand quantum phenomena. It is the quantum (information) potential that welds subject and object together. This feature of 'wholeness' also appears in Bohr's answer to the Einstein, Podolsky and Rosen objection. He points out that it is not a mechanical force that acts between the two spatially separated entangled systems, but more an "influence on the very conditions which define the possible types of predictions regarding the future behaviour of the system." In the Bohm approach it is the quantum potential that carries the information of the experimental conditions in both of these examples. My claim is that the approach I am proposing here actually offers an explanation of why the quantum phenomena are the way they are but to make sense of this explanation we must introduce the notion of active information.

But surely the Bohm picture was an attempt to return to reductionism. Originally that may have been the motivation but such a position cannot be maintained consistently. What the Bohm approach does is to separate *logically* the system from the measuring apparatus but we find that the quantum (information) potential actually unites the two systems into an indivisible whole. It does this not by mechanical interactions but by a pool of common information that is shared by the particles finding themselves in entangled states. It is only when this potential vanishes that the classical world appears. Thus the quantum (information) potential gives rise to the meaning of the relation between the measuring apparatus and the system under investigation. Putting it in a different language, we can say that the system (soma 2) has a significance within the context defined by the apparatus (soma 1), the link between them being the active information.

Finally I want to point out that we have concentrated on the organisation of particles in the above discussion only for simplicity. The global processes in the brain seem to be organised in terms of fields, action potentials and so on. But note that field theories can also be treated in exactly the same way. We find that the fields, both scalar and electromagnetic, are organised by information (quantum) potentials (see Bohm et al. 1987; Holland 1988 and Kaloyerou 1994). Everything we have said using particles can therefore be generalised to fields. Here it is the fields that are organised by a super-quantum potential, so the generalisation is straightforward in principle. The problem now is to carry this programme further by identifying suitable processes in the brain. This is a very important problem but space does not allow me to discuss this problem in more detail here.

Acknowledgements

I should like to thank Karl Pribram and Paavo Pylkkänen for many helpful discussions on the work reported in this paper. I also am deeply indebted to the Institute for Scientific Interchange Foundation for financial help.

Note

1. This is the origin of Bohr's 'indivisibility of the quantum of action'.

References

Arecchi, F. T. (2003). Chaotic neuron dynamics, synchronisation, and feature binding: quantum aspects. Presented at Aspects of Mind-Matter Research, organized by I. Antoniou and H. Atmansapcher. Wildbad Kreuth-Germany, 21–27 June 2003; to be published in Mind/Matter.

Binz, E., Pods, S., & Schempp, W. J. (2003). Heisenberg Groups-a unifying structure of signal theory, holography and quantum information theory. *J. App. Math. and Comp.*, *11*, 1–57.

Bohm, D. (1994). Soma-Significance: a new notion of the relationship between the physical and the mental. *Psycoscience*, *1*, 6–26.

Bohm, D., Hiley, B. J., & Stuart, A. E. G. (1970). On a New Mode of Description in Physics. *Int. J. Theor. Phys.*, *3*, 171–183.

Bohm, D. & Hiley, B. J. (1981). On a Quantum Algebraic Approach to a Generalised Phase Space. *Found. Phys.*, *11*, 179–203.

Bohm, D. & Hiley, B. J. (1983). Relativistic Phase Space Arising out of the Dirac Algebra. In A. van der Merwe (Ed.), *Old and New Questions in Physics and Theoretical Biology* (pp. 67–76). London: Plenum.

Bohm, D. & Hiley, B. J. (1987). An Ontological Basis for Quantum Theory: I – Non-relativistic Particle Systems. *Phys. Rep.*, *144*, 323–348.

Bohm, D., Hiley, B. J., & Kaloyerou, P. N. (1987). An Ontological Basis for Quantum Theory: II – A Causal Interpretation of Quantum Fields. *Phys. Rep.*, *144*, 349–375.

Bohm, D. & Hiley, B. J. (1993). *The Undivided Universe; an Ontological Interpretation of Quantum Mechanics.* London: Routledge.

Bohr, N. (1961). *Atomic Physics and Human Knowledge.* New York: Science Editions.

Brillouin, L. (1956). *Science and Information Theory.* New York: Academic Press.

Cushing, J. (1994). *Quantum Mechanics: Historical contingency and the Copenhagen hegemony.* Chicago: Chicago Uni. Press.

Denbigh, K. G. & Denbigh, J. S. (1985). *Entropy in Relation to Complete Knowledge.* Cambridge: Cambridge University Press.

Gabor, D. (1946). Theory of Communication. *J. Inst. Elect. Engineers*, *93*, 429–441.

Gosson, M. de (2001). *The Principles of Newtonian and Quantum Mechanics.* London: Imperial College Press.

Guillemin, V. & Sternberg, S. (1984). *Symplectic Techniques in Physics.* Cambridge: Cambridge University Press.

Heisenberg, W. (1976). The nature of elementary particles. *Phys. Today*, 32–39.

Hiley, B. J. (1971). A Note on Discreteness, Phase Space and Cohomology Theory. In T. Bastin (Ed.), *Quantum Theory and Beyond* (pp. 181–190). Cambridge: Cambridge University Press.

Hiley, B. J. (1995a). The Conceptual Structure of the Bohm Interpretation of Quantum Mechanics. In K. V. Laurikainen, C. Montonen, & K. Sunnarborg (Ed.), *Symposium on the Foundation of Modern Physics: 70 Years of Matter Waves* (pp. 99–118). Gif-sur-Yvette: Editions Frontiers.

Hiley, B. J. (1995b). Quantum Mechanics, The Implicate Order and The Relationship between Mind and Matter in New Directions in Cognitive Science. In P. Pylkkänen & P. Pylkkö (Eds.), *Proc. Int. Symp. Saariselka 4–9 Aug.* (pp. 44–55). Helsinki: Hakapaino.

Hiley, B. J. (1997). Information and Quantum Theory. In A. M. Fedoric & P. J. Marcer (Eds.), *The Outer Limits of Computing, Proceedings of the Fifth Annual Greenwich Symposium on the Frontiers of Quantum Computing* (pp. 25–42). Greenwich: University of Greenwich Press.

Hiley, B. J. (2001). Non-Commutative Geometry, the Bohm Interpretation and the Mind-Matter Relationship. In D. M. Dubois (Ed.), *Computing Anticipatory Systems: CASYS 2000 – Fourth International Conference* (pp. 77–88). New York: AIP.

Hiley, B. J. (2003). Phase Space Descriptions of Quantum Phenomena. To appear in *Proc. Int. Conf. Quantum Theory: Reconsideration of Foundations 2.* Växjö, Sweden.

Hiley, B. J. & Pylkkänen, P. (1997). Active Information and Cognitive Science – A Reply to Kieseppä. In P. Pylkkänen, P. Pylkkö, & A. Hautamäki (Eds.), *Brain, Mind and Physics* (pp. 64–85). Amsterdam: IOS Press.

Hiley, B. J. & Pylkkänen, P. (2001). Naturalizing the Mind in a Quantum Framework. In P. Pylkkännen & T. Vaden (Eds.), *Dimensions of Conscious Experience* (pp. 119–144). Amsterdam: John Benjamins.

Holland, P. R. (1988). Causal Interpretation of Fermi Fields. *Phys. Lett., 128A*, 9–18.

Kaloyerou, P. N. (1994). The Causal Interpretation of the Electromagnetic field. *Phys. Rep.*, *244*, 287–358.

Leff, H. S. & Rex, A. F. (1990). *Maxwell's Demon. Entropy, information and computing.* Adam Hilger.

MacKay, D. M. (1969). *Information mechanism and meaning.* Cambridge, MA: MIT Press.

Maroney, O. J. E. (2001). *Information and entropy in quantum theory.* PhD Thesis, London.

Moyal, J. E. (1949). Quantum Mechanics as a Statistical Theory. *Proc. Camb. Phil. Soc., 45*, 99–123.

Schempp, W. (1986). *Harmonic analysis on the Heisenberg nilpotent Lie group, with applications to signal theory.* London: Longman Scientific & Technical Press.

Schempp, W. (1993). Analog VLSI Network Models, Cortical Linking Neural Network Models, and Quantum Holographic Neural Technology. In K. Pribram (Ed.), *Rethinking Neural Networks: Quantum Fields and Biological Data.* Proceedings of the First Appalachian Conference on Behavioral Neurodynamics (pp. 233–297). New Jersey: Lawrence Erlaum Associates.

Shannon, C. E. (1948). *Bell Syst. Tech. J., 27*, 379 & 623.

Thom, R. (1975). *Structural Stability and Morphogenesis.* Reading, MA: Benjamin.

Vitiello, G. (2001). *My Double Unveiled – the dissipative quantum model of the brain.* Amsterdam: John Benjamins.

Wheeler, J. A. (1990). Information, Physics, Quantum: the Search for Links. In W. H. Zurek (Ed.), *Complexity, Entropy and the Physics of Information* (pp. 3–28). Redwood City: Addison-Wesley.

Wigner, E. P. (1972). On the Time-Energy Uncertainty Relation. In A. Salam & E. P. Wolf (Eds.), *Aspects of Quantum Theory.* Cambridge: Cambridge Uni. Press.

Wigner, E. P. (1932).On the Quantum Correction for Thermodynamic Equilibrium. *Phys. Rev., 40*, 749–759.

Chapter 12

Brain and mathematics

Karl H. Pribram
Georgetown University / George Mason University, USA

In describing the results of mapping brain processes I became aware of the fact that the same mathematical formulations apply to a variety of data bases. Not only brain processes but information measurement, thermodynamics and quantum physics share similar and often identical formulations. The relationships between the formulations can be portrayed by recourse to the Fourier transformation. When this is done insights into prespace-time, into the mind/matter dependency and into the dual of information processing and meaning emerge.

The fundamental connecting link between mathematics and theoretical physics is the pattern recognition capabilities of the human brain.

George Chapline *Physics Reports 315* (1999: 95–105)

It sometimes appears that the resistance to accepting the evidence that cortical cells are responding to the two dimensional Fourier components of stimuli [is due] to a general unease about positing that a complex mathematical operation similar to Fourier analysis might take place in a biological structure like cortical cells. It is almost as if this evoked for some, a specter of a little man sitting in a corner of the cell huddled over a calculator. Nothing of the sort is of course implied: the cells carry out their processing by summation and inhibition and other physiological interactions within their receptive fields. There is no more contradiction between a functional description of some electronic component being a multiplier and its being made up of transistors and wired in a certain fashion. The one level describes the process, the other states the mechanism.

DeValois & DeValois (1988: 288)

The fact that the formalism describing the brain microprocess is identical with the physical microprocess allows two interpretations: (a) The neural microprocess is in fact based on relations among microphysical quantum events, and (b) that the laws describing quantum physics are applicable to certain macrophysical interactions when these attain some special characteristics (p. 270).

> The formalism referred to describes the receptive fields of sensory neurons in the brain cortex. These were mapped in terms of Gabor wavelets or more generally, four dimensional information hyperspaces based on Jacobi functions (Atick & Redlich 1989) or Wigner distributions (Wechsler 1991).
>
> Pribram (1991) *Epilogue*

A personal road of discovery

The story of how, as a non-mathematician, my interest was engaged in Gabor-like mathematics is worthwhile repeating. Why would I follow such a path, when so many neurophysiologists and experimental psychologists shun, with the exception of statistical analyses, mathematical expressions (one could say, mathematical metaphors) in attempts to understand brain/mind transactions?

The story begins in the late 1930s, working in Ralph Gerard's laboratory at the University of Chicago. Gerard showed us that a cut separating two parts of the brain cortex did not abolish transmission of an electrical stimulus across the separation as long as the parts were in some sort of contact. Meanwhile, I discussed these observations with my physics professor. I argued with both Gerard and the physicist that such large scale phenomena could not account for the brain processes that allowed us to perceive, think and act. Gerard, of course, agreed but insisted that more than simple neuronal connections were important in understanding brain function. My physics professor also agreed but had nothing to offer. He may have mentioned quantum physics but was not versed in it.

At about the same time, Walter Miles, Lloyd Beck and I were pondering the neural substrate of vision. I was writing an undergraduate thesis on retinal processing in color sensation under the supervision of Polyak, making the point that beyond the receptors, the bipolar cells seemed to differentiate the three color bands to which the receptors were sensitive into a greater number of more restricted bandwidths. We bemoaned our inability to come up with some similar understanding for form vision. I distinctly recall saying: "wouldn't it be wonderful if we had a spectral explanation for brain processing of black and white patterns."

By 1948 I had my own laboratory at Yale University and began a collaboration with Wolfgang Koehler who told me of his Direct Current hypothesis as the basis for cortical processing in vision. He and demonstrated to me and my laboratory PhD students, Mort Mishkin and Larry Weiskrantz just how the anatomy of the auditory system would explain how the scalp auditory at the apex of the skull was transmitted by the brain's tissue: no neural connections needed. Shades of my experience with Gerard.

This time I set to work to test Koehler's hypothesis. We worked together with monkeys and humans displaying a white cardboard in front of their eyes and recorded from their visual cortex. (It was easy in those days to do such experiments with awake humans with their permission. Surgery had been done for clinical purposes with local anesthesia of the scalp – touching the brain itself is not felt by the patient.) Indeed we found a Direct Current (DC) shift during the display. One of my students and I then repeated the experiment using auditory stimulation in monkeys and obtained the same result in recording from the auditory cortex. (See Pribram 1971 Lecture 6 for review.)

In addition, I created minute epileptogenic foci in the visual cortex of monkeys and tested for their ability to distinguish very fine horizontal from vertical lines. Once electrical seizures commenced as shown by electrical recordings from their visual cortex I expected their ability to distinguish the lines to be impaired and even totally lost. The recordings showed large slow waves and total disruption of the normally patterned electroencephalogram (EEG).

Contrary to expectation, the monkeys performed the task without any deficiency. Koehler exclaimed: "Now that you have disproved not only my theory of cortical function in perception but everyone else's, as well, what are you going to do?" I answered: "I'll keep my mouth shut". In fact, I refused to teach a course on brain mechanisms in sensation and perception when I transferred to Stanford University (in 1958) shortly thereafter.

I did not come up empty-handed, however. What did occur was that the epileptic seizures delayed the monkeys' learning of the task some seven fold. This led to another series of experiments in which we imposed a DC current across the cortex from surface to depth and found that a cathodal current delayed learning while an anodal current enhanced it. There is more to this story but that has to wait for another occasion.

Once at Stanford I turned to other experiments that demonstrated cortical control of sensory input in the visual and auditory systems, feedback processes that were important to the conceptions Miller, Galanter and I had put forward in "Plans and the Structure of Behavior" (1960).

Some years into my tenure at Stanford, Ernest Hilgard and I were discussing an update of his introductory psychology text when he asked me about the status of our knowledge regarding brain physiology in perception. I answered that I was dissatisfied with what we knew: I and others had disproved Koehler's (1958) suggestion that perception could be ascribed to direct current brain electrical fields shaped like (isomorphic with) envisioned patterns. Hubel and Wiesel (1968) had just shown that elongated stimuli such as lines and edges were the best shapes to stimulate neurons in the primary visual receiving cortex – and that perception followed from putting together something like stick figures from these elementary sensitivities. As much of our perception depends on shadings and texture, the stick figure approach failed for this and other reasons to be a satisfactory. I was stumped. Hilgard, ordinarily a very kind and patient person seemed peeved and declared on a second encounter, that he did not have the luxury of procrastination as he had to have something to say in the text. So he asked once again to come up with some viable alternative to the ones I had so summarily dismissed.

I took the problem to my laboratory group and told them about Hilgard's problem and my dissatisfaction with the two extant proposals. I added that there was one other suggestion that had been offered which had the advantage that neither I nor anyone else knew how it might work either neurologically or with regard to perception: Lashley (1942) had proposed that interference patterns among wave fronts in brain electrical activity could serve as the substrate of perception and memory as well. This suited my earlier intuitions, but Lashley and I had discussed this alternative repeatedly, without coming up with any idea what wave fronts would look like in the brain. Nor could we figure out how, if they were there, how they could account for anything at the behavioral level. These discussions taking place between 1946 and 1948 became somewhat uncomfortable in regard to Don Hebb's book (1948) that he was writing at the time we were all together in the Yerkes Laboratory for Primate Biology in Florida. Lashley didn't like Hebb's formulation but could not express his reasons for this opinion: "Hebb is correct in all his details but he's just oh so wrong".

Within a few days of my second encounter with Hilgard, Nico Spinelli a postdoctoral fellow in my laboratory, brought in a paper written by John Eccles (Scientific American, 1958) in which he stated that although we could only examine synapses one by one, presynaptic branching axons set up synaptic wavefronts. Functionally it is these wavefronts that must be taken into consideration. I immediately realized (see Fig. 1–14, Languages of the Brain 1971) that axons entering the synaptic domain from different directions would set up interference patterns. (It was one of these occasions when one feels an utter fool. The answer to Lashly's and my first question as to where were the waves in the brain, had been staring us in the face and we did not have the wit to see it during all those years of discussion.)

Within another few days I received my current edition of Scientific American in which Emmet Leith and J. Upatnicks (1965) describe how recording of interference patterns on film tremendously enhanced storage and processing capability. Images could readily be recovered from the store by appropriate procedures that had been described by Dennis Gabor (1946) almost two decades earlier. Gabor called his mathematical formulation a hologram.

Using the mathematical holographic process as a metaphor seemed like a miraculous answer to Hilgard's question. Shading, detail, texture, everything in a pattern that we perceive can be accomplished with ease. Russell and Karen DeValois (1988) book on "Spatial Vision" and my (1991) book "Brain and Perception" provide detailed reviews of experimental results that support the conjecture that holography is a useful metaphor in coming to understand the brain/mind relation with regard to perception. Here I want to explore some further thoughts engendered by this use of a mathematical formulation to understand the brain/mind relation.

Some years later, in Paris, during a conference sponsored by UNESCO where both Gabor and I were speakers, we had a wonderful dinner together. I told him about the holographic metaphor for brain processing and we discussed its Fourier basis. Gabor was pleased in general but stated that "brain processing [of the kind we were discussing] was Fourier-like but not exactly Fourier." I asked, what then might such a relation look like and Gabor had no answer. Rather we got onto a step-wise process that could compose the Fourier – an explanation that I later used to trace the development of the brain process from retina to cortex. Gabor never then nor later told me about his 1946 contribution to communication theory and practice: that he had developed a formalism to determine the maximum compressibility of a telephone message that renders it still intelligible. He used the same mathematics that Heisenberg had used to describe processes in quantum physics and therefore called his "unit" a quantum of information. It took me several years to locate this contribution which is referred to in Likleiter's article on acoustics in Stevens (1951) Handbook of Experimental Psychology.

Does this application indicate that the formalism of quantum physics applies more generally to other scales of inquiry? Alternatively, for brain function, at what scale do actual quantum physical processing take place? At what anatomical scale(s) do we find quantum coherence and at what scale does decoherence occur? What relevance does this scale have for our experience and behavior?

To summarize: The formalisms that describe the holographic process and those that describe quanta of information apparently DO extend to scales other than the quantum. Today we use quantum holography to produce images with the technique of functional Magnetic Resonance (fMRI). The quantities described by terms of the formalisms such as Planck's constant will, of course, vary but the formulations will to a large extent be self-similar. The important philosophical implications for the brain/mind issue have been addressed in depth by Henry Stapp on several occasions (e.g. 2003, "The Mindful Universe") as well as by many others including myself (e.g. Pribram 1997, What is mind that the brain may order it?).

Scale

Deep and surface processing scales

Brain, being material, has at some scale a quantum physical composition. The issue is whether the grain of this scale is pertinent to providing insights into those brain processes that organize experience and behavior. In my book "Languages of the Brain" (1971) I identify two very different scales at which brain systems operate. One such scale, familiar to most students of the nervous system, is composed of circuits made up of large fibers usually called axons. These circuits operate by virtue of nerve impulses that are propagated along the fibers by neighborhood depolarization of their membranes.

But other, less well popularized, operations take place in the fine branches of neurons. The connections between neurons (synapses) take place for the most part within these fine fibers. Pre-synaptically, the fine fibers are the terminal branches of axons that used to be called teledendrons. Both their existence and their name have more recently been largely ignored. Postsynaptically, the fine fibers are dendrites that compose a feltwork within which connections (synapses and electrical ephapses) are made in every direction. This feltwork acts as a processing web.

The mathematical descriptions of processing in the brain's circuits needs to be different from the descriptions that describe processing in fine fibers. The problem that needs to be addressed with regards to circuits is that the connecting fibers are of different lengths and diameters that can distort the conduction of a pattern. The problem that needs addressing with regards to fine fiber processing is that, practically speaking, there are no propagated impulses within them so conduction has to be accomplished passively. Roberto Llinas (2000; Pellionitz & Llinas 1979, 1985) has provided a tensor theory that addresses the propagation in circuits and my holonomic (quantum holographic) theory models processing in the fine fibered web.

For me it has been useful to compare Llinas theory with mine to be able to detail their complementarity. The primary difference between the theories rests on the difference between the neural basis each refers to: Llinas is modeling neural circuits, what I (Pribram 1997; Pribram & Bradley 1998) have called a surface processing structure. Holonomic theory models what is going on in the fine fibered parts of these circuits, what I have referred to as deep processing. (The terms were borrowed from Noam Chomsky's analysis of linguistic structure and may, perhaps be able to provide a neurological account of these aspects of linguistic processing.)

Despite the different scales of these anatomical substrates, both Llinas and I emphasize that the processing spacetime in the brain is not the same as the spacetime within which we ordinarily get about. Llinas developed a tensor theory that begins, as does holonomic theory with oscillators made up of groups of neurons or their fine fibered parts. Next both theories delineate frames of reference that can be described in terms of vectors. Llinas uses the covariance (and contravarience) among vectors to describe tensor matrices where the holonomic theory uses vectors in Hilbert phase space to express the covariance. Llinas' tensor metric is not limited to orthogonal coordinates as is holonomic theory. (Llinas indicates that if the frame of reference is thought to be orthogonal, proof must be provided. I have provided such evidence in "Brain and Perception" and indicated when orthogonality must be abandoned in favor of non-linearity.)

In keeping with his caveat, Llinas does use the Fourier transform to describe covariation for the input, that is the sensory driven vectors: "[There are] two different kinds of vectorial expressions both assigned to one and the same physical location P, an invariant. The components v/i of the input vector are covariant (they are obtained by the orthogonal projection method) while the components v\j of the output vector are contravariant (obtained by the parallelogram method)" (Pellionitz & Llinas 1985: 2953). As in the holonomic theory, the tensor theory needs to establish entities and targets and it does this (as in the holonomic theory (see Pribram 1991, Lectures 5 and 6) by using the *motor* output to create contravariant vectors. The covariant-contravariant relationship is combined into a higher level invariant tensor metric.

Thus Llinas states that "sensory systems in the CNS are using expressions of covariant type while motor systems use components of a contravariant type" (p. 2953). This is similar to the use of motor systems in "Brain and Perception" to form Lie groups to produce the perception of invariants basic to object perception. Llinas' theory is more specific in that it spells out contravariant properties of the motor process. On the other hand, Holonomic theory is more specific in specifying the neural substrate produced by nystagmoid and other such oscillating movements (that result in co-ordination of pixels moving together against a background of more randomly moving pixels).

Another advantage of the holonomic theory is that it can explain the fact that the processes that form the experiencing of objects, project them away from the processing medium. "Projection" can be experienced by viewing a transmission hologram. Georg von Bekesy (1967) demonstrated this attribute of visual and auditory processing by arranging a set of vibrators on the skin of the forearm. Changing the phase relations among the vibrators resulted in feeling a point stimulus moving up and down the skin. Bekesy then placed two such arrays of vibrators, one on each forearm. Now, with appropriate adjustments of phase, the sensation obtained was a point in space in front of and between the arms. A similar phenomenon occurs in stereophonic sound: adjusting the phase of the sound coming out of the two or more speakers projects the sound away from the speakers (and, of course the receiver where the processing is actually occurring).

There is more to the rich yield obtained by comparing the Tensor theory to the Holonomic theory. For instance, Pellionisz and Llinas develop a look-ahead module via Taylor-assemblies that are practically the same as the anticipatory functions based on Fourier series (Pribram 1997).

The two theories also converge as Tensor Theory is based on "a coincidence of events in which both the target and interceptor merge into a single event point. This is an invariant known in physical sciences as a four dimensional Minkowski-point or world-point" (Pellionitz & Llinas, p. 2950). Holonomic Theory also requires a high-dimensional position-time manifold. "As originally implied by Hoffman (1996) and elaborated by Caelli et al. (1978), the perceptual representation of motion should be subject to laws resembling the Lorenz transformations of relativity theory." This means that the Poincare group (Dirac 1930; Wigner 1939) is relevant, requiring a manifold of as many as ten dimensions. In the context of modeling the brain process involved in the perception of Shepard figures, what needs to be accomplished "is replacing the Euclidian group [that ordinarily describes geodesics] with the Poincare group of space time isometries, the relativistic analogues of geodesics –" (Pribram 1991: 117).

Both theories handle the fundamental issue as to "how can coordinates be assigned to an entity which is, by its nature, invariant to coordinate systems" (Pellionez & Llinas, p. 2950). The very term "holonomy" was chosen to portray this issue.

It is fitting that surface structure tensor circuit theory uses insights from relativity theory while deep structure holonomy regards quantum-like processing. As physicists struggle to tie together relativity and quantum field theory in terms of quantum gravity, perhaps further insights will be obtained for understanding brain processing (Hameroff & Penrose 1995; Smolin 2004; Ostriker & Steinhardt 2001).

The main practical difference between the theories is that in the Tensor Theory, time synchrony among brain systems (which means correlation of their amplitudes) is all that is required. Holonomic theory indicates that a richer yield is obtained when phase coherence is manifest. Principle component analysis will get you correlations but it takes Independent Component Analysis (equivalent to 4th order statistics) to capture the detail (e.g. texture) represented in the phase of a signal (King, Xie, Zheng, & Pribram 2000).

Some of the relationships between the theories are being implemented in the production of functional Magnetic Resonance Imaging (fMRI). Heisenberg matrices (representations of the Heisenberg group) are used and combine in what is called quantum holography (that is, holonomy) with the tensor geometry of relativity (Schempp 2000).

Llinas, in a book called the "i of the vortex" (2001) spells out in detail the primacy of the Motor Systems not only in generating behavior but also in thinking (conceptualized as internal movement) and the experience of the self. This is an important perspective for the psychological and neurosciences (see e.g. Pribram in press) but addresses issues beyond the scope of this essay.

Quantum Brain Dynamics

Henry Stapp in two excellent articles (Stapp 1997a and b) reviews the development of quantum theory and outlines how it is essential to understanding the mind/brain relationship. Stapp sets up the issue as follows. "Brain process is essentially a search process: the brain, conditioned by earlier experience, searches for a satisfactory response to the new situation that the organism faces. It is reasonable to suppose that a satisfactory response will be programmed by a template for action that will be implemented by a carefully tuned pattern of firings of some collection of neurons. The executive pattern would be a quasi-stable vibration that would commandeer certain energy resources, and then dissipate its energy into the initiation of the action that it represents." Patterns of firings and quasi-stable vibrations are, what I have termed the surface and deep structures of processing that are represented by Llinas' Tensor and my Holonomic Theories respectively.

Stapp goes on to note that "If the programmed action is complex and refined then this executive pattern must contain a great deal of information and must, accordingly, be confined to a small region of phase space." Holonomic theory indicates that spread functions such as those that compose holography, do indeed make it possible to contain a great deal of information within a small region (patches of dendritic receptive fields) of phase space. Stapp further notes that "the relative timing of the impulses moving along the various neurons, or groups of neurons, will have to conform to certain ideals to within very fine levels of tolerance. How does the hot, wet brain, which is being buffeted around by all sorts of thermal and chaotic disturbances find its way to such a tiny region in a timely manner?" Llinas' Tensor Theory deals with the timing issue.

Further: "How in 3n dimensional space (where n represents some huge number of degrees of freedom of the brain) does a point that is moving in a potential well that blocks out those brain states that are not good solutions to the problem – but does not block the way to good solutions find its way in a short time to a good solution under chaotic initial conditions?" Stapp notes that classical solutions to this problem won't work and that "the quantum system [will work as it] has the advantage of being able to explore simultaneously (because the quantum state corresponds to a superposition of) all allowed possibilities." Stapp provides a viable metaphor in a glob or cloud of water acting together rather than as a collection of independently moving droplets. "The motion of each point in the cloud is influenced by its neighbors."

However, classical holography will also do just this. But the advantage of holonomy, that is quantum holography, is that it windows the holographic space providing a "cellular" phase space structure, in patches of dendritic fields thus enhancing the alternatives and speed with which the process can operate. In short, though the information within a patch is entangled, cooperative processing between patches can continue to cohere or de-coherence can "localize" the process.

With regard to evidence regarding the scale at which quantum processes are actually occurring, a number of publications have reported that quantum coherence characterizes the oscillations of ions within neural tissue channels. (e.g. see Stapp 1997; and Jibu et al. 1994; Jibu & Yasue in this volume). The question immediately arises as to whether decoherence occurs when the channels communicate with each other and if so, how. Stapp notes that "phase relationships, which are essential to interference phenomena, get diffused into the environment, and are difficult to retrieve. These decoherence effects will have a tendency to reduce, in a system such as the brain, the distances over which the idea of a simple quantum system holds."

Hameroff and Penrose (1995) have also dealt with the limited range over which quantum coherence can operate. These authors suggest that excitation at the microtubular scale follows quantum principles but that decoherence self-organizes towards the end of an axon or dendrite. Davydov (see Jibu this vol.) Ross Adey (1987) and I (1991) have independently proposed that microtubular quantum coherence provides saltatory coupling between dendritic spines and saltatory conduction in between nodes of Ranvier in axons via soliton waves (see also, Jibu & Yasue this volume). Soliton waves would thus provide a longer range over which coherence can be maintained.

An additional mechanism for coherent channel interaction has been proposed by Jibu, Pribram and Yasue (1996). This proposal focuses on the phospholipd bilayer that composes the membrane within which the channels occur. The phosphate parts of the molecule are hydrophilic capturing water as in a swamp. The water in such a region can become ordered into a super-liquid form that, by way of boson condensation, can act as a superconductor. Channels become connected over a limited distance by a transitional process that is quantum-like at a somewhat larger scale than the channels per se.

Thus, at the neural systems scale, there are two quantum-like fields, one pre-synaptic composed by the fine branching of axons as they approach the synapse; the other post-synaptic composed of the fine branches of dendrites. Hiroomi Umezawa and his collaborators (Stuart, Takahashi, & Umezawa 1979) pointed out that not only quantum but "classical" processing can be derived from quantum field theory. The relevance of all this to the brain/mind issue is that both Umezawa and Giuseppe Vitiello (2004) have, on the basis of mathematical insights, proposed that interactions among these two quantum brain fields is necessary for self-reflective consciousness to occur. Hiley notes: "this is part of a bigger mathematical structure of bi-algebras that Umezawa and Vitiello are exploring. The doubling arises from a natural duality." I add, could this doubling arise from the nature of the Fourier relationship? The Fourier transformation results in a complex number that represents both a real and a virtual line, a built-in duality.

My question is not an idle one. Our optical system performs a Fourier transform that results in the dual of real and virtual. One of these must be repressed in getting about in the space-time world. But the repression is incomplete. Experiments using glasses that invert the optical image to make the world look upside down, have shown that actively moving about re-inverts the image so that the world again looks "normal". Re-reversal takes place over time when the glasses are removed. Vitiello's "double" is thus twice unveiled.

To return to the topic of "scale": In the brain, at what scale does decoherence initially occur? There are two types of processes that are excellent candidates. The local chemical activities, constituted of neuro-transmitters, neuro-modulators and neuro-regulators appear, at present, not to share properties that are best described in quantum terms. Their operation transforms the entangled quantum processes into larger scale influences on neural circuitry especially at synaptic sites. A second locus for decoherence is the region of the axon hillock. It is here that the passive conduction of dendritic activity influences the spontaneous generation of the discrete impulses that transmit the results of processing at one location to another location via neural circuitry.

Formalisms

The quantum formalism

The initial quotation introducing this essay is from the ending of an excellent paper by George Chapline (1999) entitled "Is theoretical physics the same thing as mathematics". Chapline's provocative title employs a bit of poetic license. Nonetheless the paper provides considerable insight as to the applicability of the quantum formalism to other scales of inquiry. Chapline shows that quantum theory "can be interpreted as a canonical method for solving pattern recognition problems" (p. 95). In the paper he relates pattern recognition to the Wigner-Moyal formulation of quantum theory stating that this "would be a good place to start looking for a far reaching interpretation of quantum mechanics as a theory of pattern recognition" (p. 97). In a generalization of the Wigner- Moyal phase space he gives the physical dimensions as the Weyl quantization of a complete holographic representation of the surface. He replaces the classical variable of position within an electromagnetic field with ordinary creation and annihilation operators. He shows that "representing a Riemann surface holographically amounts to a pedestrian version of a mathematically elegant characterization of a Riemann surface in terms of its Jacobian variety and associated theta functions" (p. 98). This representation is equivalent "to using the well known generalized coherent states for an SU(n) Lie algebra" (p. 98). This is the formalism employed in "Brain and Percption" (Pribram 1991) to handle the formation of invariances that describe entities and objects.

There is much more in Chaplin's paper that resonates with the holonomic, quantum holographic formulations that describe the data presented in "Brain and Perception". These formulations are based on quantum-like wavelets, Gabor and Wigner phase spaces. Whether these particular formulations will be found to be the most accurate is not the issue: rather it is that such formalisms can be attempted due to the fact that the "fundamental connecting link between mathematics and physics is the pattern recognition of the human brain" (p. 104).

As an example of the utility of these insights, Chapline indicates how we might map the co-ordination of processing in the central nervous system. He notes that "the general idea [is] that a quantum mechanical theory of information flow can be looked upon as a model for the type of distributed information processing carried out in the brain." He continues, "one of the fundamental heuristics of distributed information processing networks is that minimization of energy consumption requires the use of time division multiplexing for communication between processors, and it would be natural to identify the local internal time in such networks as quantum phase" (p. 104). The caveat is, as noted, that quantum phase is fragile in extent and must be supplemented by the processes described in comparing holonomic (quantum holographic) theory with the tensor theory of Llinas (which applies to neural circuitry rather than to the fine fibered quantum holographic processing per se).

Bohm and Hiley (1981) had also undertaken a topological approach to quantum mechanics based on a Wigner-Moyal cellular structure of phase space. In the current volume, Hiley (this volume) carries the approach further by relating it to Gabor's handling of signal transmission (communication) with what we now call a Gabor function (he called it a quantum of information) which is the centerpiece of the Holonomic Brain theory presented in "Brain and Perception" (1991). Hiley is able to introduce a phase space distribution function that allows calculation of quantum probabilities without having to resort to non-commuting dynamic variables. This makes easier the transition to the commuting aspects of groups.

It thus shows the intimate connection with the Heisenberg group as used by Schempp in describing the fMRI process.

Hiley goes on to note that underlying the Wigner-Moyal distribution is the simplectic group. (Note that Chapline has focused on an SU(n) Lie group. The simplectic group is mathematically the more general.) "The simplectic group is in turn covered by a metaplectic group that underlies Schroedinger's equation, as well as Hamilton's equation of motion and the classical ray formulation of optics. The metaplectic 'double' covers the symplectic group in the same way that the spin group SU(2) double covers the rotation group SU(3).

The importance of these insights is that "we have a mathematical structure that is basic to both classical mechanics and quantum mechanics. At this level there is no basic difference between the dynamical equations of classical and quantum mechanics. The difference arises once one asserts there is a minimum value for this action and equates this value to Planck's constant h" (Hiley this volume).

> What this means is that certain results – may look as if they are quantum in origin but in fact have nothing to do with quantum mechanics per se but arise from the group structure that is common to both forms of mechanics. For example the Fourier transformation is common to both classical and quantum situations. Indeed the Fourier transformation is at the heart of of Gabor's discussion of information transfer. Thus any results that emerge from an analysis of either the Wigner-Moyal approach or the Bohm approach may not necessarily have to do with quantum phenomena per se, and for that reason I would like to call the emerging dynamics that I will discuss below 'information dynamics'. (Hiley this volume)

Observables, observations, and measurement

Just what is the specific role of the brain in helping to organize our conscious relatedness? A historical approach helps sort out the issues. The Matter/Mind relationship has been formulated in terms of cuts. In the 17th century the initial cut was made by Renee Descartes (1662/1972) who argued for a basic difference in kind between the material substance composing the body and its brain and conscious processes such as thinking. With the advent of quantum physics in the 20th century Descartes' cut became untenable. Werner Heisenberg (1930) noted a limitation in simultaneously measuring the moment (rotational momentum) and location of a (*material*) mass. Dennis Gabor (1946) found a similar limit to our understanding of communication, that is, *minding*, because of a limitation in simultaneously measuring the spectral composition of the communication and its duration.

These indeterminacies place limits to our observations of both matter and mind and thus the location of the matter/mind cut. Heisenberg (1930) and also Wigner (1972) argued that the cut should come between our conscious observations and the elusive "matter" we are trying to observe. Niels Bohr (1961) argued more practically that the cut should come between the instruments of observations and the data that result from their use.

In keeping with Bohr's view, these differences in interpretion come about as a consequence of differences in focus provided by instrumentation (telescopes, microscopes, atom smashers, and chemical analyzers). Measurements made with these instruments render a synopsis of aspects of our experience as we observe the world we live in.

The Fourier relationship

The formalisms found to be important in quantum measurement as it relates to the brain/mind issue is the Fourier (1802) relationship. This relationship states that any space-time pattern can be transformed into the spectral domain characterized by a set of waveforms that encode amplitude, frequency and phase. Inverting the transform realizes the original space-time configuration. The transform domain is "spectral" not just "frequency" because the Fourier transformation encodes both the cosine and sine of a waveform allowing the interference between the 90 degree phase separation to be encoded discretely as coefficients.

The advantage gained by transforming into the spectral domain is that a great variety of transformed patterns can be readily convolved (multiplied) so that by performing the inverse transform the patterns have become correlated. This advantage is enhanced in quantum holography (which I have called Holonomy). Chapline (2002) in a paper entitled: "Entangled states, holography, and quantum surfaces" argues that the simplest way to encode "fundamental objects – may be as multi-qubit entangled states" (p. 809). I suggest that, impractical as it may currently seem, it would be more productive to encode "qulets", wavelet transformations, to preserve phase. As noted, Lie group theory can be used to describe how, by way of co-variation, various perspectives (images) of an object can form an invariant entity (Pribram 1991). Image processing as in tomography such as PET scans and fMRI are prime examples of the utility of such encoding.

The diagram below provides one summary of what these measurements indicate both at the quantum and cosmic scale. The diagram is based on a presentation made by Jeff Chew at a conference sponsored by a Buddhist enclave in the San Francisco Bay area. I had known about the Fourier transformation in terms of its role in holography. But I had never appreciated the Fourier-based fundamental conceptualizations portrayed below. I asked Chew where I might find more about this and he noted that he'd got it from his colleague Henry Stapp who in turn had obtained it from Dirac. (Eloise Carlton a mathematician working with me and I had had monthly meetings with Chew and Stapp for almost a decade and I am indebted to them and to David Bohm and Basil Hiley for guiding me through the labyrinth of quantum thinking.)

The diagram has two axes, a top-down and a right-left. The top-down axis distinguishes change from inertia. Change is defined in terms of energy and entropy. Energy is measured as the amount of actual or potential work necessary to change a structured system and entropy is a measure of how efficiently that

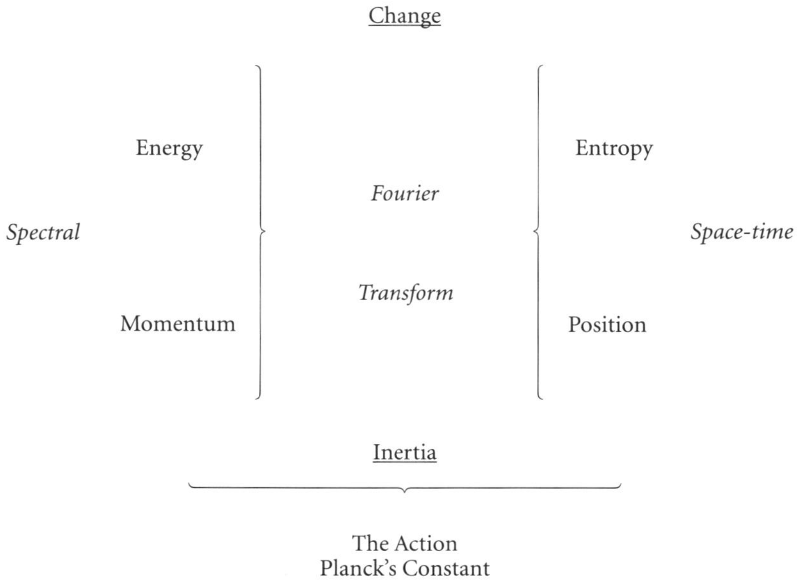

Figure 1. The wave/particle dichotomy is orthogonal to the above distinction

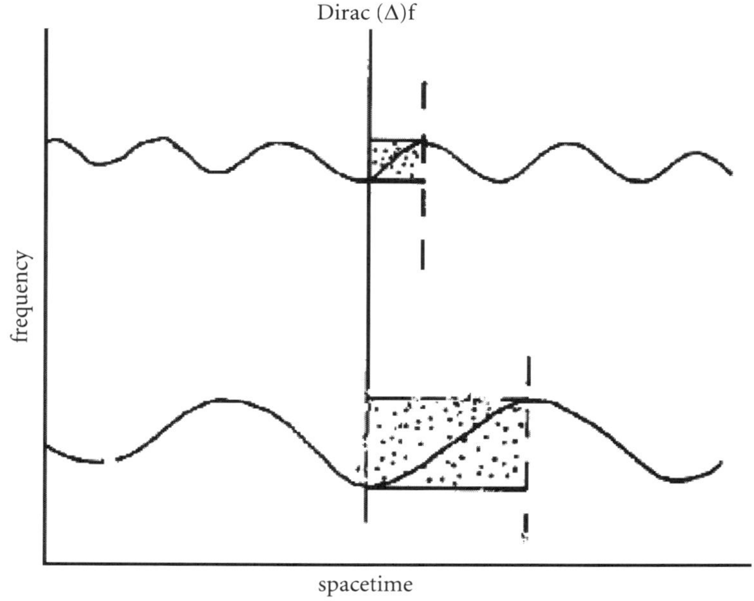

Figure 2. Logons, Gabor elementary functions: Quanta of information

change is brought about. Structure is classically described in numbers and their relations to one another. Inertia is defined as moment, the rotational analogue of mass. Location is indicated by its spatial coordinates, i.e. by the geometry of the shape of that location.

The right-left axis distinguishes between measurements made in the spectral domain and those made in spacetime. Spectra are composed of interference patterns where fluctuations intersect to reinforce or cancel. Holograms are examples of the spectral domain. I have called this pre-spacetime domain a potential reality because we navigate the actually experienced reality in spacetime.

The up-down axis relates mind to matter by way of sampling theory (Barrett 1993). Choices need to be made as to what aspect of matter we are to "attend". The brain systems coordinate with sampling have been delineated and brain systems that impose contextual constraints on sampling have been identified (Pribram 1959, 1971). The down-up axis indicates the emergence of in-formation from patterns of brain processes.

My claim is that the basis function from which both matter and mind are "formed" is the potential reality, the flux (or holo-flux, see Hiley 1996). This flux provides the ontological roots from which conscious experiences regarding matter as well as mind (psychological processes) become actualized in spacetime. To Illuminate this claim, let me begin with a story I experienced: Once, Eugene Wigner remarked that in quantum physics we no longer have observables, (invariants) but only observations. Tongue in cheek I asked whether that meant that quantum physics is really psychology, expecting a gruff reply to my sassiness. Instead, Wigner beamed a happy smile of understanding and replied, "yes, yes, that's exactly correct". If indeed one wants to take the reductive path, one ends up with psychology, not particles. In fact, it is a psychological process, mathematics, that describes the relationships that organize matter. *In a non-trivial sense current physics is rooted in both matter and mind* (Chapline 2000, "Is physics and mathematics the same thing?").

Conversely, communication ordinarily occurs by way of a material medium Bertrand Russell (1948) addressed the issue that the form of the medium is largely irrelevant to the form of the communication. In terms of today's functionalism it is the communicated sample of a pattern that is of concern, not whether it is conveyed by a cell phone, a computer or a brain and human body. *But not to be ignored is the fact that communication depends on being embodied, instantiated in some sort of material medium.* This convergence of matter on mind, and of mind on matter, gives credence to their common ontological root (Pribram 1986, 1998). My claim is that this root, though constrained by measures in spacetime, needs a more fundamental order, a potential that underlies and transcends spacetime. The spectral basis of the quantal nature of both matter and of communication portray this claim.

Of matter and mind

One way of interpreting the "Fourier" diagram is that it indicates *matter to be an "ex-formation", an externalized (extruded, palpable, concentrated) form of flux. By contrast, thinking and its communication (minding) are the consequence of an "internalized" (neg-entropic) forming of flux, its in-formation.*

Hiley (this volume) comes to a similar perspective in that he stresses the formative aspect of in-formation. As noted, in discussing Bohm's quantum potential, Hiley begins with the Wigner-Moyal approach to the Schroedinger wave function. The real part of the equation describes what, in my formulation, is ex-formation. The virtual part of the equation describes the quantum potential: "it has no external source in the sense that the electric field has its source in a distribution of charges. Thus it does not emerge from an interaction Hamiltonian as does classical force.... In this sense it cannot be thought to act as an efficient cause. It is more like a formative cause that shapes the development of the process.... Thus we can think of the information as active from within giving shape to the whole process and this shape depends on the environment [the material context] in key ways." In the Fourier diagram this formative cause is labeled action (after Feinman).

Flux, measured as spectral density, is here defined (see Pribram & Bradley 1998) as change or lack thereof, basic to both energy (the amount of actual or potential work involved in altering structural patterns) and inertia (measured as the rotational momentum of mass). David Bohm (1973) had a concept similar to flux in mind which he called a holomovement. He felt that my use of the term "flux" had connotations for him that he did not want to buy into. I, on the other hand, felt holomovement to be vague in the sense of asking "what is moving?" We are dealing with fluctuations, and in the nervous system with oscillating hyper- and depolarizations characterized by the field potentials we can map from the fine fibered parts of the system.

Quantum physics is a science of matter. In quantum physics the Fourier transformation is primarily applied in relating the position in space of a mass to its rotational momentum (spin). Much has been written regarding the indeterminacy of this relationship at the lower limit of measurement, that is, that at the limit it is impossible to accurately measure both position and moment. This is also known as Heisenberg's uncertainty principle.

In the physics of matter the terms moment and position refer to a stable status: "moment" to the inertia of a mass and "position" to its location. By contrast, "energy" and "entropy" in thermodynamics refer to change measured as a quantitative amount of work necessary to effect the change and the efficiency with which the change is carried out. Both moment (rotational momentum) and energy are measured in terms of frequency (or spectral density) (times Planck's constant). Position is measured with respect to location, entropy as it evolves over duration for instance as power, the amount of work per unit time).

The Fourier relation envisions the waveforms involved in measuring frequency not as a linear continuum but rather as a clock-face-like circle – thus one can triangulate and obtain the cosine and sine of the waveform to produce their interference and measure phase in the spectral domain. This was Fourier's definitive insight (or was it that of the mathematicians in Egypt with whom he discoursed during Napoleon's expedition?) that has made his theorem "probably the most far reaching principle of mathematical physics" as Feynman has declared it. Thus, the Fourier energy-time relation becomes, in a sense, "spatialized".

In quantum physics very little has been made of the uncertainty involved in relating energy and time. Dirac and especially Wigner (1972) called attention to this indeterminacy in discussing the delta function, but for the most part quantum physicists (e.g. Bohr) have focused on the relationship between energy and mass as in Einstein's equation: $E = mc^{*}$. By squaring c, the constant representing the speed of light, a linear measure of time becomes "spatialized" into an area-like concept, Minkowsky's space-time. I will return to a discussion of this version of time when considering brain processes. In short, much of the thinking that has permeated theories describing matter has been grounded in space-time, not the spectral aspects addressed by the Fourier transformation. For quantum physicists interested in the composition of matter, the Einstein/Minkowsky spatialization of time and energy comes naturally.

For brain function, Dirac's and Wigner's indeterminacy in the relation between energy and time is the more cogent. As noted, during the 1970s and 1980s the maps of dendritic receptive fields of neurons in the primary visual and other sensory cortexes were described by a space-time constrained Fourier relation, the Gabor elementary function, a windowed Fourier transform, essentially a sinc function, a kind of wavelet in phase (Hilbert) space. Gabor had used the same mathematics that Heisenberg had used; he therefore called his unit a "quantum of information" warning that by this he meant only to indicate the formal identity of the formulation, not a substantive one.

Gabor had undertaken his mathematical enterprise to determine the minimum uncertainty, the maximum compressibility, with which a telephone message could be transmitted across the Atlantic cable without any loss in intelligibility. He later (1954) related this minimum uncertainty to Shannon's BIT, the measure of a reduction of uncertainty. In turn, Shannon had related his measure of uncertainty to Gibbs' and Boltzman's measure of entropy. The stage was set for the issues of current concern in this part of the essay: a set of identical formalisms that refer to widely different substantive and theoretical bodies of knowledge.

Thermodynamics

Contrast the referents of the formulations in classical, relativity and quantum physics to those in thermodynamics: First there are no references to the momentum and position of a mass. Second, the emphasis is on energy as measured not as a pseudo-spatial quantity but as dynamic, often "free" energy. The utility of energy for structured work (as in a steam engine) is of concern in thermodynamics; its efficiency in structured use or rather, its inefficiency as dissipation into unstructured heat is measured as entropy. In the diagram of the Fourier relation, thermodynamics focuses on the upper part of the relationship (the dynamics of energy and time) just as physics focuses on the lower part (the statics of momentum and location of a mass or particle).

The distinction devolves on the conception of time. As noted, time in relativistic and quantum physics has been spatialized as clock time, the Kronos of the ancient Greeks. Time in thermodynamics is a measure of process, how quickly energy is expended. This amount of time, its duration, may vary with circumstance. It is the "Duree" of Bergson, the Kairos of an "Algebraic Deformation in Inequivalent Vacuum States" (Correlations, ed. K. G. Bowden, *Proc. ANPA, 23*, 104–134, 2001).

Brain processes partake of both aspects of time. In the posterior parts of the brain, the processes described by the Fourier transform domain, by virtue of movement, form symmetry groups that describe invariance, that is, objects in space and in Kronos, clock time. Alternatively, in the frontal and limbic portions of the brain the processes described result in the experience of Kairos, the duration of an episode. The evidence for these statements is reviewed in detail in Lecture 10, "Brain and Perception".

Meaning

Shannon (1948; Shannon & Weaver 1949) insisted that his measure of the amount of information as the amount of reduction of uncertainty did not provide a measure of meaning: "One has the vague feeling that information and meaning may prove to be something like a pair of conjugate variables in quantum theory, they being subject to some joint restriction that condemns a person to the sacrifice of the one as he insists on having much of the other". Looking at the Fourier diagram, we can ask, which of the conjugate relationships are appropriate to serving Shannon's intuition with regard to meaning? My answer is that it is the relationship between Shannon's and Gabor's measures of information as negentropy and the location (the placement, the sampling) of a mass on the right side of the diagram.

Meaning is, in a nontrivial sense, the instantiation in matter of information. We might say, meaning matters. Bohm noted that his "active information" did something, had an influence on the course of the quantum material relationship. Charles Pearce stated: "What I mean by meaning is what I mean to do." Doing acts on the material world we live in.

This returns us to the statements made by Stapp: "Brain process is essentially a search process – the brain searches for a satisfactory response – and then dissipates [increases the entropy of] its energy in the initiation of the action that it represents". Llinas also emphasizes the primacy of the motor systems in implementing thought and in the experiencing of the self. A "satisfactory" response is a meaningful one. "Implementation" involves acting on the world we live in.

With regard to language, meaning is the semantic relationship between linguistic "informative" patterns that ultimately lead to the deictic, "the pointing to the lived-in material world" to which that pattern refers (Pribram 1975).

But there is another meaning to meaning, the meaning in music and in the pragmatics (the rhetoric) of language (Pribram 1982). This meaning of meaning does not involve doing. Rather it is evocative, it engages not the striped muscular system of the body but the smooth muscles and endocrines. What is needed to account for this form of meaning is an addition to Pearce's "what I mean to do". This addition is: "What I mean by meaning is what I mean to experience." When I walk into a concert hall I am prepared to experience a familiar or not so familiar rendition of a repertoire. When Marc Antony addressed the crowd at Caesar's funeral he proclaimed: "I come to *bury* Caesar, not to praise him". The prosodics of this declamation as well as the semantics play into the expected experience of the audience. Prosody is a right hemisphere, semantics a left hemisphere process.

The time is ripe for untangling patterns of information from patterns of meaning. The proposal presented here stems directly from the other analyses undertaken. I continue to be amazed and awed by the power of mathematical conceptualizations in understanding the roots of brain function. These roots grow in the soil of the pattern processing of the brain, patterns we call information and meaning.

To summarize: The formal, mathematical descriptions of our subjective experiences (our theories) of observations in the quantum, thermodynamic and communications domains are non-trivially coordinate with each other. They are also coordinate with brain processes that, by way of projection, unify the experiential with the physical. By this I mean that the experiences of observations (measurements) in quantum physics, in thermodynamics and in communication appear to us to be "real", that is, extra-personal. Adaptation to living in the world makes it likely that this coordination of mathematical descriptions thus represents the useful reality within which we operate.

Acknowledgements

As noted in the text, I am deeply indebted to David Bohm and to Basil Hiley for inspiration and corrective management of my course of theorizing. Additionally I have learned much in my association with Henry Stapp and Geoffrey Chew and more recently from Sisir Roy. I hope this manuscript will challenge them to continue to critique what often I feel they think of as my wayward ways.

References

Adey, W. Ross (1987). Electromagnetic fields, the modulation of brain tissue functions – a possible paradigm shift in biology. In G. Adelman (Ed.), *The International Encyclopedia of Neuroscience*, Vol. 2.

Barrett, T. W. (1993). Is Quantum Physics a Branch of Sampling Theory? In C. Cormier-Delanous, G. Lochak, & P. Lochak (Eds.), *Courants, Amers, Ecueilsen Microphysique*. Paris: Fondation Louis DeBroglie.

Bekesey, G. (1967). *Sensory Inhibition*. Princeton, NJ: Princeton University Press.

Bohm, D. (1973). Quantum Theory as an indication of a new order in physics. Part B. Implicate and Explicate Order in physical law. *Foundations of Physics, 3*, 139–168.

Bohm, D. & Hiley, B. (1993). *The Undivided Universe: An Ontological Interpretation of Quantum Theory*. London: Rutledge.

Bohr, N. (1961). *Atomic Physics and Human Knowledge*. New York: Science Editions.

Chapline, G. (1999). Is theoretical physics the same thing as mathematics? *Physical Reports, 315*, 95–105.

Chapline, G. (2002). Entangled states, holography, and quantum surfaces. *Chaos, Solitons and Fractals, 14*, 809–816.
DeValois, R. L. & DeValois, K. K. (1988). *Spatial Vision* (Oxford Psychology Series # 14). New York: Oxford University Press.
Descartes, R. (1972/1662). *Treatise on man* (T. S. Hall, Trans.). Cambridge: Harvard University Press.
Eccles, J. C. (1958). The physiology of imagination. *Scientific American, 199*, 135–146.
Fourier, J. (1807). Sine and Cosine Series for an Arbitrary Function. In *Joseph Fourier 1768–1830*. Ed. and annotated by I. Grattan-Guinness. Cambridge, MA: The MIT Press.
Gabor, D. (1946). Theory of communication. *Journal of the institute of Electrical Engineers, 93*, 429–441.
Gabor, D. (1948). A new microscopic principle. *Nature, 161*, 777–778.
Hameroff, S. & Penrose, R. (1995). Orchestrated reduction of quantum coherence in brain microtubules: a model for consciousness. In J. S. King & K. H. Pribram (Eds.), *Is the brain too important to be left to specialists to study?* The third Appalachian Conference on Behavioral Neurodynamics. Mahwah, NJ: Lawrence Erlbaum Associates.
Hebb, D. O. (1949). *The Organization of Behavior: A Neuropsychological Theory*. New York: Wiley.
Heisenberg, W. (1930). *The Physical Principles of the Quantum theory*. London: Dover Publications.
Hiley, B. J. (1996). Mind and Matter: Aspects of the Implicate Order Described through Algebra. In K. H. Pribram & J. S. King (Eds.), *Learning as Self Organization: Proceedings of the Third Appalachian Conference on Behavioral Neurodynamics*. Mahwah, NJ: Lawrence Erlbaum Associates.
Hiley, B. J. (2001). Towards a Dynamics of Moments: The Role of Algebraic Deformation and Inequivalent Vacuum States. In K. G. Bowden (Ed.), *Correlations, Proc. ANPA, 23* (pp. 104–134).
Hubel, D. N. & Wiesel, T. N. (1968). Receptive fields and functional architecture of monkey striate cortex. *Journal of Physiology, 195*, 215–243.
Jibu, M., Pribram, K. H., & Yasue, K. (1996). From Conscious experience to memory storage and retrieval: the role of quantum brain dynamics and boson condensation of evanescent photons. *International Journal of Modern Physics B, 10*(13 & 14), 1735–1754.
King, J. S., Min Xie, Bibo Zheng, & Pribram, K. H. (2000). Maps of the Surface Distributions of Electrical Activity in Spectrally Derived Receptive Fields of the Rat's Somatosensory Cortex. *Brain and Mind, 1*, 327–349.
Koehler, W. (1958). The present situation in brain physiology. *Am. Psychologist, 13*, 150–156.
Koehler, W. & Held, R. (1949). The cortical correlate of pattern vision. *Science, 110*, 414–419.
Lashley, K. S. (1942). The problem of cerebral organization in vision. In *Biological Symposia, VII: Visual Mechanisms* (pp. 301–322). Lancaster: Jaques Cattel Press.
Leith, E. N. (1976). White light holograms. *Scientific American, 235*(4), 80–87.
Leith, E. N. & Upatnicks, J. (1965). Photography by Laser. *Scientific American, 212*, 24–35.
Mach, E. The Analysis of the Sensations. *The Monist, 1*, 48–68.
Llinas, R. R. (2001). *I of the Vortex: From Neurons to Self*. Cambridge: MIT Press.
Miller, G. A., Galanter, E., & Pribram, K. H. (1960). *Plans and the Structure of Behavior*. New York: Henry Holt. (Russian trans.; also in Japanese, German, Spanish, Italian.)

Moyal, J. E. (1949). Quantum Mechanics as Statistical Theory. *Proc. Camb. Philosophical Soc., 45*, 99–123.

Peirce, C. S. (1934). *Collected Papers*, Vol. V. Cambridge: Harvard University Press.

Pellionisz, A. & Llinas, R. (1979). Brain modeling by tensor network theory and computer simulation. The cerebellum: Distributed processor for predictive coordination. *Neuroscience, 4*, 323–348.

Pellionisz, A. & Llinas, R. (1985). Tensor network theory of the metaorganization of functional geometries in the CNS. *Neuroscience, 16*, 245–273.

Pribram, K. H. (1959). On the neurology of thinking. *Behavioral Science, 4*, 265–287.

Pribram, K. H. (1971). *Languages of the Brain: Experimental Paradoxes and Principles in Neuropsychology.* Englewood Cliffs, NJ: Prentice-Hall; Monterey, CA: Brooks/Cole, 1977; New York: Brandon House, 1982. (Translations in Russian, Japanese, Italian, Spanish.)

Pribram, K. H. (1975). Neurolinguistics: The study of brain organization in grammar and meaning. *TOTUS HOMO, 6*, 20–30.

Pribram, K. H. (1982). Brain mechanisms in music. Prolegomenon for a theory of the meaning of meaning. In M. Clynes (Ed.), *Music, Mind and Brain* (pp. 21–35). New York: Plenum Press.

Pribram, K. H. (1991). *Brain and Perception: Holonomy and Structure in Figural Processing.* New Jersey: Lawrence Erlbaum Associates, Inc.

Pribram, K. H. (1997). What is Mind that the Brain May Order It? In V. Mandrekar & P. R. Masani (Eds.), *Proceedings of Symposia in Applied Mathematics, Vol. 2: Proceedings of the Norbert Wiener Centenary Congress, 1994* (pp. 301–329). Providence, RI: American Mathematical Society. Reprinted: *The Noetic Journal*, Vol. 1 (pp. 2–5), June 1997.

Pribram, K. H. (1997). The Deep and Surface Structure of Memory and Conscious Learning: Toward a 21st Century Model. In Robert L. Solso (Ed.), *Mind and Brain Sciences in the 21st Century*. Cambridge: MIT Press.

Pribram, K. H. & Bradley, R. (1998). The Brain, the Me, and the I. In M. Ferrari & R. J. Sternberg (Eds.), *Self-Awareness: Its Nature and Development* (pp. 273–307). New York: The Guilford Press.

Pribram, K. H., Xie, Zheng, Santa Maria, Hovis, Shan, & King (2004). Accepted for publication. Tokyo: Forma, Scripress.

Pribram, K. H. (2004). Consciousness Reassessed. Accepted for publication: Journal of Mind and Matter.

Russell, B. (1948). Human Knowledge, Its Scope and Limits. New York: Simon and Schuster.

Salam, A. & Woolf, P. E. (Eds.). (1972). *Aspects of Quantum Theory*.

Schempp, W. (1986). *Harmonic Analysis on the Heisenberg nilpotent Lie group, with applications to signal theory*. London: Longman Scientific and Technical Press.

Schempp, W. (1993). Analog VLSI Network Models, Cortical Linking Neural Network Models and Quantum Holographic Neural Technology. In K. H. Pribram (Ed.), *Rethinking Neuraql networks: Quantum Fields and Biological Data*. Proceedings of the First Appalachian Conference on Behavioral Neurodynamics (pp. 233–237). Hillsdale, NJ: Lawrence Erlbaum Associates.

Shannon, C. E. (1948). *Bell Syst. Tech. J., 27*, 379 & 623.

Shannon, C. E. & Weaver, W. (1949). *The mathematical theory of communications* (p. 117). Urbana: The University of Illinois Press.

Smolin, Lee (2004). Atoms of Space and Time. *Scientific American*, Vol. 290 #1.

Ostriker, J. P. & Steinhardt, P. J. (2001). The Quintessential Universe. *Scientific American*, Jan.

Stapp, H. P. (1997/1972). The Copenhagen Interpretation. *American Journal of Physics, 40*(8), 1098–1116.

Stapp, H. P. (1997). *The Journal of Mind and Behavior, 18*(2 & 3), 171–194.

Stuart, C. I. J. M., Takahashi, Y., & Umezawa, H. (1979). Mixed-system brain dynamics: Neural memory as a macroscopic order state. *Foundations of Physics, 9*, 301–327.

Vitiello, G. (2001). *My Double Unveiled – the dissipative quantum model of the brain.* Amsterdam: John Benjamins.

Wigner, E. P. (1967). *Symmetries and Reflections*. Bloomington, IN: Indiana University Press.

Wigner, E. P. (1972). On the time-energy Uncertainty relation. In A. Salam & P. E. Woolf (Eds.), *Aspects of Quantum Theory*.

Chapter 13

Searching for the biophysics of an elementary system

Martin Fleischmann
Salisbury, UK

In this contribution I describe research on the passage of ions through pores in an extremely thin membrane consisting of a black lipid bilayer. Such systems are *in vitro* realisations of some of the *in vivo* processes in living organisms. Indeed, developments of the research might well provide a means of exploring living matter. The analysis shows that we need to answer the question whether the purely atomistic view is sufficient for the description of Nature, or this requires collective dynamics (field dynamics in the physical jargon) for the analysis of the seemingly individual phenomena (in the present case, the passage of ions through pores). We believe that such a change in paradigm is a necessary precondition for building a "Quantum" bridge between matter and mind.

Research carried out during the last century has established the consensus that Quantum Theory is the correct framework for exploring the world of atoms and molecules. We would expect therefore that this should be equally true for any attempt to gain an understanding of the emergence of mind from matter. This topic would appear to be a promising field for future research at the present time. It is true, however, that Quantum Theory is still perceived as a peculiar recipe for the description of microscopic objects each considered as an individual, separate species. This view is shared both by biologists as well as by philosophers and research workers engaged in the complexities of the understanding of mind. Such a prejudice forbids the emergence of a truly holistic description which does not involve localised events but, instead, emerges from the totality of the organism.

It will be apparent also from the papers in this volume that, during the second half of the last century, scientists became increasingly aware that the description provided by Quantum Mechanics is incomplete and that the be-

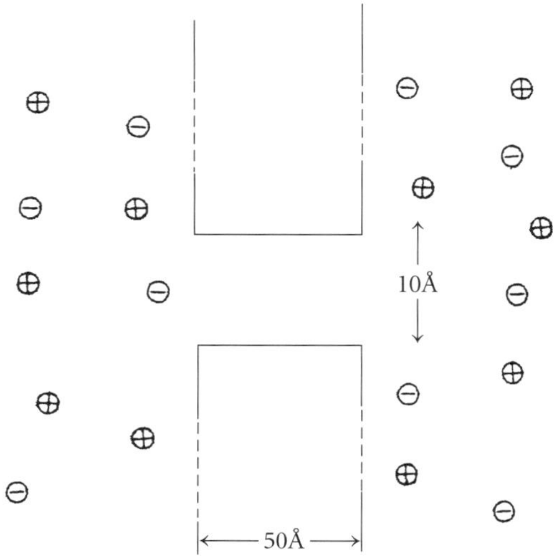

$$P + F = C + 2$$
$$3 + 2 \neq 2 + 2$$

Figure 1. The transmembrane ion conduction across pores in black lipid bilayers; violation of the Second Law of Thermodynamics

haviour of condensed matter has to be understood and modelled in terms of Field Theory (or Quantum Electrodynamics, Q.E.D., for the case of ordinary matter) e.g. see Preparata (1995). Notwithstanding the rather self-evident truth of this statement, scientists have been singularly resistant to the introduction of this particular paradigm shift. Thus the validity of the applications of Q.E.D. are challenged even though such objections have been shown to be incorrect; the publications outlining these objections do not cite the relevant literature so that the subject area (the discussion of the applicable paradigms) does not advance.

In this contribution I will describe research on one particular topic, the passage of ions through pores in an extremely thin membrane consisting of a black lipid bilayer. Figure 1 gives the approximate dimensions of such a pore and it will be clear that such systems allow us to select a small number of ions (in the limit of just one ion). It will also be clear that such systems are *in vitro* realisations of some of the *in vivo* processes in living organisms. Indeed, developments of the research might well provide a means of exploring living matter.

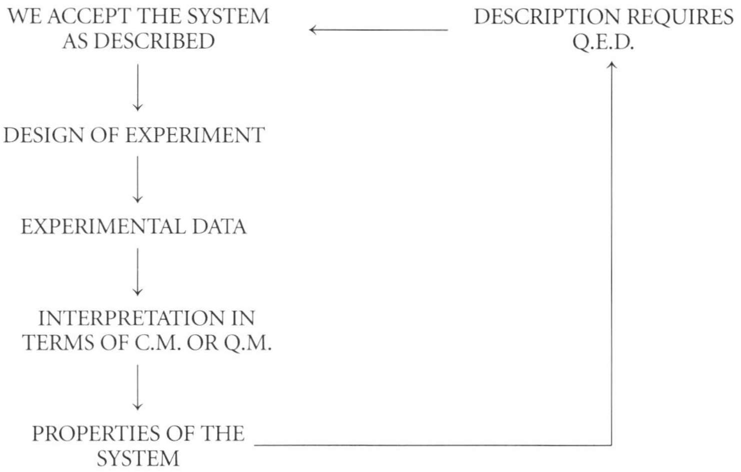

Figure 2. The "hidden agendas" of the research projects

Research on this particular topic was part of a programme of work aimed at exploring the validity of the Q.E.D. paradigm. However, this programme had a somewhat different aim to that of the development of a complete mathematical theory[1] namely, the question of whether one can devise experiments which cannot be adequately explained by means of the Classical or Quantum Mechanical Paradigms but which could find complete interpretations within the framework of Q.E.D. However, in view of the many objections to the Q.E.D. of condensed phase systems, we conducted these programmes within hidden agendas illustrated by Figure 2.

Thus, using such agendas, the systems were investigated within the accepted models of Classical and Quantum Mechanics: the influence of many-body effects and of Q.E.D. had to emerge from the interpretation of the results. It will be evident that this particular strategy had the advantage of avoiding the presupposition of the importance of Q.E.D.: for example, it might well have turned out that the effects were unimportant or not measurable. This strategy also had the advantage of avoiding the premature criticisms of the multitude of scientists opposed to the application of Q.E.D. to condensed phase systems.[2] The overall aim has been the illustration of the development of inconsistencies arising from the purely atomistic view of Quantum Physics. We need to answer the question whether the description of Nature requires collective dynamics (*field* dynamics in the physical jargon) for the analysis of the seemingly individual phenomena (in the present case, the passage of ions through pores). We believe that such a change in paradigm is a necessary precondition for building a "Quantum" bridge between matter and mind.

As has already been noted, the experiment design, Figure 1, selects the behaviour and properties of a small number of ions or molecules (ideally, just one species, the Elementary System of the object being investigated). It will be apparent from the papers presented at this Meeting that Consciousness must be a consequence of the fact that Field Theory applies to the operation of the assembly of Elementary Systems (i.e. the Compound System). It is, therefore, an article of faith that such ideas should apply also to the operation of the elementary systems.[3] However, if this hypothesis is correct, then it will always be much more simple to derive the properties of the elementary systems (especially of the dynamics) by direct observations rather than to attempt to do this by deconvoluting the properties of the Compound System.

In this paper we focus attention on the Biophysics of a single system, the voltage-gated insertion of the polypeptide pore former alamethicin into black lipid bilayers (Fleischmann 1980), illustrated in Figure 1. It has been known for some time (Mueller 1967; Gordon 1976) that this voltage-gated insertion leads to the creation of pores in the lipid bilayer and that the processes of the aggregation and deaggregation of the pore by the polypeptide pore former leads to steps in the transmembrane ion current, e.g. see Figure 3, taken from our own data (Fleischmann 1980) (the system has been studied by numerous research groups following the initial measurements (Mueller 1967; Gordon 1976)). It will be evident that this *in vitro* system is a reasonable model for switching processes taking place in *in vivo* systems on the biomembranes of living organisms.

It will also be apparent that observations on such an elementary system provide ideal opportunities for investigating both the energetics and dynamics of the aggregation/deaggregation processes. Thus the number of transitions between adjacent states give direct access to the kinetics at the molecular level while the dwell-times in each level give the energetics of that state (via the Maxwell-Boltzmann distribution). The application of the usual equations of electrochemical kinetics (relating the rates of the processes to the free energy difference between the initial and final states) lead to the model for pore formation/removal illustrated in Figure 4. The details of these particular processes are not of any special interest with regard to the material presented in this paper; on the other hand, the form of the energy versus number state diagram, Figure 5, has such a special importance.

We can see that the parabolic relationship shows that the free energy is positive at all number states (the deviations at high number states are due to the

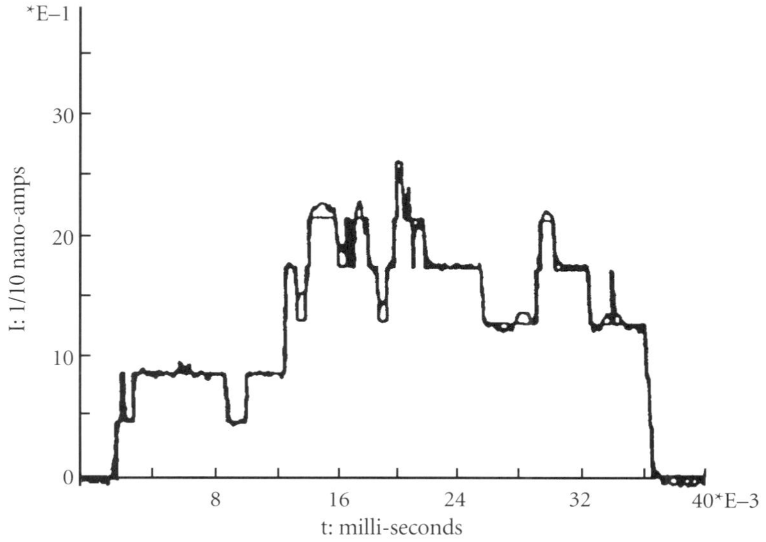

Figure 3. Section of a current (I) – time (t) record of the voltage gated transmembrane ion current induced by alamethicin in a black lipid membrane

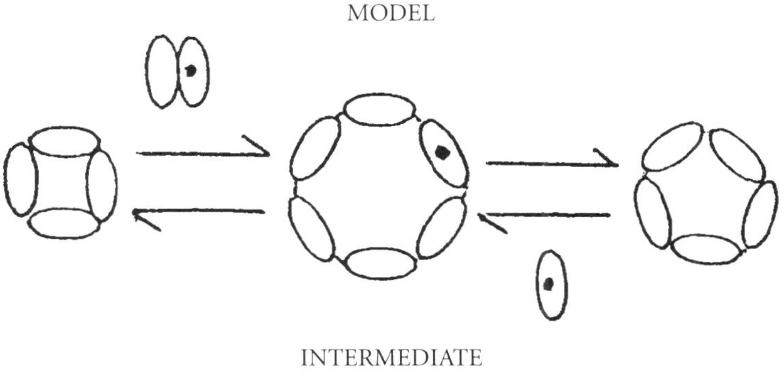

Figure 4. The model for the formation/removal of a pore in the black lipid layers

contributions of several pores, see below). The form of the relationship must be contrasted with the usual form for two-dimensional nucleation shown in the lower part of Figure 5. In this case a positive edge energy must be combined with a negative bulk energy leading to a maximum in the total free energy. In consequence this maximum in the free energy acts as a free energy of activation for the two-dimensional nucleation processes; the negative values of the free

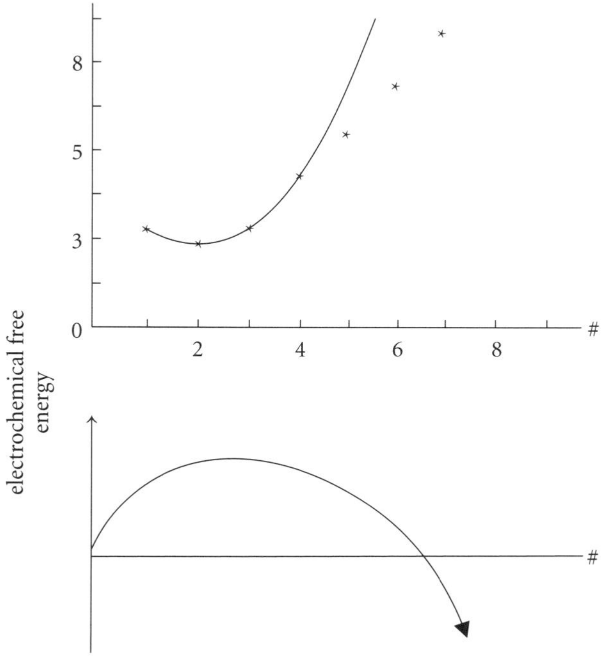

Figure 5. Comparison of the energetics of pore formation with the usual behaviour for two-dimensional nucleation

energy at high number states show that two-dimensional nucleation is subject to an upper absorbing state leading to eventual spontaneous growth. On the other hand, pore formation in the lipid bilayer is subject to an upper reflecting state so that the pore must again close at sufficiently long times (see Figure 3). It is clear that we can only have this behaviour if the process of pore formation is subject to negative edge and positive bulk free energies. It is of some interest that the signs of these free energy terms could have been predicted from the very fact that membranes observed *in vivo* are stable!

We naturally have to ask also: how can it be that the bulk free energy of the electrolyte filling the pore is positive with respect to the free energy of the macroscopic electrolyte? It is this fact which requires interpretation in terms of the Q.E.D. of water (Arani 1995) and of electrolyte solutions (Del Giudice 2000). This interpretation shows that water is divided into highly structured "Coherence Domains" and into "Incoherent Regions" which have a positive free energy with respect to the bulk of the liquid; electrolyte is expelled from the Coherent into the Incoherent Regions where it forms a Second Coherent

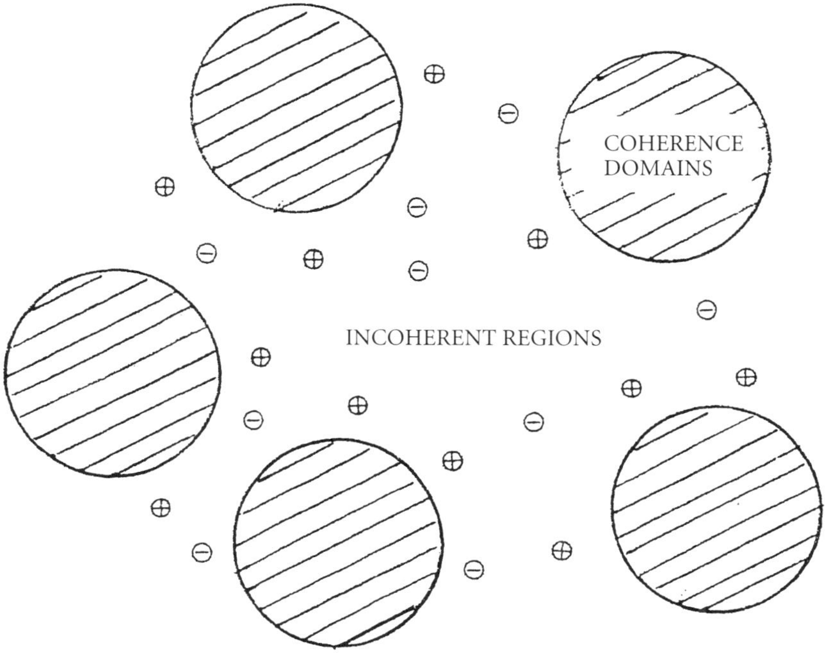

Figure 6. The model of electrolyte solutions based on the principle of Q.E.D. Coherence (Arani 1995; Del Giudice 2000)

system. We can now see the reason for the positive free energy of the "bulk" liquid in the pores within the membrane: these pores are filled by an Incoherent Region of the electrolyte solution which has a positive free energy with respect to the bulk of the solution.[4]

We can see that the phenomenology of the voltage-gated transmembrane ion currents can be quite adequately explained in the context of the left-hand-side of Figure 1 but that it is the detailed description of the system which requires us to invoke the consequences of Q.E.D. We note also that we need to invoke Q.E.D. to explain the very existence of the black lipid bilayers. Thus, we can apply the Phase Rule (an expression of the Second Law of Thermodynamics) to the existence of the bilayers. We can see that the complete system is divided into three geometrically distinct phases, i.e. $P = 3$, by two components (the bulk phases and the lipid) i.e. $C = 2$. We would expect, therefore, that the system should conform to the usual demands of the Phase Rule:

$$P + F = C + 2$$

where the system has two degrees of freedom. However, the description given above shows that

$$3 + 2 \neq 2 + 2$$

so that we evidently need a further actor, the consequences of Q.E.D. (compare Del Giudice 1995). We see that the structure of the system is dictated by the demands of this paradigm just as is the case for the structure of electrolyte solutions.

We can see that the detailed discussion of the behaviour of the Elementary System indicates that this consists of a single domain of the Incoherent Region of the electrolyte. We would expect, therefore, that the dynamics of the system should also reflect this aspect. We note that the investigation outlined in Fleischmann (1980) showed that the system conforms strictly to a Birth-and-Death process, i.e. that transitions only take place between adjacent states. At the same time we note that the volume of the pores is so small that the pores must be free of electrolyte for an appreciable fraction of the experiment time.[5] If we therefore attempt to model the system in terms of the left-hand-side of Figure 1, we would need to conclude that the conductivity of the pore has to drop to zero when the pore does not contain electrolyte (as the current has to drop to zero during these periods). The behaviour should therefore be as indicated in the lower half of Figure 7, showing that the system cannot conform to a "Birth-and-Death" process.

This leads us to the question; "how can it be that the system conforms to a Birth-and Death process yet we can show that it cannot be so?" The answer is that the "noise level" in the system is too low if we attempt to explain it's behaviour according to the left-hand-side of Figure 2, a phenomenon also indicated by other characteristics of the system which will not be discussed here. Evidently, we therefore have a system characterised by quantum noise or, at least, dominated by quantum noise,[6] see Figure 8.

It should be noted that we have not produced a complete mathematical description of the behaviour of the system. Such a complete description would require the theory of stochastic processes subject to quantum noise, a topic which was expected to raise profound difficulties (the existing theory of stochastic processes is semi-classical: Quantum Mechanics is only taken into account to the extent of recognising the existence of individual Quantum States). Our interpretation rests instead on the application of "hidden agendas", Figure 2. These "hidden agendas" indicate the nature of the reasoning required to complete the interpretation: we find that the behaviour of the ions in the pores must be governed by the dictates of Q.E.D.

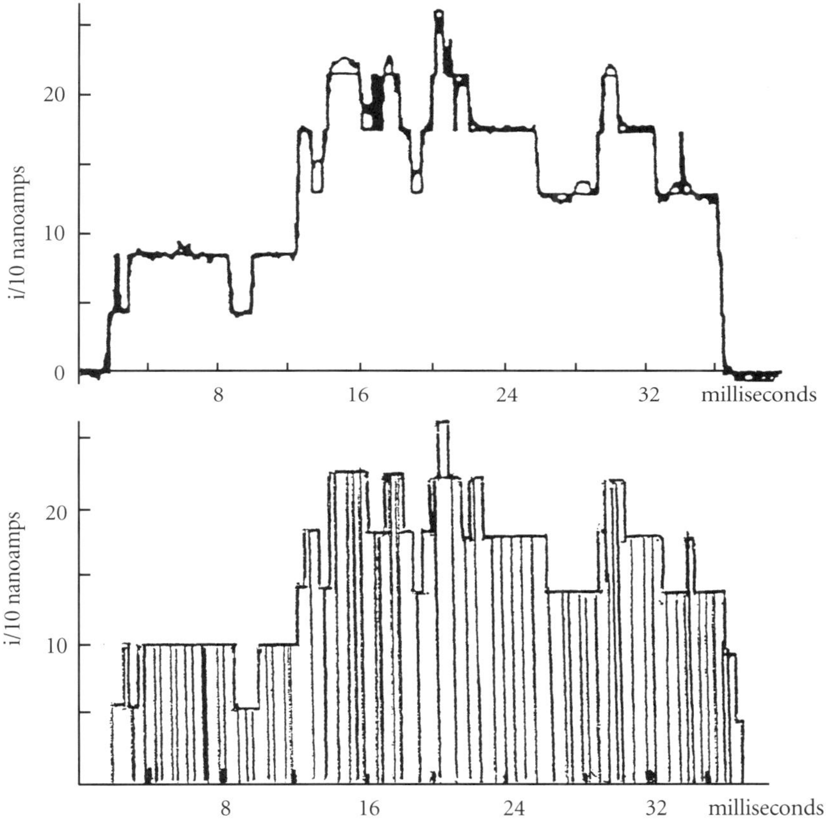

Figure 7. Comparison of the experimentally observed current-time series with that predicted from elementary considerations of the ion transduction process. Schematic diagram.

Towards the future

By the early 1980's we had reached a watershed in the development of this topic. It was clear that we had developed a methodology which could be developed much further for *in vitro* investigations[7] and it was also clear that we could take steps to apply the methodology *in vivo*. Unfortunately, we then ran into the difficulties associated with a different research project (the topic now known as "Cold Fusion", a topic also predicated on the application of the Q.E.D. paradigm). These difficulties have continued in the intervening years

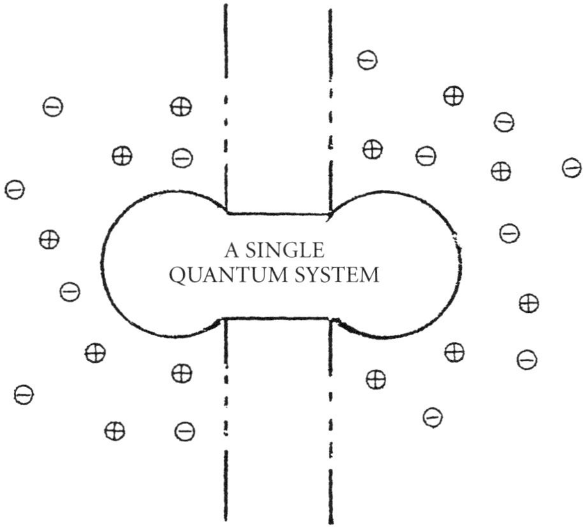

Figure 8. The model of the pore revealed by the hidden agenda

and we have been unable to develop further the topic of the voltage-gated transmembrane ion currents.

It is important here to draw attention both to a limitation of the research carried out and to the way in which this limitation might be overcome. We believe that we had reached an adequate illustration of the fact that one elementary system of Biophysics has to be modelled in terms of Q.E.D. This in turn leads us to two important questions; firstly, what other illustrations can we find in Biophysics of the need to invoke this paradigm and, secondly, if we stay within the confines of the voltage-gated transmembrane ion currents, then how exactly is the Elementary System (the single pore) related to the behaviour of the Macroscopic System (an assembly of pores). We will consider briefly this second question.

We have drawn attention to the fact that the electrochemical free energy of the system, Figure 5, shows that we have contributions from more than one pore at high number states. The development of the mathematical analysis (Fleischmann 1986a) allowed us to deconvolute the experimental current-time series up to # 20 to give the behaviour of the single pores making up the compound system (Fleischmann 1986b). This was a somewhat pyrrhic victory because it answered a question which was unimportant in the further development of the subject. The question we need to answer is how the behaviour of

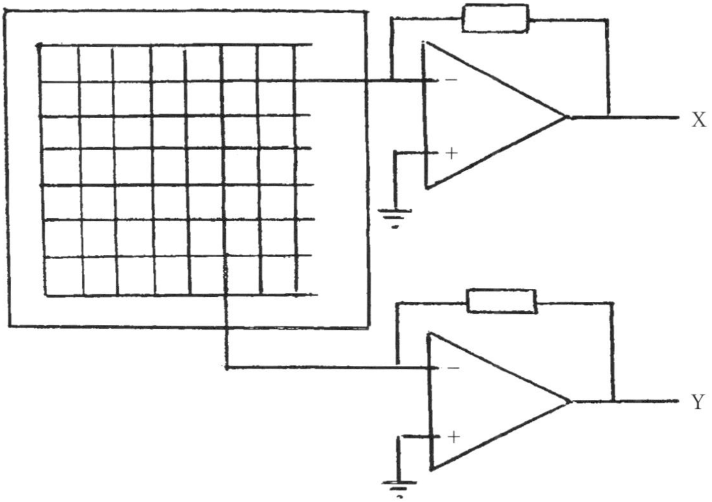

Figure 9. Current measurements with an array of wire microelectrode counterelectrodes

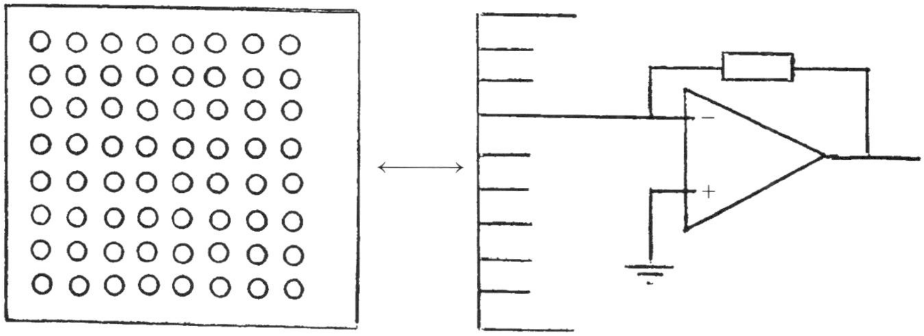

Figure 10. Current measurements with an array of circular microelectrode counter-electrodes

a single elementary system determines the development in space and time of further elementary systems.[8] Figures 9 and 10 show how we might attempt to answer this question. In the device of Figure 9 we would look for coincidences in the currents of the lines in the *x* and *y* directions. Although such a device would be less costly than the array detector, Figure 10, we believed that it was such an array detector which would be required and which would have extensive applications in other fields of electrochemistry.[9] Our work was confined to the proof of concept using a multiplexed 8 × 8 array; see Figure 11; the full im-

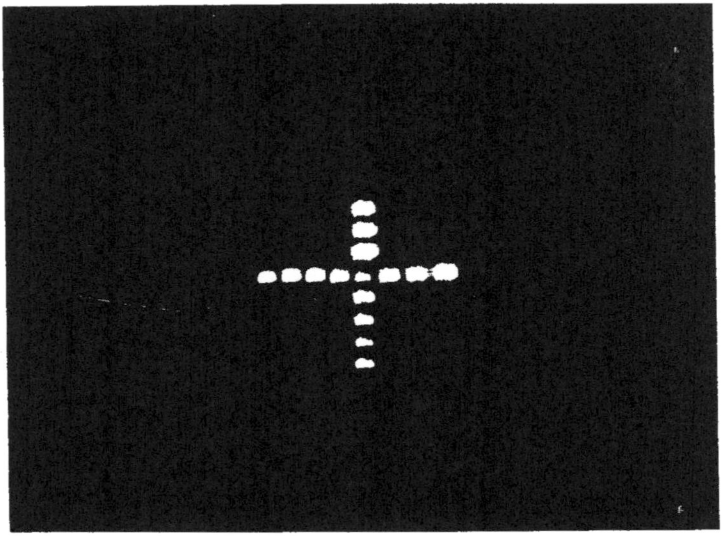

Figure 11. Image of a cross of Cu wires undergoing anodic dissolution in 0.5 M $CuSO_4$ + 0.1 M H_2SO_4 as seen by an 8 × 8 array of Cu microelectrodes. The output from the array (after suitable discrimination) was used to apply a *Z* – modulation to the beam of a CRT display.

plementation of the project using arrays in the range 100 × 100 to 1000 × 1000 was beyond our reach.[10]

It will be clear that the space-time correlations of the generation of a multiplicity of pores would give access to the Q.E.D. of this particular many-body system (i.e. the compound system consisting of a multiplicity of elementary systems). Devices of this type would therefore provide a means of exploring the link between field theory and matter.

Notes

1. The basis of the discussion of the appropriate paradigm rests on the development of a complete mathematical theory.

2. This description is not completely correct from an historical point of view. We adopted this strategy following our first investigation of the kinetics of fast reactions in solution at very short space-times (in the range 2×10^{-22}cm.s to 3×10^{-12}cm.s). The reactions to our attempts to interpret the results in terms of memory propagators and the development

of structures with a typical size $\sim 10^{-6}$cm were uniformly negative and, sometimes, quite violently so. This convinced us that the effects of Q.E.D. had to be "hidden".

3. It should be observed that the operation of the Compound Systems might be governed by Field Theory while the elementary systems are governed by Classical and/or Quantum Mechanics (and the converse is equally possible!). The hypothesis which we have advanced must therefore be verified. The material presented in this paper shows that such a verification should be relatively straightforward. See further below.

4. The division of the electrolyte into Coherent and Incoherent regions also gives a natural explanation of a number of other puzzling phenomena such as that of Reverse Osmosis. We have to ask: "how can it be that the larger pores of reverse osmosis membranes only allow the passage of the solvent?" The answer must be that the pores in these membranes select the Coherent Regions of the Electrolyte (Del Giudice 2001).

5. The emptying of the electrolyte from the pores will be enhanced by the extremely short transit time of ions in the system (say 2–6ns)

6. Quantum Systems do not generate the noise which we associate with semi-classical behaviour; they only generate the noise due to quantum fluctuations.

7. Such developments included improvements in the experiment design (especially the enhancement of the frequency response of the current measurements) and the theory of processes subject to quantum noise (see above). See also the main text for a further important aspect.

8. The technical description is that we need to find the Lagrangian for the development of a multiplicity of pores.

9. It has been customary in electrochemistry to study a working electrode when using a single counter electrode. It has become apparent that a considerable amount of new information on the space-time correlations of processes on working electrodes would become available by using such arrays of counter electrodes.

10. However, we should note that elementary quantum switches would not necessarily dictate that the triggering of further switches would be governed by Q.E.D. In the *in-vivo* system the elementary switch may merely be a response to phenomena which do not generate transmembrane ion currents.

References

Arani R., I. Bono, E. Del Giudice, & G. Preparata (1995). Q.E.D. Coherence and the Thermodynamics of Water. *International Journal of Modern Physics*, *B9*, 1813–1841.

Del Giudice, E. & G. Preparata (1995). Coherent Dynamics in Water as a Possible Explanation of Biological Membranes Formation. *Journal of Biological Physics*, *20*, 105–116.

Del Giudice, E., G. Preparata, & M. Fleischmann (2000). Q.E.D. Coherence and Electrolyte Solutions. *Journal of Electroanalytical Chemistry*, *482*, 110–116.

Del Giudice, E. & M. Fleischmann (2001). Unpublished work.

Fleischmann M., C. Gabrielli, M. T. G. Labram, A. I. Mc. Mullen, & T. H. Wilmshurst (1980). Alamethicin-Induced Conductances in Lipid Bilayers: I, Data analysis and Simple Steady-State Model. *Journal of Membrane Biology*, *55*, 9–27.

Fleischmann M., C. Gabrielli, M. T. G. Labram, & T. Markvart (1986a). Alamethicin Induced Conductances in Lipid Bilayers: II, Decomposition of the Observed (Compound) Process. *Journal of Electroanalytical Chemistry*, *214*, 427–440.

Fleischmann M., C. Gabrielli, & M. T. G. Labram (1986b). Alamethicin Induced Conductances in Lipid Bilayers: III, Derivation of the Properties of the Elementary Process form the (Compound) Observed Process. *Journal of Electroanalytical Chemistry*, *214*, 441–457.

Gordon, L. G. M. & D. A. Haydon (1976). Kinetics and Stability of Alamethicin Conducting Channels in Lipid Bilayers. *Biochimica et Biophysica Acta*, *436*, 541–556.

Mueller, P. & D. A. Rudin (1967). Action Potential Phenomena in Experimental Biomolecular Lipid Membranes. *Nature*, *213*, 603–604.

Preparata, G. (1995). *Q.E.D. Coherence in Matter.* Singapore: World Scientific.

CHAPTER 14

Brain and physics of many-body problems* **

Luigi Maria Ricciardi[a,b] and Hiroomi Umezawa[a]
[a]Istituto di Fisica Teorica, Napoli, Italy
[b]Gruppo Nazionale di Cibernetica del CNR

On the basis of a recent physical theory of many-body problems developed in our Institute,[1] a model of the brain is formulated, and it is shown how some of its typical features, such as *learning* and *memory processes*, find therein a natural and simple explanation. In the Appendix a short surview of the necessary mathematical formalism is finally given.

1. Introduction

Although nowadays the study of natural and artificial brain plays the main role among those researches which were indicated by N. Wiener as *Cybernetics*, it seems that a very few concrete results have been obtained, in the sense that the question *how the brain works out the information received from the outside, and which is the logic on which the operations performed by the brain are based* is still far from a satisfactory solution. Several mathematical models of brains have been formulated (McCulloch & Pitts 1943; Caianiello 1961; Harth 1966), and all of them, although apparently more or less different, are based upon the same assumption, partially demonstrated by electrophysiological evidence, that significant parameters in the description of the brain are the activity or inactivity states of its neurons, the values of their thresholds, and the magnitude of their coupling coefficients. Neurons are looked on as binary elements, whose output can be described by a two-valued function which, after obvious normalization, is usually represented by the Heaviside unit step-function. In this way, it is possible (Caianiello 1961) to write a system of N nonlinear algebraic equations (where N is the number of neurons) which once solved after specification of couplings and thresholds, furnish the instantaneous behaviour of the nerve net.

Models of this type are particularly useful in order to design machines, whose fundamental elements are artificial neurons, able to perform preassigned operations, as for instance *pattern recognition*; they also contain the possibility of introducing *learning processes*, and of explaining memory as due to the time variation of the coupling coefficients and to the formation of periodical activity patterns in the firing of neurons (Caianiello 1961). It seems very appropriate to indicate these models as *simulators of neuronal networks*. Being a very drastic idealization of the reality, they represent a powerful tool of investigation of the neuron dynamics, and moreover they offer undeniable advantages in *machine-designing* problems, i.e. in designing *active networks* characterized by certain specific features. However, in the case of natural brains, it might be pure optimism to hope to determine the numerical values for the coupling coefficients and the thresholds of all neurons by means of anatomical or physiological methods. Moreover, as soon as one asks the question whether or not these models can be looked on as models of the true brain, many questions immediately arise. First of all, at which level should the brain be studied and described? In other words, is it essential to know the behaviour in time of any single neuron in order to understand the behaviour of natural brains? Probably the answer is negative. The behaviour of any single neuron should not be significant for the functioning of the whole brain, otherwise higher and higher degree of malfunctioning should be observed, unless to assume the existence of "*special*" neurons, characterized by an exceptionally long half life: or to postulate a huge redundancy in the circuitry of the brain. However, up to our knowledge, there have been no evidences which show the existence of such "*special*" neurons, and to invoke the redundancy is not the best way to answer the question.

Another possibility is the following one: the activity of any single neuron is not significant, but rather the patterns of activity of clusters of them; what is important is only a "quantity" somehow related to the activity of the whole cluster, which does not change appreciably as function of the number of *alive* neurons belonging to that cluster. However, apart from the difficulties offered by the latter hypothesis, it is easily seen that it also contains the same difficulty as the one present in the former. Moreover, any effort to support any hypothesis of this type by means of anatomo-physiological evidence, for the time being, does not seem to be successful.

Another remark which seems rather important is that neurons do not seem to be the only fundamental units of the brain; or, at least, together with them some other elements, as for instance *glia cells*, may play important role (Arbib 1963; Schmitt 1966).

From all the previous considerations, it again emerges that the proposed simulators of brain may be useful only in order to design "thinking machines" able to perform preassigned tasks, and although all of them stemmed from some anatomical and physiological evidence, they cannot be looked on as describing the true dynamics of the brain.

Then, another possible approach to the *brain-dynamics* naturally arises. The point of departure is that even though the elementary entities like neurons and others could be the most fundamental, their dynamical behaviour may be so complicated that many of their original features may become hidden under the macroscopic dynamics of the brain as a whole. This hypothesis has been recently supported by some beautiful experimental results obtained by John (1966); he clearly demonstrated that the existence of similar and almost simultaneous responses in several regions of the brain (a kind of *long range correlation*) to a particular stimulation technique does not find any explanation in terms of activity of the single nerve cells: new *non-classical* mechanisms have to be looked for. It then arises naturally a possibility: since one usually ignores the mechanism according to which the brain performs *intelligent* operations, but only hypothesis (even though interesting and sometimes useful) can be formulated, one could try to give a *more general* description of the brain dynamics; as we shall see later, from a phenomenological point of view it is strongly suggestive of a *quantum* model. In other terms, one can try to look for specific dynamical mechanisms (already known in physics of many degrees of freedom) which can satisfy the essential requirements of the observed functioning of the brain.

In the following, we shall first examine which are the essential requirements to be satisfied by a model of the brain, and we shall later see how an already existing physical theory can be used in order to give a quantum description of the brain, such to take into account well recognized features of the living beings, as the capacity of learning and of remembering; moreover, the existence of *long range correlations* automatically finds an explanation in the theory itself.

2. Some necessary requirements to be satisfied by a model of brain

One of the most typical features of the brain is the capacity of remembering, that is of storing information patterns for a more or less long time interval. More precisely, there seems a great deal of evidences for the existence of essentially two types of memories, i.e., a long-term and a short-term memory (Arbib

1964; Schmitt 1966), and it appears that we have to retain an idea for quite a while in a short term memory before it is transferred into long-term memory.

How can we account for memory? The brain is estimated to contain 10^{10} neurons and ten times that many glia cells, which closely invest the neurons and the blood vessels. Many neurons receive hundreds or thousands of fiber connections from other neurons (Braitenberg et al. 1965). The resulting neuronal nets are too complicated to be explored with the microscope, and experts estimate that only a few per cent of the connections have actually been traced (Schmitt 1966), and these myriad "nets-within-net" are particularly complex and numerous in the brain cortex, the region most intimately associated with higher mental functioning.

Especially long term memory, learning and other higher nervous functions have been thought to result from impulses traversing such complex neuronal nets, and individual long term memories could have their physical explanation in ceaseless reverberatory passage of impulses, each in its particular net (Schmitt 1966).

However, there seems to be some serious objections. Bioelectrical waves in the brain can be stopped by treatment with cold, electric shock, or drugs, without loss of memory after recovery, and moreover, memory is not lost after many *ablation* experiments or when a brain is sliced in many directions so that certainly some pre-existent networks are destroyed.

These facts suggest that memories are not "wired" into individual neuronal nets, but are instead *diffused in the brain* (Schmitt 1966). The first net that corresponds to an incoming message has its information transferred to other nets often far from the original location of the stimuli carrying the message.

In this framework, it automatically emerges the possibility of observing some kind of correlations between the activities of neurons (at present neurons are much more easily observed than other systems or "entities") even though they are situated in places far from each other, and this has been supported by experimental evidence (Braitenberg 1965; Ricciardi 1967).

Another remark has finally to be made: it seems that no conscious simultaneous recall of several memorized *informations* is possible; more precisely, there are good reasons to believe that the stored information can be recalled according to serial processes rather than in parallel, and, as it is well known, often it happens that as soon as something has been remembered, further and sometimes unrelated informations are recalled, too. Also this last remark, as it will soon be seen, finds its natural explanation in the model that we are going to propose.

3. Outline of the model

In this section we will be concerned with the essential requirements to be satisfied by a model which describes the brain as a collection of very many mutually interacting units. As it will clearly appear in the following, the model is firmly based on the experimental evidence which was mentioned in the previous sections. In the foregoing, while exposing the logical steps to be done in order to construct our model, we will see how naturally our argument follows the line of an already well founded physical theory; from the latter, in order to give a deeper development of our model, the necessary mathematical tools can be derived.

Let us look at the brain as a *system* in interaction with the external world; from the latter it receives *stimuli*, and as a consequence of it the system is put into particular *states*. A first requirement is that these stimuli should be in some way *coded* and their effects should be represented into the brain also after they have ceased; this implies that the brain, as effect of the received stimulations, changes its state; the states may be described by a suitable set of variables, which in the following according to the physical terminology we will call "dynamical variables". In the usual models of brain these variables may be for instance identified with the binary variables describing the activity or inactivity states of the neurons. However, for the previously mentioned reasons, here we do not intend to consider necessarily the neurons as the fundamental units in the brain; therefore, our variables should be thought as related to a certain (very large) number of entities, essential for producing the different states of the brain; their nature does not here need to be analysed. To be precise, our variables must be such to describe the *stationary* or *quasi-stationary states* of the *isolated* brain. According to the usual terminology of Quantum Mechanics, such states can be classified in terms of *quantum numbers* associated with the invariance of the system under certain transformations.[2] We can then assume the specification of the quantum number as the code itself; moreover only those variables which do not change under invariant transformations will be responsible for the storing of the information represented by the corresponding code.

These variables must be classified into sets (henceforth called independent sets) which transform independently under a given invariant transformation; these sets, in turn, can be labelled by the quantum numbers.[3] Moreover the classification of the independent sets depends on the choice of the invariant transformations, since in general there may exist several *non-commutative* invariant transformations.[4]

According to different stimuli, these variables are changed to create excited states of the quantum numbers associated to them; already at this stage the information coming to the system from the outside is coded into the brain. Since the first requirement for memory processes is the *stability*, the code should be later on transferred to the ground state of the system. This can be achieved by means of condensation into the ground state of the coded information already held by the excited states. Such a condensation can then give account of learning processes.

The previously invoked condensation finds its explanation in the *Bose-Einstein* condensation phenomenon, well known in the physics of many-body problems. More in detail, the condensation of Bose particles with a specific quantum number transfers to the ground state the quantum number itself (for instance, the Bose condensation of spin wave in the case of ferromagnetism creates the polarized ground state). This argument requires the existence in the brain of variables obeying the Bose-statistics. Such variables should also posses long range property in order to supply the code to the brain system itself and to maintain it. In other words, the interaction of such a complex system consisting of a huge number of entities, has as its effect the creation of the long range Boson correlation which regulates the brain dynamics.

In the terminology of physics, what the present model proposes is that the brain function is a manifestation of spontaneous break-down of various symmetries in brain dynamics and the long range Bosons mentioned above correspond to the so called Goldstone quanta.

By means of the previous description, long term memory is related to the ground state; short-term memories can instead be related to the existence of *meta-stable* excited states.

The description of the brain as a system of an enormous number of interacting entities, offers also another advantage: the gross properties of the system should not change appreciably under destruction of a considerable number of units, which, instead, as we mentioned before, has a catastrophic effect for the usual brain simulators.

A final remark is that from our argument one should expect correlations also among the neurons' activity, which, as we said before, recently has been already observed.

4. Concluding remarks

Although the outlined model only represents the start point of further theoretical researches, already at this stage some additional remarks, which spontaneously emerge from the previously sketched theory, are to be done.

The existence of long range correlation, by means of which fundamental properties of the brain, as memory processes, find an explanation, seems to be supported by experimental evidence (John 1966).[5] Moreover, it also suggests the existence of different types of short-term memory processes. In the previous section, indeed, we interpreted short term memory as related to transitions between excited states. Such an interpretation has the advantage of giving reason for the so often observed "association of ideas"; moreover, the fact that memory is associated with the states of the brain system, leads us to predict that several codes cannot be transferred to the brain at the same time, and it would be useful to recall here that psychological experiments seem to show that several memory processes may not take place simultaneously.

Also another possibility is offered, indeed, for the explanation of the short term memory. As a matter of fact, we may consider the brain as consisting of several regions; our previous argument may then be applied to each of these regions, considered as *physically* disjoined. However, because of the existence of long range correlation in our model, these regions should not be *functionally* separated, and therefore the code held by one of them may be destroyed or modified because of the influence of the other regions.

Another remark is the following: as we said in Section 3, the classification of the independent sets depends on the choice of the invariant transformations, because of the possible existence of several non commutative invariant transformations. This situation, which is an obvious property of some descriptions of many-body problems, provides in this context an explanation for the often observed "many-to-one" type response of living beings under given stimulations, and at the same time it accounts for the destructive or associative correlations in memory processes.

To conclude, we wish to remark that our present knowledge of the brain and of the microscopic processes therein involved does not yet permit us to guess which types of invariance may be present, even though a high number of them should certainly exist; therefore, for the moment we prefer not to anticipate anything. However we want to point out that often in many-body problems the possible types of invariance are suggested just by inspection of the properties of the long range correlation; and also in the case of the brain we feel that following this way we will eventually obtain encouraging results.

Appendix[6]

Let us consider a set $\{X_i(\vec{x},t)\}$ of dynamical variables describing stationary or quasi-stationary oscillations; by $\vec{x}$ and t we denote the position and the time respectively. According to the terminology of Quantum Mechanics, stationary oscillations can be described by harmonic oscillators, while the quasi-stationary oscillations are represented by damped oscillators. These oscillators are described by the set of real variables $\{X_i(\vec{x},t)\}$ already introduced, together with their canonical conjugates $\{\chi_i(\vec{x},t)\}$; index i may in general assume both discrete and continuous values.

Let us now indicate by n_i ($n_i = 0, 1, 2, \ldots$) the principal quantum number; the corresponding energy E_i is then of the form:

$$E_i = n_i\omega_i \qquad (n_i = 0, 1, 2, \ldots)$$

where ω_i denotes the energy of the oscillation. Since we are concerned only with long half-life quasi-stationary oscillations, we will consider them, in a first approximation, as stationary oscillations.

Let us now consider the variables X_i as obeying the Bose Statistics;[7] this implies that the following commutation relations hold:

$$\left[\chi_i(\vec{x},t), X_j(\vec{y},t)\right] = -i\hbar\delta_{ij}\delta(\vec{x}-\vec{y}).$$

Let us now construct the new variables ξ_i defined as:[8]

$$\mathrm{X}_i = \frac{X_j + i\chi_j}{2}.$$

Their hermitic conjugate ξ_j^+ are then given by

$$\xi_j^\dagger = \frac{X_j - i\chi_j}{2}$$

and between the ξ_i 's and $\xi_j^\dagger$'s the following commutation relations hold:

$$\left[\xi_i(\vec{x},t), \xi_j^\dagger(\vec{y},t)\right] = i\hbar\delta_{ij}\delta(\vec{x}-\vec{y}). \tag{1}$$

As it is well known, the principal quantum number n_j is eigenvalue of the operator

$$\int d^3x\,\xi_j^\dagger\xi_j$$

and ξ_i depends on t only through an exponential factor of the form

$$\exp\left[-i\omega_j t\right].$$

Denoting now by Φ_0 the ground state, whose energy is regarded as zero, the states $\xi_j^\dagger \Phi_0$ represent excited states whose correspondent energies are $z\omega_j$. Proceeding in this way we may construct all the excited states.

The Bose-Einstein condensation appears provided there exist certain variables $\tilde{\xi}_j$ such that their ground state expectation values are non-zero, that is

$$\left\langle \tilde{\xi}_j \right\rangle = \eta_j \neq 0 \tag{2}$$

These variables $\tilde{\xi}_j$ can then always be written in the form

$$\tilde{\xi}_j = \xi_j + \eta_j \tag{3}$$

where ξ_j are the already introduced operators.

Let us note that since ξ_j 's obey the Bose statistics, the variables $\tilde{\xi}_j$ satisfy the commutation relations (1), and therefore they also are canonical.

The dynamical variables may now be classified into *independent* sets which independently transform under given invariant transformations, and each of which is specified by its quantum number. Let us recall that the invariant transformations are those under which the dynamical system does not change at all.

Let us now focus our attention on the independent set containing the variables $\tilde{\xi}_j$ which satisfy (2). Now, since $\tilde{\xi}_j$ carries a quantum number, because of (3) also η_j does, and finally in virtue of (2) this quantum number is given to the ground state. Intuitively speaking, the condensation of bosons carrying a quantum number gives origin to a ground state which maintains the same quantum number.

The quantum number transmitted to the ground state has to be identified with the *code* introduced in Section 3.

It is obvious that any ground state of non-trivial quantum number is not a state which is invariant under the invariant transformations. Still, such a ground state can appear without conflicting with the invariant nature of the system, when the energy spectra (ω_i) of the bosons (ξ_i) are gapless.[9] Here an energy spectrum is said to be gapless when its minimum value coincides with the energy of the ground state. The condensation of bosons can take place without any external energy supply when the boson energy spectra are gapless. Since any invariant transformation can be performed without any energy supply, the condensation in the ground state can be regulated by the invariant transformations. It must be noted that the gapless nature of the boson energy spectra is the origin of *the long range property of the boson effects* (i.e. long range correlation).[10]

The excited states $\xi_i^\dagger \Phi_0$, $\xi_i^\dagger \xi_j^\dagger \Phi_0$, etc. appear under the effect of *external stimuli*, and since ξ_j carries a quantum number, the excited states carry additional quantum numbers: these represent the code of the short-term memory.

Let us note that the excited states may decay into lower states either because of their intrinsic instability or because of the interaction of a particular subsystem with the neighbouring regions.

Let us finally recall that the invariance of the considered dynamics under any continuous transformation gives origin to locally conserving currents carrying specific quantum numbers, and this is the origin if the long range correlation. Noteworthy is the fact that the communication of the code (i.e. of the quantum number) is performed only by conserving currents, and this circumstance automatically ensures the conservation of the code, in the sense that it cannot spontaneously disappear.

Acknowledgments

The first-named author (L. M. R.) would like to thank Prof. F. O. Schmitt for hospitality of the Neurosciences Research Program, Intensive Study Program held at Boulder, Colorado from July 18 through August 12, where some of the work was performed.

Warm thanks are due by the second-named author (H. U.) to Prof. E. R. Caianiello for hospitality at the Istituto di Fisica Teorica in Naples.

Notes

* Reprinted with kind permission from *Kybernetik*, Vol. 4 (1967), pp. 44–48.
A brief outline of the model presented in this paper was informally given by one of the authors (L. M. R.) at the I.S.P. Neurosciences Research Program; Boulder Colorado, in July 1966.

** This work has been sponsored in part by: Air Force Avionics Lab., Research and Technology Div., Wright Patterson Air Force, Air Force System Commend-U.S.A.F.-Government of United States of America. Contract No. AF 33(615)-2786.

1. Umezawa (1965, 1966); Leplae, Sen, and Umezawa (1967).

2. The reader is reminded, for instance, of the angular momentum associated with spherical symmetry.

3. For example the spin doublet has 1/2 spin quantum number, associated with the spin rotation.

4. For instance, this happens in the case of rotations in many-fold spaces.

5. An extremely stable type of memory, as it is well known, seems to be linked with the genetic code contained in the structure of DNA-molecules. The proposed model can be utilized also in order to describe the genetic mechanisms, and again one may account for stability by means of the condensation of the long range correlation.

6. In this Appendix we present a very brief sketch of the mathematics on which our model is based. The interested reader, however, will find a complete treatment of the theories of the type utilized in the text in Umezawa (1966) (which is easy readable also by the non-specialists), Leplae et al. (1967) and Umezawa (1965). Detailed mathematical considerations on the Bose condensation will be found in Araki and Woods (1963) and Ezawa (1965). Finally, a simple presentation of the many-body problems in terms of our framework can be found, for instance, in Henley and Thirring (1962).

7. By this we do not mean that the Bosons represent original constituents of the brain, but rather we assume the existence of a mapping between the original variables and the X_i here introduced. Moreover, let us stress that Bosons do not represent a pure speculative fiction; in fact from their observed properties we must be able to identify them as realistic entities. In this respect a very urgent and significant problem is to measure the energy spectrum of the long range correlation.

8. When no risk of ambiguity arises, for simplicity we will drop the argument $(\vec{x}, t)$.

9. Remember, for example, the crystals and ferromagnets. In the crystals the interactions are invariant under spatial translations even though they create the lattice structure. In the polarized magnets the interactions are invariant under the spin-rotation. In both cases there exist certain gapless bosons.

10. When there exists a long range potential of the Coulomb type, the boson energy spectra need not to be gapless. In such a case the long range property of the boson effects comes from that of the potential.

References

Araki, H. & E. J. Woods (1963). *J. Math. and Phys., 4*, 637.

Arbib, M. (1964). *Brains Machines and Mathematics*. New York, Toronto & London: McGraw-Hill.

Braitenberg, V., G. Gambardella, G. Ghigo, & U. Vota (1965). *Kybernetik, 2*, 197.

Caianiello, E. R. (1961). *J. Theor. Biol., 1*, 204.

Ezawa, H. (1965). *J. Math. and Phys., 6*, 380.

Harth, E. M. (1966). In Caianiello (Ed.), p. 201. London: Academic Press.

Henley, E. M. & W. Thirring (1962). *Elementary quantum field theory*. New York, Toronto & London: McGraw-Hill.

John, E. R. (1966). *Proc. Neurosc. Res. Progr. I.S.P.*

Leplae, L., R. N. Sen, & H. Umezawa (1967). *Nuovo Cimento, 1*, 1.

McCulloch, W. S. & W. Pitts (1943). *Bull. Math. Biophys., 5*, 115.

Ricciardi, L. M. (1967). *Exp. Brain Res., 3*, 1.

Schmitt. F. O. (1966). *New Scientist, 23*, 643.
Umezawa, H. (1965). *Acta, Phys. Hung., 19*, 9.
Umezawa, H. (1966). *Festschrift of Progr. Theor. Phys.* (*Kyoto*), *Supplement, 37, 38*, 585.

Chapter 15

Quantum Brain Dynamics and Quantum Field Theory

Mari Jibu and Kunio Yasue*
Notre Dame Seishin University, Japan
*Okayama Institute for Quantum Physics, Japan

An introductory exposition of Quantum Brain Dynamics (QBD) is presented in which the fundamental physical process of the brain can be described within the realm of quantum field theory. QBD is nothing else but Quantum Electrodynamics (QED) of the electric dipole field of dipolar solitons and water molecules with a symmetry property under the dipole rotation. The highly systematized functioning of the brain is found to be realized by the spontaneous symmetry breaking phenomena. Memory printing, recall and decay processes are represented by the fundamental physical processes standing for the phase transition process, the symmetry restoring process and the quantum tunneling process, respectively.

1. Motivation

The problem of understanding the mechanism of memory in terms of fundamental physical processes providing the brain tissue with a highly systematized functioning (i.e., cybernetics) has been of particular interest from both physical and physiological points of view. Since only a restricted number of modern physicists have been interested in the problem, the conventional approach to it has remained of phenomenological and macroscopic nature. Indeed, the usual way for physicists and physiologists to understand the memory mechanism has been at most to consider the so-called neuronal network dynamics of transmembrane ionic transfer phenomena. The Waldeyer-Sherrington neuron doctrine has been believed widely without recourse to any further deep consideration. There, emphasis has been put on a belief that pathway conduction of neural impulses is the sole basis for signal transfer and information processing in the brain. Neural impulses are macroscopic membrane potential differences arising from the transmembrane ionic transfer phenomena.

Such a macroscopic and phenomenological approach to the problem has been shown increasingly to fail in several directions, and a further theoretical approach from microscopic point of view has been expected. In a series of papers Umezawa and coworkers (Ricciardi & Umezawa 1967; Stuart et al. 1978, 1979) proposed a completely new framework of brain dynamics within the realm of quantum field theory. They considered the brain as a mixed physical system composed of the macroscopic neuron system and an additional microscopic system. The former consists of pathway conduction of neural impulses. The latter is assumed to be a quantum mechanical many-body system interacting with the macroscopic neuron system. There, they emphasized the importance to notice a fact that the long-range correlation of Goldstone mode type plays an essential role in making brain dynamics highly ordered or systematized. Indeed, they developed a quantum field theoretical model of unlocalized memory in the brain in which memory storage is nothing but a vacuum state of spontaneous symmetry breaking type. Such a vacuum state is realized as a spatial domain in which all the ingredients manifest a strongly correlated homogeneous configuration up to the quantum fluctuation. In other words, a memory is stored in an unlocalized domain of a quantum field theoretical vacuum state just as the Bose-Einstein condensate in the superconducting media. It manifests, therefore, long-range correlation and long-term stability. These very characteristics of the quantum vacuum state well explain both nonlocal presence and long-term stability of the memory.

The theory of the mixed physical system proposed by Umezawa and coworkers seems to be the first approach to develop a physical framework describing the fundamental process of the brain information processing on the basis of conventional physical principles in quantum field theory. Therefore, it may be allowed to call the theory "Quantum Brain Dynamics" and abbreviate it by "QBD."

QBD is nothing but a quantum dynamics of a macroscopic many-body system of two different types of ingredients interacting with the macroscopic neuron system. The first ingredient is called a "corticon" and assumed to be the fundamental dynamical element of the microscopic system. Corticons are not spatially confined within each neuron. They are distributing both inside and outside of neuronal membrane and constitute rather a fuzzy region of dynamical exchange points. In other words, corticons manifest a global distribution in the whole cell assembly of the brain tissue. The second ingredient is an exchange boson. Corticons show a global distribution in the whole cell assembly of the brain tissue.

Umezawa and coworkers have presented an interesting model of memory printing and recalling processes in terms of a long-range ordered vacuum state of the microscopic system of corticons and exchange bosons. Roughly speaking, the memory printing process is nothing but a phase transition process of the QBD vacuum state which is assumed to be of spontaneous symmetry breaking type. The recalling process is, then, merely a creation process of Goldstone mode (or equivalently Goldstone bosons), that is, long-range correlation waves with zero energy requirement.

Within the realm of QBD, it is also shown that the memory stored as a vacuum state of spontaneous symmetry breaking type is rendered unstable due to the quantum tunnel effect. The decay process of the memory is then described by the notion of the instanton just as the recall process is by that of Goldstone boson.

The purpose of the present article is to provide the researcher in medical, biological and computer sciences with an introductory exposition of QBD. We avoid, therefore, the use of mathematical notations and equations, and try to make the conceptual aspect of QBD clearer.

2. Why Quantum Field Theory?

Any matter in macroscopic scale is made of as many atomic ingredients as the Avogadro constant. It is, therefore, difficult to perform straightforward physical analyses starting from the most fundamental principle of dynamics for atomic ingredients, that is, quantum mechanics. Simpler systems like the harmonic oscillator, the hydrogen atom and the helium plus ion can be solved completely by means of the direct use of quantum mechanics. However, complex systems with more atomic ingredients cannot be solved. Of course, this does not mean incompleteness of quantum mechanics any more. Rather, it is our problem that we cannot find a proper mathematical way to solve the fundamental equation of quantum mechanics, that is, Schroedinger equation or Dirac equation for complex systems.

This difficulty in question has forced physicists to find appropriate approximations so that certain physical aspects of macroscopic matter may be well understood qualitatively. Knowing or guessing a typical physical aspect of macroscopic matter, physicists look for an approximative way to solve the fundamental equation of quantum mechanics or to describe the typical physical aspect of macroscopic matter by certain phenomenological or heuristic equations which are easy to handle. For example, most of the non-living matter in macroscopic scale have a common physical aspect to be described phenomenologically by good old thermodynamics. This typical physical aspect is said to be thermal equilibrium. Then, a systematic approximation adequate for describing or analyzing complex systems of atomic ingredients in thermal equilibrium has been obtained. It is called "quantum statistical mechanics" or "equilibrium quantum statistical mechanics." Most of the theoretical results on physical analyses of non-living matter in macroscopic scale are given within the realm of quantum statistical mechanics. In other words, quantum statistical mechanics has been highly successful in understanding the typical physical aspect of non-living matter in macroscopic scale.

It seems important to notice here what the success of quantum statistical mechanics means. The basic assumption of quantum statistical mechanics is to incorporate an approximative viewpoint that typical physical characteristics of macroscopic matter in thermal equilibrium are same as those of less complex systems of ideal disordered atomic ingredients without mutual correlation. The fact that quantum statistical mechanics has had much success in understanding the typical physical aspect of non-living matter in macroscopic scale claims simply the absence of mutual correlation in such matter in actual circumstance. In other words, macroscopic matter in thermal equilibrium can be thought of as a complex system of atomic ingredients manifesting completely disordered (i.e., uncorrelated or thermalized) dynamics so that quantum statistical mechanics happens to give appropriate approximations.

Encouraged by the success, quantum statistical mechanics has been a little enlarged so that it may cover complex systems out of but near thermal equilibrium. Then, chemical and biochemical reactions of atomic ingredients have come to fall into the competition of modern physical analysis from which recent molecular biology has been developed extensively. Namely, nowadays medical and biological sciences find their physical and theoretical foundation in quantum statistical mechanics, and provide us with detailed microscopic understanding of certain biochemical processes taking part in living matter. Since few physicists mention explicitly that quantum statistical mechanics is an approximative framework limiting its validity to complex systems of atomic ingredients without mutual correlation, the researcher in medical and biological sciences believes it to be the most fundamental physical law and also reliable in the analysis of living matter from the microscopic point of view of quantum mechanics.

When the famous mathematician von Neumann saw scientists investigating dynamics of water by using the Euler equation for a perfect fluid without viscosity, he laughed and said it was dynamics of "dry" water. Similarly, we should keep it in mind that the picture of living matter investigated by molecular biology and biochemistry in terms of quantum statistical mechanics is not a real one but that of "dry" living matter. The meaning of this warning is that living matter is not a complex system of atomic ingredients without mutual correlation but those with strong mutual correlation. As it has been pointed out by Schroedinger in his famous monumental lecture (Schroedinger 1944), living matter is characterized by production of negentropy without recourse to incoming information gain. In other words, living matter decreases entropy and increases order even under the supply of energy in completely thermalized and disordered form, whereas non-living matter in thermal equilibrium is with maximum entropy (i.e., disorder). To investigate the correct physical aspect of living matter, therefore, another approximative framework is needed in which complex systems of atomic ingredients with strong mutual correlation can be well described from the fundamental point of view of quantum mechanics. It may be opposite, in a sense, to quantum statistical mechanics in which mutual correlation of atomic ingredients is completely neglected.

The above quite important fact was first emphasized by a physicist H. Umezawa in the early 1960's. He claimed the necessity of emergence of quantum field theory in describing and investigating the typical physical aspect of living matter, because complex systems of atomic ingredients with strong mutual correlation like living matter can not be treated by quantum statistical mechanics but only by quantum field theory. In 1967 he published a monumental paper with one of his Italian colleagues, L. M. Ricciardi in Kybernetik (Ricciardi & Umezawa 1967). There, focusing on the highly systematized functioning of the brain tissue as the most elaborated example of typical aspects of living matter, emphasis is put on the role of collective mode and Goldstone mode in quantum field theory of complex systems of atomic ingredients with strong mutual correlation. Just as the thermal equilibrium is the key concept in approximative physical analyses of complex systems of atomic ingredients with vanishing mutual correlation by means of quantum statistical mechanics, collective mode and Goldstone mode are the key concepts in those with strong mutual correlation.

3. A brief history of QBD

Before proceeding to the conceptual exposition of QBD in the succeeding sections, we give here a little bit of history after the original proposal of Umezawa (Ricciardi & Umezawa 1967). Slightly later, another physicist H. Froehlich pointed out also that collective mode or long-range coherent dynamics of atomic ingredients in macroscopic scale may play an essential role in energy storage of biological systems (Froehlich 1968). Coherent dipolar wave propagation is shown to exist in the cytoskeletal structure of the biological cell and exchange energy with the external electromagnetic field. Such a coherent dipolar wave propagation is nothing but a collective mode of many dipolar oscillations maintained by nonlocalized electrons trapped in the one-dimensional chains of protein molecules as well as hydrogen bonds recurring therein. Due to Froehlich's estimation, the coherent dipolar wave propagation realizes a branch of longitudinal electric modes in a frequency region between 10^{11} and 10^{12} sec^{-1}, called the "Froehlich frequency." Then, it is concluded that energy larger than Froehlich frequency times Planck's constant supplied to the cytoskeletal structure of biological cells is not completely thermalized but stored in a highly ordered fashion. Notice that thermalized energy or heat is the form of energy characteristic to complex systems of atomic ingredients with vanishing mutual correlation.

In the 1970's, Umezawa proceeded to developing his original idea to describe the typical physical aspect of living matter in terms of collective mode and Goldstone mode in quantum field theory so that the memory printing and recalling process in the brain can be well investigated with the use of known facts in quantum field theory. In two succeeding papers with his colleagues, C. I. J. M. Stuart and Y. Takahashi, he presented an interesting physical picture of fundamental processes of memory printing and recalling (Stuart et al. 1978, 1979).

Memory printing is maintained by a physical process of phase transition from disordered dynamics to ordered one of the brain tissue (i.e., cortex) seen as a complex system of atomic ingredients called "corticons." Memory recalling is maintained by a physical process of symmetry restoring typical in the ordered dynamics which can be considered as a creation process of Goldstone mode or Goldstone boson. Phase transition from disordered dynamics to an ordered one as well as the symmetry restoring process are both central concepts to understand the collective mode and Goldstone mode in quantum field theory.

Much inspired by those results of incorporating quantum field theoretical method of physical analysis into the investigation of living matter initiated by Umezawa and Froehlich in 1960's, several physicists began to consider in the 1980's biological systems of living matter as complex systems of atomic ingredients with strong mutual correlation in which energy transfer does not suffer from Maxwell's demon, that is, thermalization. They have continued to apply quantum field theory to physical analyses of various fundamental processes taking part in biological systems of living matter and ensured Umezawa's and Froehlich's original ideas (Davydov 1982; Del Giudice et al. 1982, 1985, 1986, 1988, 1992; Sivakami & Srinivasan 1983; Jibu & Yasue 1993, 1995, 1997a, 1997b; Jibu 2001). Thanks to those extensive research activities, we are nowadays in the best position to understand various typical physical aspects of living matter from the very fundamental point of view of quantum field theory. QBD is one of many examples of understanding such aspects.

4. Fundamental and metabolizing systems of living matter

Most of the living matter are known to manifest a specific form of existence, that is, the biological cell. It may be understood as the most fundamental element of biological systems. In other words, most of the biological systems including of course the human body are made of biological cells. For the purpose of making our discussion as general as possible, we will consider a typical form of biological system as a macroscopic assembly of biological cells and call it a "cell assembly." A biological cell is also a macroscopic structure of living matter which can be considered as a complex system of quite a large number (i.e., the Avogadro constant) of atomic ingredients with strong mutual correlation.

From a macroscopic point of view, the biological cell can be regarded essentially as living matter called "cytoplasm" confined within a spatial region of macroscopic scale by other living matter called the "cell membrane." Of course, there exist other macroscopic structures embedded in cytoplasm such as the cell nucleus, Golgi apparatus and mitochondria. However, they may be thought of as other biological cells cooperating inside the biological cell in question. From a microscopic point of view, cytoplasm is a complex system of water molecules and protein molecules, and cell membrane is that of lipid molecules and protein molecules. The former manifests a dense and dynamical three-dimensional network structure of protein filaments (i.e., one-dimensional chain of protein molecules) called the "cytoskeletal structure," surrounded by water molecules, and the latter manifesting a double layered two-dimensional surface structure of lipid molecules patched from both inside and outside by a dynamical two-dimensional network structure of protein filaments. There are many protein molecules embedded in the lipid bilayer of the cell membrane which plays the role of active gates for the ionic metabolism of the biological cell. The functioning of those protein gates has been investigated intensively in molecular biology. Nowadays, they are known to control the diffusion process of various ions between the inside and outside of the biological cell.

Such a diffusion process of ions is a typical example of completely thermalized (i.e., disordered or incoherent) dynamics of atomic ingredients well described by quantum statistical mechanics. It has been supposed, therefore, to be the most fundamental physical process of living matter in medical and biological sciences. This may be the limit of conventional analyses of biochemistry and molecular biology. For further investigation of the typical physical aspect of living matter such as the cell assembly, the truly most fundamental physical process should be considered from the fundamental point of view of quantum mechanics.

Since we are not interested in completely thermalized dynamics of the ionic diffusion process as the fundamental physical process of living matter, we look for another possibility starting from the original ideas of Umezawa and Froehlich. Let us neglect all the atomic ingredients taking part in the completely thermalized dynamics of cell assembly, that is, the lipid bilayer of cell membrane and protein molecules embedded therein. Then, a living matter such as the cell assembly comes to manifest a rather simpler microscopic structure of a huge and dense three-dimensional network of protein filaments surrounded by and interacting with water molecules. We call this structure the "fundamental system of living matter." Comparatively, we call the molecular biologically well-known structure maintaining the completely thermalized dynamics of cell assembly the "metabolizing system of living matter." Thus, the living matter is a mixed system composed of the fundamental system and the metabolizing system. The latter manifests completely thermalized (i.e., disordered and incoherent) dynamics of atomic ingredients playing the role of heat engine to supply energy such as the ATP cyclic process, and well described by quantum statistical mechanics. The former manifests conversely non-thermalized (i.e., ordered and coherent) dynamics of atomic ingredients with strong and long-range mutual correlation, the role and characteristic of which is the scope of quantum field theory. The idea that the living matter is from a fundamental point of view of physics a mixed system of coherent dynamics and incoherent dynamics is due to Umezawa (Ricciardi & Umezawa 1967; Stuart et al. 1978, 1979). There, the fundamental system of the brain tissue as typical living matter is called the "microscopic system," and the metabolizing system is called the "macroscopic (physiologically nonclassical) neuron system."

It seems worthwhile to notice here again that both fundamental and metabolizing systems of living matter are made of as many atomic ingredients as the Avogadro constant. In this sense, they are both macroscopic systems of extremely many microscopic (i.e., atomic) ingredients. The difference between them appears only in their dynamical characteristics. Namely, the fundamental system manifests ordered dynamics of atomic ingredients with strong and long-range mutual correlation which may be well described by quantum field theory. The metabolizing system manifests disordered dynamics of atomic ingredients with vanishing mutual correlation such as ionic diffusions which has been well investigated by quantum statistical mechanics.

5. Physical picture of the fundamental system of living matter

Let us investigate the essential characteristic of the fundamental system of living matter. First, we visualize relevant degrees of freedom of the fundamental system of living matter, that is, a huge and dense three-dimensional network of protein filaments surrounded by and interacting with water molecules. Although the network of protein filaments changes its form and connectivity due to spatial motion of protein filaments, such degrees of freedom belong no longer to the fundamental system. This is simply because such a dynamical change of the network structure is driven by disordered dynamics of protein filaments and belongs to the metabolizing system of living matter. Indeed, the dynamical change of the network structure of protein filaments (i.e., cytoskeletal structure) results in protoplasmic streaming. We may be allowed, therefore, to regard the network structure of protein filaments as a mere background for the coherent and dynamics of the fundamental system. Thus, the first degree of freedom we are looking for in the fundamental system of living matter may be found as an internal degree of freedom of the background three-dimensional network structure of protein filaments free from thermalization (i.e., Maxwell's demon).

In 1979 such a degree of freedom was found by Davydov as a coherent dipolar solitary wave propagation along the one-dimensional chain of protein molecules such as the protein filament (Davydov 1979). In quantum field theory, a coherent solitary wave propagation is considered as a localized degree of freedom maintaining and carrying energy without loss due to thermalization, and it is called the "Davydov soliton" or "dipolar soliton." Namely, energy incoming from the metabolizing system of living matter through the ATP cyclic process to the fundamental system of living matter induces first dipolar solitons localized in each protein filament. As a specific character of soliton in quantum field theory, energy stored in soliton form is kept free from thermalization and belongs to the fundamental system of living matter, though creation of soliton is triggered by incoherent and disordered interaction with the metabolizing system. In other words, the creation and annihilation process of dipolar solitons plays the role of a gateway between metabolizing and fundamental systems. The dipolar soliton is a collective mode of many dipolar oscillations maintained by nonlocalized electrons trapped in the one-dimensional chain of protein molecules and may be regarded as the first degree of freedom of the fundamental system of living matter. It is a quantum mechanical degree of freedom representing electric dipole moment localized in each background protein filament. In the case of brain tissue, Stuart et al. (1978, 1979) called it the "corticon." In the general case of cell assembly, we call it simply "dipolar soliton." Thus the first degree of freedom of the fundamental system of living matter is found to be the dipolar soliton localized in each protein filament of the background three-dimensional network structure. It may be well visualized from physical point of view by a quantum mechanical variable representing nonvanishing electric dipole moment localized in each protein filament.

Let us look for the second degree of freedom of the fundamental system of living matter. From materialistic point of view, the fundamental system of living matter is composed of a huge and dense three-dimensional network of protein filaments and enormously large number (i.e., the Avogadro constant) of water molecules surrounding it. The former has been thought of as a mere background structure for the fundamental system and plays the role of supporting the existence of the first quantum mechanical degree of freedom, that is, dipolar solitons. The latter is of purely quantum mechanical nature and an enormously large number of atomic ingredients, that is, water molecules, forces us to rely on quantum field theory. The water molecule, H_2O, is a typical molecule simple in its form but rich in its physical characteristics. The origin of richness can be found, however, in simpleness of its form. Namely, due to the spatial geometric configuration of two hydrogen atoms relative to one oxygen atom, the water molecule manifests nonvanishing electric dipole moment. Thus, the totality of enormously large number of water molecules can be well described from a physical point of view by a quantum mechanical degree of freedom of electric dipole moment moving and rotating freely. This is the second degree of freedom of the fundamental system of living matter. We call it the "water dipole moment."

Finally, we have obtained a physical picture of the fundamental system of living matter. It is essentially a quantum mechanical many-body system described by two different degrees of freedom interacting with each other, that is, dipolar solitons localized in the background three-dimensional network structure of protein filaments and water dipole moments surrounding them.

6. What is Quantum Brain Dynamics?

Having obtained a physical picture of the fundamental system of living matter, we are now in the best position to make a conceptual exposition of quantum brain dynamics, QBD. Identifying the corticon in Umezawa's original viewpoint with the dipolar soliton of the fundamental system of brain tissue as typical living matter, we may visualize the most fundamental physical process providing the brain tissue with a highly systematized functioning within the realm of quantum field theory. Thus, QBD is a completely new theoretical framework to describe the fundamental physical process of the brain dynamics that makes man human on the basis of quantum field theoretical analysis of the fundamental system of brain tissue.

Let us start from the physical picture of the fundamental system of brain tissue with two different degrees of freedom standing for dipolar solitons localized in the background three-dimensional network structure of protein filaments and dipole moments of surrounding water molecules. The first degree of freedom, that is, the dipolar soliton arises from a coherent solitary wave propagation of nonlocalized electron along each protein filament. The dipolar soliton is created at the end of each protein filament by energy gain from the metabolizing system through, for example, the ATP cyclic process.

If we focus on the mere electric dipole moment of the dipolar soliton, we may consider an electric dipole field confined on each protein filament. Once created, such a confined electric dipole field is kept conserved there as long as there exists no other electric dipole moments nearby. However, in the fundamental system of brain tissue, we have the second degree of freedom, that is, electric dipole moments of water molecules surrounding the protein filaments. Then, it is most likely that the electric dipole field on each protein filament is no longer kept conserved and confined thereon but propagates into the spatial region occupied by water molecules.

This suggests to us the following fact: As long as we look at the fundamental system of the brain tissue in terms of the electric dipole field, both dipolar solitons and water dipole moments can no longer be different degrees of freedom.

In other words, the fundamental system of brain tissue can be well described by a single degree of freedom of electric dipole field spanning the spatial volume of the brain tissue.

Astonishingly, QBD is now translated into QED (i.e., quantum electrodynamics) of the electric dipole field which may be easily supposed to fall into the competition of quantum field theory.

It is found that the principal degree of freedom of QBD, that is, the corticon is not merely the dipolar soliton as was expected at first sight but also the water dipole moment surrounding the protein filaments. The corticon in QBD is now fully described by the electric dipole field (of both dipolar solitons and water dipole moments) spanning the spatial volume of the brain tissue. In this sense, we may call the fundamental system of brain tissue simply as the "system of corticons," hereafter. Considering the physical background of the electric dipole field as those of dipolar solitons and water dipole moments, we may assume that the electric dipole field manifests symmetry under rotation. Namely, even if the electric dipole field on each position is rotated by any spatial angle, the total energy of the system of corticons is kept invariant. In quantum field theory, the total energy of the system of any field quantity plays an important role in specifying dynamics of the field, and it is usually called the "Hamiltonian." So, we refer to the total energy of the system of corticons as the Hamiltonian of the system of corticons or equivalently the Hamiltonian of QBD. Then, we obtain the following invariant or symmetry property: The system of corticons in QBD manifests a symmetry under the rotation of the electric dipole field in a sense that the Hamiltonian of QBD is invariant.

At this point it seems easy to answer the question "What is QBD?" tentatively. QBD is nothing but QED of the electric dipole field with symmetry under the dipole rotation.

7. Collective mode and Goldstone mode

This section is devoted to an exposition of two basic concepts in quantum field theory, that is, the collective mode and the Goldstone mode which play the most important roles in understanding the fundamental physical process of QBD providing the brain tissue with the highly systematized functioning.

Let us consider the system of corticons in QBD. Recall that it is nothing else but the totality of as many atomic ingredients with electric dipole moments as the Avogadro constant. Then, a naïve question may arise naturally:

We may have certainly non-living matter made of as many atomic ingredients with electric dipole moments as the Avogadro constant. It is essentially living and thinking about the beginning of our universe?

The answer is, of course, negative. It is neither living nor thinking about. As we have emphasized in the preceding Section 2, living matter such as brain tissue is the totality of atomic ingredients with strong mutual correlation, whereas the non-living matter is that with vanishing mutual correlation. In other words, the non-living matter made of as many atomic ingredients with electric dipole moments as the Avogadro constant has a typical physical aspect that it manifests disordered dynamics well described by quantum statistical mechanics. There, each atomic ingredient with electric dipole moment manifests a time evolution irrespective of the precise time evolutions of neighboring ones but only respective to the energy transfer between them. Thus, the neighboring electric dipole moments cancel out mutually, obtaining no net electric dipole field any more. Conversely, the living matter made of as many atomic ingredients with electric dipole moments as the Avogadro constant has a different typical physical aspect that it manifests ordered dynamics well described by quantum field theory. There, each atomic ingredient with electric dipole moment manifests a time evolution strongly correlated to the precise time evolutions of neighboring ones. In such a peculiar situation, the neighboring electric dipole moments no longer cancel out, but accumulate to their collective value. Such a collective value of strongly correlated neighboring electric dipole moments given in each position provides us with the electric dipole field which is nothing but the principal degree of freedom of the system of corticons in QBD. This is the very reason why we should rely on quantum field theory to understand the fundamental physical process taking part in the fundamental system of living matter such as the system of corticons.

Let us investigate the fundamental physical process of the system of corticons by means of the electric dipole field arising from the ordered collective dynamics of atomic ingredients with electric dipole moments within the general framework of quantum field theory. We refer to Stuart et al. (1979) and Del Giudice et al. (1985), though the essential framework had been originally developed by Ricciardi and Umezawa (1967).

In quantum field theory, the simplest way to describe the electric dipole field is given by a spinor field. It is a two-component complex field usually described by a two by one matrix form. The spinor field itself does not correspond directly to the collective value of electric dipole moments. In the spinor representation, dynamical variables providing us with the electric dipole moment are the Pauli spin matrices. The electric dipole moment of the electric dipole field is then given by first multiplying the spinor field by the Pauli spin matrices and then multiplying the result by the conjugate spinor field. Thus the spinor field does not stand for the electric dipole moment, but it represents physically the field of molecular vibrations of protein filaments and water molecules in the brain tissue from which the electric dipole field arises (Del Giudice et al. 1985). Therefore, the spinor field of molecular vibrational field is more essential than the electric dipole moment itself.

The corticon in QBD described by the electric dipole field of both dipolar solitons and water dipole moments is thus found to be essentially described by the spinor field of molecular vibrations of both protein filaments and water molecules spanning the spatial volume of the brain tissue. Therefore, we may be allowed to call the spinor field the corticon field. Recall that any physical quantity such as the electric dipole moment is given by an expectation of a certain function of field variables over possible field configurations characterized by a certain proper value of the Hamiltonian (i.e., the total energy function) in quantum field theory. In the present case of the corticon field, the electric dipole moment is the expectation of a quantity given by first multiplying the spinor field by the Pauli spin matrices and then multiplying the result by the conjugate spinor field. Then, the invariant or symmetry property of the system of corticons in QBD under the rotation of the electric dipole field may be translated into that of the corticon field. For this aim, we present in what follows the quantum field theoretical framework of the system of corticons in terms of the corticon field.

The system of corticons in QBD is represented by the corticon field. The Hamiltonian, that is, the total energy of the system of corticons is then given by a certain functional of the corticon field, possibly depending on both its spatial and temporal derivative. The symmetry property of the system of corticons in QBD implies that the Hamiltonian is invariant under the group of dipole rotation. In terms of the corticon field, such a rotation group can be well represented by the SU(2) group of complex two by two unitary matrices with determinants equal to unity. Since we are mainly interested in the symmetry structure of the system of corticons in QBD, we will not specify the explicit form of the Hamiltonian as a functional of the corticon field, but concentrate on its invariant property under the SU(2) group. In other words, the present investigation of QBD remains general and independent of the details of the dynamics. Those who are interested in the explicit form of the Hamiltonian are invited to see the appendices of Stuart et al. (1979).

Let us see the typical physical aspect of the system of corticons in QBD which may explain the highly systematized functioning of the brain tissue. For this aim, we focus on the dynamics of corticons corresponding to the minimum proper value of the Hamiltonian. Dynamics of corticons with larger proper value of the Hamiltonian will be easily investigated on the basis of that with the minimum one. In quantum field theory, such a dynamical state of the system is said to be the lowest energy state or the vacuum state. The vacuum state of the system of corticons (i.e., the corticon field) in QBD violates its original dynamical symmetry structure under the SU(2) group of dipole rotations. Such a vacuum state is said to be of spontaneous symmetry breaking type. There, corticons are all fallen into a uniform configuration of their electric dipole moments up to the quantum fluctuation, and so the original dynamical symmetry under dipole rotations is broken.

The vacuum state of the corticon field may be well characterized by the presence of nonvanishing uniform electric dipole moment. We call this uniform electric dipole moment characteristic of the vacuum state of the corticon field the "vacuum polarization." It represents the mean value of the uniform electric dipole moment of every dipolar soliton and water molecule aligned along one and the same direction. In this sense, the system of corticons in the vacuum state of spontaneous symmetry breaking type manifests a typical physical aspect that it creates a long-range (i.e., large-scale) order in its dynamics. Namely, there exists a long-range order so that the corticon field is systematized globally to realize the uniform configuration of electric dipole moment. Umezawa and coworkers used the expression "macroscopic ordered state" to refer to large-scale phenomena of order creation whose occurrence cannot be explained without recourse to the action of specific quantum field theoretical mechanisms responsible for the spontaneous rearrangement of the symmetry attributes of the system. Thus, the vacuum state of the corticon field in QBD is a typical macroscopic ordered state.

It is known from the Goldstone theorem in quantum field theory that in any macroscopic ordered state, cooperative excitations of the symmetry attributes appear as long-range correlation waves and behave as bosons (i.e., quanta obeying the Bose-Einstein statistics) whose minimum energy is zero. They are called "Goldstone bosons" or "Goldstone modes." Since the Goldstone boson manifests a continuous energy spectrum above zero, it is also called a "gapless mode" because there exists no energy gap in the spectrum. It is nothing else but the action of the corticon field in the vacuum state accounting for the rearrangement of symmetry attributes leading to the creation of a macroscopic order with long-term stability and nonlocal presence.

Dynamics of the corticon field in the vacuum state of spontaneous symmetry breaking type is a typical example of peculiar physical aspect called the "collective mode" in quantum field theory. Namely, the corticon field manifests highly ordered dynamics as aligning all the electric dipole moments of corticons along one and the same direction up to the quantum fluctuation. It is a static collective mode of the corticon field which does not change in time. In general, the collective mode of a certain field stands for a cooperative dynamics of the field variable which is highly systematized and synchronized. The collective mode provides us with an appropriate approximative framework to describe the typical physical aspect of a complex system of atomic ingredients with strong mutual correlation such as living matter. Thus, the collective mode of the corticon field is a typical dynamical aspect of the system of corticons with strong mutual correlation in which every corticon manifests one and the same dynamical configuration. If we translate by analogy this quantum field theoretical concept of the collective mode into our familiar concept, it may be compared with the synchronized swimming. Just as you find a collective movement of a team of synchronized swimmers as a single large-scale ordered state of the team, the collective mode of the corticon field behaves as a distinct physical entity and the precise dynamical factors of the atomic ingredients on which its existence depends can be neglected.

In QBD the vacuum state of the corticon field violating the original SU(2) symmetry of the dipole rotation appears as a macroscopic ordered state and behaves itself as a distinct physical entity. It may be seen as a macroscopic spatial domain of the brain tissue in which nonvanishing electric dipole moment distributes uniformly up to the quantum fluctuation. Such a domain of the vacuum state cannot be arbitrarily small because a considerable number of atomic ingredients must take part in forming the macroscopic ordered state. Any collective mode requires the strong mutual correlation of a large number of atomic ingredients, and it cannot appear when the domain of organization is too small. The minimum linear dimension (i.e., size) of the domain for realizing the macroscopic ordered state such as the vacuum state is called the "coherence length."

It seems worthwhile to notice that consequently the vacuum state of QBD can only appear in a spatial domain with linear dimension larger than the coherence length. Umezawa emphasized this fact as saying that the order is intrinsically diffused. In other words, the fundamental system of the brain tissue manifests a macroscopic dynamical structure made of nonoverlapping large-scale spatial domains in which the nonvanishing vacuum polarizations exist. Therefore, the fundamental physical process of the fundamental system of the brain tissue may be well investigated by taking the dynamical action of the vacuum state of the corticon field against the energy supply into account. This is the very reason why the concept of Goldstone mode becomes so important in QBD. The Goldstone mode is nothing but a wave of defects of the corticon field from the vacuum polarization created by any amount of the energy supply from the metabolizing system of the brain tissue. It appears as a new degree of freedom carrying the original symmetry property of the corticon field with respect to the SU(2) group of dipolar rotations. Thus, taking the symmetry property of Goldstone modes, the system of corticons in QBD insures again the original symmetry even though its vacuum state remains of spontaneous symmetry breaking type. The broken symmetry is now restored by the creation of Goldstone bosons.

8. Memory as vacuum state of the corticon system

Having seen the two basic concepts in quantum field theory, that is, the collective mode and the Goldstone mode, we proceed to investigating the fundamental physical process of QBD which provides the brain tissue with the highly systematized functioning. More precisely, we introduce an interesting scheme of the information processing of the brain in terms of the phase transition from a disordered state to an ordered one of the system of corticons in QBD.

A new mechanism for memory has been presented by Umezawa and coworkers in which the two systems are necessarily coupled in order to achieve memory printing and recall (Ricciardi & Umezawa 1967; Stuart et al. 1978, 1979). The coupling is given in QBD by the creation of dipolar solitons in the microscopic protein filaments triggered by the energy supply from the ATP cyclic process which protect memory in stable form against excitations with the communication mode.

We consider in what follows how the mechanism proposed by Umezawa can account for the highly systematized functioning of the brain associated with memory:

Let us start with the fundamental system of the brain, that is, the system of corticons where memory of a specific stimulus from the external world is yet to be printed. The brain tissue is here exposed to stray signals including unattached perception of external events as well as activity associated with general physiological events such as motor activity and the like. Those stray signals are organized and transmitted to the system of corticons through the metabolizing system (i.e., the neuronal network). Namely, they can create corticons indirectly in the fundamental system, and various spatial domains of the macroscopic ordered states are formed provided that the created corticons manifest a long-range correlation whose spatial extent is larger than the coherence length. Thus, there exist apparent thresholds for the incoming energy from the metabolizing system to the fundamental one to create ordered domains. The stray signals to the brain tissue transmitted with energy slightly exceeding the threshold value are all tacitly coded in the dynamical domain structure of macroscopic ordered states in the small domains. Notice that in such a dynamical structure the direction of electric dipole moments of the corticon field can randomly vary from domain to domain and there exists no hierarchy among the coded signals.

The learning process can be identified with the phase transition of the system of corticons from the less ordered state of many but small ordered domains to the more ordered state of a few but large ordered domains. Such a phase transition is induced by an external stimulus which supplies the system of corticons with enough energy to break domain boundaries of many small ordered domains, thus aligning the electric dipole moments of the corticon field in much larger domains. Of course, the notion of external stimulus denotes an energy flow organized and transmitted through the metabolizing system. We obtained quite an interesting point of view of the fundamental physical process of QBD that may be understood as the learning process of a typical signal from the external world. It seems worthwhile to notice here that the phase transition requires the onset of the aligning process of the electric dipole moments which goes through an interaction between the external stimulus and the Goldstone boson. This implies a fact that the external stimulus can induce the phase transition from the less ordered state to the more ordered one only when it can interact with the Goldstone bosons and the interaction energy is sufficiently large to break the domain boundaries. In other words, there exists a filter such as the selection rule for the coupling of transmitted signals to the system of corticons in which stray signals have been coded in the domain structure. Stuart et al. (1978) saw there inherent limitations on learnability for the brain tissue.

By such a learning process, a typical external stimulus can be printed in the system of corticons of the fundamental system as a stable macroscopic ordered state. There, the external stimulus is coded into the vacuum state of the corticon field manifesting a large-scale uniform alignment of the electric dipole moments along one and the same direction. The stability of printed memory comes from the very fact that it is coded into the vacuum state of the corticon field the stability of which is a consequence of quantum field theory. Apparently, printed memory manifests nonlocal (i.e., diffused) existence.

Once memory of a typical external stimulus is printed in the macroscopic ordered state of the system of corticons, it can be recalled quite easily thanks to the Goldstone bosons. Namely, when the system of corticons receive even a weak signal of a nature similar to that used in the learning process, it can excite the gapless Goldstone mode of the vacuum state corresponding to printed memory. The recalling process is nothing but a creation process of Goldstone bosons (i.e., long-range correlation waves) with almost no energy requirement. They play the role of a replication signal of the original external signal. In this way, the existence of printed memory can be taken into account by consciousness. Consciousness in QBD means the quantum field theoretical dynamics of the system of corticons itself. Thus, any weak external signal can recall printed memory, and it take part in consciousness. It is implicit to this view that recall involves the stimulation of those parts of the metabolizing system of the brain that were once excited to organize and transmit external signals during the original learning process. It seems also worthwhile to point out that a weak external stimulus might interact with two or more kinds of Goldstone bosons, thereby recalling more than one stored code. This may explain association in the recall process.

9. Quantum decay process of memory

The most important concept in QBD is the vacuum state of spontaneous symmetry breaking type as we have seen in the preceding section. In quantum field theory, a system with the Hamiltonian which is invariant under some continuous transformation group like the system of corticons is known to have infinitely many vacuum states. Furthermore, those vacuum states are not the true vacuum state which must be the unique proper state of the Hamiltonian with the lowest proper value of energy. Each of them is an approximative vacuum state of the system in which the field variable manifests the smallest deviation from a classical minimum energy solution. In the terminology of quantum field theory, they are coherent states around classical field configurations corresponding to different energy minima. Each vacuum state can be transformed into another by the symmetry transformation. Therefore, those vacuum states are in general not invariant under the original symmetry transformation of the Hamiltonian. They are of spontaneous symmetry breaking type.

As Umezawa and coworkers have shown, each vacuum state of the system of corticons plays the role of memory storage in QBD. It is important to see the fact that the vacuum state of spontaneous symmetry breaking type is capable of creating Goldstone bosons with zero energy requirement. This means that the stored memory code can be easily excited during the recall process and it contributes to other fundamental physical processes of the fundamental system of the brain tissue by means of Goldstone bosons. In this way the existence of a memory can be "consciously" taken into account in the information processing of the brain. The fact that the system of corticons in QBD has infinitely many vacuum states of spontaneous symmetry breaking type implies an infinite capacity of memory coded and stored in the brain. The whole fundamental system of the brain tissue is divided into extremely many ordered domains of vacuum states, and Goldstone bosons corresponding to those vacuum sates take part in the quantum field theoretical dynamics of the corticon field, that is, consciousness. Thus memory coded in the vacuum state always affects the further development of consciousness because of its long-term stability. It may be concluded, therefore, that the memory codes stored in the vacuum sates of the system of corticons will never be lost, which seems not the case in the actual functioning of the brain. However, the symmetry property of the system of corticons in QBD does provide us with a more realistic feature of the memory codes as being rendered unstable due to the quantum tunnel effect. The memory code is no longer stable completely if we take the vacuum tunneling phenomena of the corticon field into account. Let us see this by restricting our discussion to a typical spatial domain of the vacuum state.

Suppose that after a certain learning process the system of corticons in this domain has fallen into a vacuum state, say vacuum state A, that breaks the original symmetry of the system spontaneously. It is a memory code which can be easily recalled by the creation process of Goldstone bosons. As we have mentioned before, it is not a true vacuum state of the corticon field from the strict point of view of quantum field theory, but an approximative one standing for a classical field configuration with minimum energy. This means that the vacuum state A is no longer invariant under the time evolution of the system of corticons driven by the Hamiltonian. As time passes, the vacuum state A is put into another state, say state B, which differs from the vacuum state A. Only the exact vacuum state remains always the same vacuum state as the time passes. Then, the probability that the actual state B of the system of corticons is found to be in another vacuum state, say vacuum state C, is given by the absolute square of the inner product of the state B and the vacuum state C in quantum field theory. Since the actual state B is not the vacuum state, the inner product in question does not vanish. Namely, the system of corticons in this domain may happen to be in another vacuum state C with nonvanishing probability. This is the quantum tunnel effect. It is also a decay process of a vacuum state storing a memory code into another which stores a meaningless code having nothing in common with the original memory code. In quantum field theory, such a quantum decay process of vacuum state due to the quantum tunnel effect can be thought of as the onset of a new quantum field theoretical mode called the "instanton." We may conclude in QBD that the memory codes stored in the ordered domain structure of vacuum states of the system of corticons are rendered unstable due to the tunnel effect and the decay process of printed memory is described by the creation process of instantons just as the recall process is by that of Goldstone bosons.

It may be of certain interest, though speculative at this moment, to interpret the virtual dynamics of instantons as the fundamental physical process corresponding to the dream. A dream is a virtual action of consciousness based on the actual memory codes but resulting in irrelevant memory codes which cannot be reached by the normal action of consciousness. The dream is not a causal event at all. This empirical fact on the typical aspect of the dream may be compared with that of virtual dynamics of instantons. Namely, dynamics of instantons is a virtual physical process starting from the actual vacuum state but ending with an irrelevant vacuum state by the quantum tunnel effect which cannot be transformed by the normal and causal physical process of QBD. In this sense, the quantum decay process of the vacuum state is said to be a virtual physical process. The virtual dynamics of instantons triggered by the quantum tunnel effect is not a causal event at all.

10. Outlook

It seems of much importance now to let the neuro- and cognitive scientists know the truth: it is necessary to rely on quantum theory even if physical phenomena of matter and light in the macroscopic scale are concerned. Of course, the brain is not the exception, and intensive researches from the very fundamental point of view of quantum theory will be respectable (Pribram 1991; Jibu & Yasue 1995; Jibu et al. 1996, 1997; Jibu 2001). It has been widely believed in neuro- and cognitive sciences that the brain can be understood as a macroscopic object governed not by quantum physics but by classical physics or at most by molecular and chemical biologies.

Recently, several ambitious theories have been proposed which aim at explaining such basic features of consciousness as "unity (binding problem)," "qualia," "non-algorithmic processing," "synchrony" and "free will" in terms of quantum theoretical concepts (Penrose 1989, 1994; Hameroff 1987; Eccles

1986). In most cases, quantum theoretical concepts are quoted from quantum mechanics, and the famous conceptual difficulties of quantum mechanics such as "nonlocality," "superposition of states," "uncertainty principle," "reduction or collapse of wave function (state)," "EPR-paradox," "non-separability," "Bell's theorem" and "measurement by abstract ego" are put into heavy use.

However, it must be warned strongly that quantum mechanical concepts are necessarily restricted to the highly idealistic cases of microscopic objects imperceptible directly by our consciousness. As most of neuro- and cognitive scientists believe, the brain is a macroscopic object open to its noisy thermal surroundings, and a naïve application of those quantum mechanical concepts to the brain might be hardly accepted. Neverthless, in the present paper, it is emphasized that incorporation of quantum theory into the investigation of brain functioning is an inevitable turning point of the consciousness research. Here, unlike those recent ambitious theories, quantum theory means not quantum mechanics but quantum field theory that provides us with the first principle of modern physics (Umezawa 1993; Jibu & Yasue 1995; Vitiello 2001).

It may be true that consciousness would not fall yet into the competition of modern scientific activity because of difficulty to elucidate any defining aspects. Among a few defining aspects of consciousness frequently focused on in neuro- and cognitive sciences, we may restrict our discussion to "unity" or "binding problem." This is because unity is the most consistently identified defining aspect of consciousness capable of being approached from the fundamental scientific framework of theoretical physics. The long-standing difficulty of understanding the origin and mechanism of unity of consciousness (as a unified self) has been called the "binding problem" in neurophysiology: What is it that controls and unifies all the physico-chemical processes taking part in the stratified society of brain cells?

We may suppose in QBD that quantum electromagnetic phenomena play essential roles in realizing unity of the stratified society of brain cells spanning hierarchically from molecular biology of neural networks to chemical biology of cytoskeletal networks and extracellular matrices. The binding problem must be solved not by introducing the idealistic quantum mechanical nonlocality but by investigating the usually neglected quantum electromagnetic phenomena taking place in the dynamically ordered regions (i.e., perimembranous regions) of intracellular and extracellular water. There, each brain cell is enfolded within a common field of macroscopic condensation of evanescent photons and all the physico-chemical processes taking part in the stratified society of brain cells are subject to the control and unification by quantum electrodynamics. Unity of consciousness thus arises from the existence of the global field of condensed evanescent photons overlapping the whole brain tissue in the cranium. The origin and mechanism of EEG (i.e., electroencephalogram, brain wave) may be explained within the same quantum electrodynamical framework of QBD (Jibu et al. 1997).

References

Davydov, A. S. (1979). Solitons in molecular systems. *Physica Scripta, 20*, 387–394.

Del Giudice, E., Doglia, S., & Milani, M. (1982). A collective dynamics in metabolically active cells. *Physics Letters, 90A*, 104–106.

Del Giudice, E., Doglia, S., Milani, M., & Vitiello, G. (1985). A quantum field theoretical approach to the collective behaviour of biological systems. *Nuclear Physics, B251*, 375–400.

Del Giudice, E., Doglia, S., Milani, M., & Vitiello, G. (1986). Electromagnetic and spontaneous symmetry breaking in biological matter. *Nuclear Physics, B275*, 185–199.

Del Giudice, E., Preparata, E., & Vitiello, G. (1988). Water as a free electric dipole laser. *Physical Review Letters, 61*, 1085–1088.

Del Giudice, E., Doglia, S., Milani, M., & Vitiello, G. (1992). A dynamical mechanism for cytoskeleton structures. In M. Bender (Ed.), *Interfacial phenomena in biological systems*. New York: Marcel Dekker.

Eccles, J. C. (1986). Do mental events cause neural events analogously to the probability fields of quantum mechanics? *Proceedings of the Royal Society of London, B277*, 411–428.

Froehlich, H. (1968). Long-range coherence and energy storage in biological systems. *International Journal of Quantum Chemistry, 2*, 641–649.

Hameroff, S. R. (1987). *Ultimate computing: Biomolecular Consciousness and NanoTechnology*. Amsterdam: North Holland.

Jibu, M. (2001). Theory of cell membrane organizers and pressure reversal of anesthesia. *Medical Hypotheses, 56*, 26–32.

Jibu, M. & Yasue, K. (1993). Intracellular quantum signal transfer in Umezawa's quantum brain dynamics. *Cybernetics and Systems: An International Journal, 24*, 1–7.

Jibu, M. & Yasue, K. (1995). *Quantum brain dynamics – An introduction*. Amsterdam: John Benjamins.

Jibu, M. & Yasue, K. (1997a). What is mind? – Quantum field theory of evanescent photons in brain as quantum theory of consciousness. *Informatica, 21*, 471–490.

Jibu, M. & Yasue, K. (1997b). Magic without magic: Meaning of quantum brain dynamics. *Journal of Mind and Behavior, 18*, 205–228.

Jibu, M., Hagan, S., Hameroff, S. R., Pribram, K. H., & Yasue, K. (1994). Quantum optical coherence in cytoskeletal microtubules: Implications for brain function. *BioSystems, 32*, 195–209.

Jibu, M., Pribram, K. H., & Yasue, K. (1996). From conscious experience to memory storage and retrieval: the role of quantum brain dynamics and boson condensation of evanescent photons. *International Journal of Modern Physics, B10*, 1745–1754.

Jibu, M., Yasue, K., & Hagan, S. (1997). Evanescent (tunneling) photon and cellular 'vision'. *BioSystems, 42*, 65–73.
Penrose, R. (1989). *The Emperer's new mind*. London: Oxford University Press.
Penrose, R. (1994). *Shadows of the mind*. London: Oxford University Press.
Pribram, K. H. (1991). *Brain and perception*. New Jersey: Lawrence Erlbaum.
Ricciardi, L. M. & Umezawa, H. (1967). Brain and physics of many-body problem. *Kybernetik, 4*, 44.
Schroedinger, E. (1944). *What is life?* Cambridge: Cambridge University Press.
Sivakami, S. & Srinivasan, V. (1983). A model for memory. *Journal of Theoretical Biology, 102*, 287–294.
Stuart, C. I. J. M., Takahashi, Y., & Umezawa, H. (1978). On the stability and non-local properties of memory. *Journal of Theoretical Biology, 71*, 605–618.
Stuart, C. I. J. M., Takahashi, Y., & Umezawa, H. (1979). Mixed-system brain dynamics: neural memory as a macroscopic ordered state. *Foundation of Physics, 9*, 301–327.
Umezawa, H. (1993). *Advanced field theory: micro, macro and thermal physics*. New York: American Institute of Physics.
Vitiello, G. (2001). *My double unveiled*. Amsterdam: John Benjamins.

Chapter 16

Brain and Quantum Field Theory

Notes on monumental discussions presenting quantum field models of brain

Yasushi Takahashi and Mari Jibu
University of Alberta, Canada / Notredame Seishin University, Japan

The theory of Quantum Brain Dynamics (QBD) is presented in its original form developed intensively by Stuart, Takahashi and Umezawa three decades ago in Edmonton.

1. Introduction

In 1967 Ricciardi and Umezawa predicted that certain physical ordering phenomena arising from spontaneous symmetry breakings play an essential role in the fundamental process of neurons (Ricciardi & Umezawa 1967). From September 12, 1975, a series of workshops were organized by Iain Stuart on Friday afternoon in the Department of Physics, University of Alberta. There, the Ricciardi-Umezawa conjecture was discussed intensively by the physicists from the Department as well as biologists and medical doctors from other Departments. Following to the lectures on the Ricciardi-Umezawa conjecture and its underlying fundamentals by Umezawa, one of the authors (YT) presented some theoretical calculations on the quantum fields inside and outside the neuron on November 10th based on the concrete form of the Hamiltonian he had proposed. This model Hamiltonian was aimed at clarifying the essential property of physical fundamental processes of memory storage and retrieval in terms of interactions between two kinds of quanta called corticon and stuarton, and resulted in a model playing the same role as the Lee model played in the early development of quantum field theory.

Those workshops resulted in the two papers published in 1978 and 1979 (Stuart et al. 1978, 1979). Then, YT shifted his research activity to the theoretical development of Thermo Field Dynamics (Takahashi & Umezawa 1975) and had no chance to go back to the quantum field model of brain. Fortunately, it was found that the physical model in which corticons are regarded as quanta of the electric dipole field of water inside and outside neurons and stuartons as photons, quanta of the electromagnetic field, has the same Hamiltonian as YT proposed (Pribram 1991; Jibu et al. 1994), and the model has been developed into a systematic theoretical framework of quantum brain dynamics – a quantum field theory of fundamental processes of the brain functioning (Jibu & Yasue 1995). Furthermore, an international conference was held in 1999 at the United Nations University in Tokyo aiming at introduction of quantum brain dynamics to the wider research communities in neurosciences, and a more specialized workshop was held in 2002 at Villa Gualino of the ISI Foundation in Turin to which the authors would like to contribute by presenting the record of the monumental workshops held in 1967.

2. On the physical-biological discontinuity

2.1 Cooperative phenomena in quantum physics and biology

As with biological organization generally, numerous features of brain organization are qualitatively reminiscent of the correlation effects called cooperative phenomena in quantum physics. In the latter field, such effects are associated with the so-called exchange force interactions which are not seen as forces like those of classical physics.

Such quasi forces are illustrated by considering two independent electrons of a single atom. A state function for independent elements would be in the form of the product of probabilities for the individual elements. The form of such a function is symmetric since the product pair is commutable. It turns out, however, that an antisymmetric form is required in order to account for the behavior of electrons in terms of their conformity to the Pauli exclusion principle. Particularly, if the probability functions for each element are equal, then the antisymmetric equation has 0 as its solution. Thus, in its simplest form, the exclusion principle says that the probability is zero for a composite state to be constructed from identical individual states.

The point at issue here is that in this instance we encounter a form of symmetry breaking which can not be explained in terms of ordinary dynamics; the phenomenon is cooperative rather than being mechanistically determined. Although the details are clearly different, it seems apparent that biological organizations, in general, cannot be understood outside the epistemology of cooperative effects. As a straight-forward example, we note the case where a large population of neurons with variable action potential frequencies will synchronize their firing rates to that of the fastest rate in the population. A probabilistic state function for such populations would be antisymmetric in form although on a priori considerations it should be symmetric. It does not seem useful to consider such forms of symmetry breaking as being explicable in terms of ordinary neurodynamics.

As a further illustration of apparent parallels between quantum physics and brain organization, we may consider the physical situation in which operators are associated with state observables. The operators are considered to generate eigenvalues and associated eigen functions. In this context we can represent by $\psi(x)$ the amplitude of the biochemical gradients associated with the development of the neuronal action potential. The growth of $\psi(x)$ may be considered as being continuous until a critical value is reached, this value generating the full action potential. We may ask if the amplitude now drops out of consideration. But at this point it seems natural to introduce considerations related to the action potential frequency. To this end we introduce an iterative operator, Q, with values $0, 1, \cdots, q, \cdots, n$. Writing q for an eigenvalue associated with Q, we can then write $Q\psi(x) = q\psi(x)$ to express the repeat rate of the action potential. This seems a quite natural way to connect the amplitude of the biochemical gradients and the action potential frequency. In particular, it makes apparent the need to determine the local events which specify Q in relation to events which specify $\psi(x)$.

It would of course be extravagant to think that the apparatus of quantum physics will apply wholesale to brain dynamics. In this regard, the disposition adopted here may be expressed along the following lines. Fundamental properties of biological organization, such as self-replication, self-repair and adaptive control, betoken cooperative phenomena precisely in the sense, though not necessarily in the specific forms, of quantum physics. That is to say, we have not been successful in explaining them in terms of principles and laws known to classical physics. This amounts to our facing the possibility of a fundamental discontinuity in nature. If we consider quantum physics as being characterizable as the physics of cooperative phenomena, then the presumed continuity of nature might appear the more discernible if biological cooperations are treated in quantum terms. The brain is an especially useful starting point for such attempts. In one direction it connects biological material with mental phenomena. In the opposite direction it invites us to ask whether a language, that of quantum physics, which applies to the fundamental substrate of all materials, the atom, will apply also to the brain considered as one of the most extreme forms of specialization among the biological materials. Metaphorically speaking, the issue is whether the grammar of the brain has important similarities with the grammar of the atom.

2.2 Biological union and partitioning

It is obviously useless to talk about qualitative parallels. Moving in the direction of quantities, we are struck by the characterization of biological materials as yielding instances of many-body problems. It has been estimated, for example, that a typical cell involves stable patterns of organization among some 10^{15} molecular species. The brain, with some 10^{10} neurons, and vast numbers of non-neuron bodies, involves a collective of molecular species of incredible magnitude. In view of the many-body character of the brain, and the associated dynamics, it would seem useful to approach specific problems of brain dynamics in terms of the numbers of elements involved in collective behavior and to regard individual elements as dropping out of consideration in favor of numerous occupied equivalence classes.

Equivalence classes provide a natural basis for discussing two operations that are fundamental to biological materials. These are the operations of union and partitioning. Union may be considered as equivalent to biological combination, and partitioning as equivalent to the processes by which individual classes of entities, including interactions, are established. Examples of union are provided in syngamy and neurological synapses; partitioning is expresses in phenomena like meiosis and mitosis as well as in the "decomposition" of chromosomes into gene loci and the "decomposition" of the brain into correlated populations of neurons. From this point on the discussion will be restricted to the topic of partitioning with the understanding that union is the inverse of partitioning.

2.3 Choice of partitioning functions

If we have a set of state variables and an aggregate whose elements can assume any of the possible values of the state variables, then the possible values define the equivalence classes of the aggregate. The number of possible arrangements of the N elements in the aggregate over the g equivalences classes is g^N and the probability for any particular arrangement is g^{-N}. Expressed in combinatorial terms, the probability function becomes

$$\Omega_{MB} = \frac{N!}{n_1!n_2! \cdots n_i!} g^{-N} \tag{1}$$

This is the classical Maxwell-Boltzmann phase space as applied, e.g., to an ideal gas; all values of the state variables are equiprobable.

In a fairly obvious sense, biological stability depends on the union of many aggregates being determined by the internal composition of the aggregates themselves. Biological variation, on the other hand is dependent on the possibility that an aggregate may be partitioned in more than one way. It is clear that there are no absolute isolates in a biological organism and in that sense statements about organization reduce to statements about interactions. The individual elements in the system need not be distinguishable, all we need to know statistically is how many elements are in the equivalence class g_i. The behavior of the aggregate is then identifiable with the behavior of the equivalence classes. This amounts to saying that interactions among the states represented by equivalence classes will be the source of behavior in the total system.

Thus, a description of state for a given system under consideration will refer to the number of states or equivalence classes and it will assign a descriptive character to each state; in addition, it will characterize (possibly on the basis of estimates) the density of the state in terms of the number of elements occupying it, and it will also describe the interactions between the individual states. In a clear sense, the more organized a system, the more definite will be restrictions on the distribution of elements into the various states or equivalence classes. It is in this sense that "disorder" is understood when Ω_{MB} is described as a disorder parameter. Thus, if we have a set of state variables appropriate for describing the behavior of elements in a large aggregate, we would say that the behavior is disorderly if it fits the Maxwell-Boltzmann distribution. All values of the state variables would be equally likely at any given moment of observation. We might wish to say that from the point of view of an observer, who has no basis for knowing the states of a system like the brain immediately before his observation nor for controlling the sensory input into the brain, all possible values of the state variables are indeed equally probable. But, as indicated above, what we are interested in is the probabilistic character of the internal organization of the system. From this point of view, Ω_{MB} represents a freedom from definite restrictions which would be difficult to reconcile with the concept of an organized system.

The disorder we have associated with a Maxwell-Boltzmann system can be expressed in terms related to the notion of symmetry discussed earlier. Suppose we have a large space occupied by an aggregate of elements each with a particular momentum. We suppose the various momenta to induce a partitioning of the aggregate into equivalence classes each defined by a particular momentum. Each class g_i will then be occupied by a particular number n_i of elements, so we can say that the total aggregate of N elements will be expressed by an array of occupancy numbers $n_1 + n_2 + \cdots + n_m = N$. Such an array results from the placement of elements into the different classes, but nothing is said about the number of possible placements g^N. Suppose all placements are equally probable, as in the Maxwell-Boltzmann system. Then an observer who has a translatable observation post will not be able to distinguish the configuration seen at an arbitrary observation point, 1, from that seen at a different point, 2. More generally, the system will be invariant with respect to an infinity of spatially translated observations. So long as the system remains in equilibrium, this invariance will apply also to observations taken at different instants in time. The various translations of the observation post amount to operations performed on the observed system. It is generally understood that a symmetry operation is one which sends a geometric body back into itself. That is to say, an entity is symmetrical with respect to a given operation if subjecting it to that operation does not change its appearance. The Maxwell-Boltzmann configuration is then seen to be symmetrical with respect to translatory observation.

It is also noteworthy that the notion of invariance is entailed by symmetry considerations. In the most general terms, we can say that for every symmetry there is an associated invariance. From this we can see that in physical situations the invariant could well be a quantity obtained from measurement procedures. The physical entity which corresponds to this quantity would thus be conserved. If an existing symmetry is broken by some spontaneous natural process, then a previously conserved quantity would no longer be conserved.

If we set Ω_{MB} as our standard disorder parameter, then a quantitative reduction in disorder will be obtained if we reduce the number of alternative possibilities for constructing the array of occupancy numbers. This would arise if we had a shift from Ω_{MB} to some Ω_i in which the expression

$$\frac{\Omega_{MB}}{\Omega_i} = x > 0 \qquad (2)$$

is a true statement. The above expression would represent an instance of symmetry breaking and so we might expect that in consequence some quantity observed under the Ω_{MB} situation is no longer conserved; we should expect that quantity now to be reduced. If we take Ω_{MB} as the disorder parameter in Boltzmann's statistical mechanics equation $S = k \log \Omega_{MB}$, where S is the entropy, we see that dividing both sides by our Ω_i necessarily forces a reduction in the entropy S. This is the same thing as saying that where the relationship given in (2) obtains then we have an entropy S_i associated with Ω_i, an original entropy S_{MB}, and a reduction in the latter defined as

$$S_j = S_{MB} - S_i. \tag{3}$$

We now have to account for the appearance of the quantity $x > 0$ in the equation (2). The quantity x represents a growth of some kind precisely in the sense that it made its appearance when we divided Ω_{MB} by Ω_i. This growth balances the reduction expressed in (3) by the arithmetic subtraction. But if Ω_{MB} is a disorder parameter, then the quantity which grows, as the magnitude of Ω lessens, will be order itself. That is to say, ordering can be no less viable as a purely physical quantity than the entropy. In particular, ordering is the reciprocal of the entropy. If we use the word 'information' to stand for the word 'ordering' as defined above, then writing I for information, we have

$$I = S^{-1} \tag{4}$$

This expression indicates that there is a definite and measurable quantity, information, whose magnitude is always the reciprocal of the entropy: $IS = 1$, and this relationship is universally constant.

The Second Law of thermodynamics is generally understood as expressing the principle of the growth of entropy. Because of this, it is often thought that biological organizations represent local and temporary suspensions of the Second Law. The expression (4), however can be understood to mean that entropy is translated whenever I increases. From a somewhat different point of view, it seems natural to regard an increase in information, ΔI, as a discrete quantity in the sense that the differential dI would have no obvious meaning. Suppose we sum all the ΔI's until we have $\Sigma \Delta I = K$ and $\Sigma \Delta_{-1} S = 0$. The second law can then be interpreted as saying that the situation where $\Sigma \Delta_{-1} S = 0$ can never be obtained, which is the same thing as saying that a completely ordered or informed universe cannot exist.

In quantum physics we find two additional phase spaces which depart in opposite directions from the classical Maxwell-Boltzmann distributions. The latter may be regarded as the limiting case for the former distributions. Fermi-Dirac statistics describe probabilities which increase in favor of those distributions in which each equivalence class has at least one element in it but where no class is packed. This kind of distribution provides a natural form of expression for the exclusion principle. Related principles appear to operate in the singularity of antigen-antibody coupling, species mating exclusion, and neuronal lateral inhibition. In general the Fermi-Dirac distributions may be considered as representing a situation like those described in the Maxwell-Boltzmann systems but where in addition some form of exclusion phenomenon operates. In a similar vein, the second non-classical statistics, the Bose-Einstein, represents situations where a Maxwell-Boltzmann system is acted on or influenced by condensation phenomena. Here the probabilities increase for distributions in which some states, or equivalence classes, are packed and other possible states are either empty or nearly so. This form of condensation may be fundamental to the specification of nucleotide occupancy patterns in genetic material, and it seems to provide a natural form of representation, at least qualitatively, for neuron recruitment and action potential rate shifts. Umezawa has proposed this model for a neurodynamics account of memory (Ricciardi & Umezawa 1967).

Thus, if the "grammars" of biological organization are to be considered as having "rules" comparable to those of the atom, then it would seem that we must assume two restraints from the outset. The first is that we are concerned with "grammars" for many-body problems. The second is that Fermi-Dirac and Bose-Einstein "grammars" are more likely to describe biological systems, when these are considered as aggregates with state equivalence classes, than is the Maxwell-Boltzmann "grammar" of unrestricted disorder.

Clearly, in comparison with Ω_{MB}, the corresponding Fermi-Dirac and Bose-Einstein quantities, Ω_{FD} and Ω_{BE}, represent symmetry breaking situations. Each of them, in comparison with the Ω_{MB} systems, reduces the number of alternative possibilities for the assignment of occupancy numbers and thereby increases the probability of a particular arrangement of the numbers. With this increase in probability we have a corresponding ΔI.

To illustrate the increase in probability, we can consider a situation in which we have $g = 5$ classes and $N = 3$ elements. The number of alternative assignments under the Bose-Einstein model is

$$\Omega_{BE} = \frac{(g+N-1)!}{(g-1)!N!} \tag{5}$$

The corresponding expression for the Fermi-Dirac model is

$$\Omega_{FD} = \frac{g!}{N!(g-N)!} \tag{6}$$

with the stipulation that each class occupancy number equals 1 or 0.

For the values $g = 5, N = 3$, we then obtain the probabilities $\Omega_{MB}^{-1} = 0.05$, $\Omega_{BE}^{-1} = 0.03$, $\Omega_{FD}^{-1} = 0.1$. The corresponding ΔI's are obtained from $0.05 \div 0.03 = 1.67$ for the Bose-Einstein case, and $0.05 \div 0.1 = 0.50$ for the Fermi-Dirac case.

We now want to see if we can find reasonably clear a-priori grounds for believing that the brain, considered as an aggregate of neurons, might have informational characteristics which can be represented by Bose-Einstein or Fermi-Dirac quantities. The subsequent discussion is limited to the case of Bose-Einstein quantities.

3. Some assumptions about the brain

The neuron is an assembly of dynamics exchange points of types p and s, the p-points being associated with intraneuron action potential dynamics and the s-points with intraneuron synapsing dynamics.

There is no definite spatial boundary for the neuron since the point types p and s are of a transmembrane character. Representationally, the neuron is thus depictable as a somewhat fuzzy circle of dots.

The number of s-points on a single neuron may be very large with some points having excitatory function $+s$ and some with inhibitory function $-s$. The total number of combinatorial sets of neurons is thus very much larger than the total number, $\overline{N}$, of neurons.

$\overline{N}$ is approximately constant across the individuals of a given species, and in the normally functioning individual the loss rate of neurons is negligible in relation to the size of $\overline{N}$.

Genetic mechanisms partition $\overline{N}$ into morphologically distinct equivalence classes. The neuron occupancy numbers for these classes, then arranged in order of their size, are described by a step function. In addition, genetic mechanisms determine the arrangement of neurons into distinct tracts which are anatomically characteristic for the species. In these tracts we find restrictions on the number of possible connections between neurons.

Relatively transient concatenations of neurons are induced as a consequence of the individual's exchanges with the environment. These are considered as non-genetically determined patterns of inter-neural excitation and inhibition.

Concatenation may be considered as a binary operation in which a left member joins a right member to form a sequence of neurons coupled by dynamic excitations. If neurons x_1 and x_2 are in concatenation, then the pair (x_1x_2) may concatenate with neuron x_3 to form the sequence $(x_1x_2)x_3$. Provided the excitation order of the elements is unchanged, concatenation thus permits the identity $(x_1x_2)x_3 = x_1(x_2x_3)$. That is to say, the operation of excited coupling is a binary composition which satisfies the algebraic association law. Thus, excited coupling of neurons has the properties of a semi-group structure.

However, we have noted that not all pairings are anatomically possible. This leaves us with the possibility of finding not only many semi-groups but also different types of semi-group in the brain. It also leaves us with the need to extend our initial apparatus.

From now on, we shall use 'joined' to indicate that a set of neurons is physically arranged so as to permit excited or inhibitory coupling, as required by the context. From the density of $+s$ points available in the total neuronal population we may assume that each neuron may be joined to a variable $k-1$ other neurons. This permits multiple coupling in more than one spatial direction instead of the simple concatenation already discussed. We also note that the coupling is a time consuming event. It will be convenient to refer to such a spatio-temporal coupling pattern of n neurons as a synergy.

For a given population of N joined neurons, the presence of multiple $+s$ points and $-s$ points, when $|s| \gg N$, ensures $g_1, g_2, \cdots, g_m$ synergies as distinct possible entities. For exactly the same reasons, we can assume that the various synergies will have variable occupancy numbers including zero. We might also note that what was previously said about the semi-group structure still obtains only the situation has become one of added complexity.

It will, however, be clear that a synergy, g, is here considered as an equivalence class where we are concerned with the distribution of occupancy numbers of neurons to a set g_i of synergies possible in a joined population of N neurons. We can represent the classes by spaces between bars, and the occupancy numbers by marks of some kind placed between bars. There will be $g+1$ bars, the sequence beginning and ending with a bar. The N marks for elements will be placed between pairs of bars in arbitrary order and with a variable number of such marks, including zero, between any pair of bars. Since the sequence must begin and end with a bar, there are thus $g-1$ bars which can be placed in arbitrary order. The number of distinguishable distributions then equals the number of ways of selecting N places out of $g+N-1$. This yields $g(g+N-1)!$. But the number of permutations of classes and elements is $g!n!$, so for distinguishable arrangements we must divide by that quantity. The total number of distinguishable arrangements is then

$$\Omega = {}_{g+N-1}C_N = \frac{g(g+N-1)!}{(g-1)!N!} = \frac{(g+N-1)!}{(g-1)!N!} \tag{7}$$

so that $\Omega = \Omega_{BE}$.

Thus, on the basis of the neurological assumptions presented, only some of which bear on the final result $\Omega = \Omega_{BE}$, we find grounds for assuming that brain organization will exhibit characteristic states which are represented by Bose-Einstein quantities.

Only static combinatorics have been considered so far, but dynamic implications will be obvious. We can see, for example, in the motor epileptic seizure a breakdown in the cooperative information implied by Ω_{BE} into a state which more closely approximates a localized Ω_{MB} condition. More generally, we can assume that a given synergy will involve several action potential rates, including zero. In which case the synergy can be represented as an arrangement of occupancy numbers for classes representing the different frequencies $f_0, f_1, \cdots, f_k$. The occupancy numbers would then correspond to the amplitude of each frequency, yielding a coupling spectrum for the synergy. The life histories of these spectra are especially important for our understanding of brain dynamics. In particular, we have to account for the spectrum induced by an immediate sensory perception becoming condensed into a highly stable spectrum necessary for the memory of the external event. Memory is especially significant because it is here that we find the basis for what we call mental phenomena. It is thus particularly noteworthy that Umezawa's account of memory, in terms of brain dynamics, is in the grammar of Bose-Einstein condensation quantities.

4. Purpose and theoretical meaning

Developing a model of fundamental processes of the brain (i.e., brain dynamics) within the realm of quantum field theory has the following three purposes:

- Propose the zeroth order approximation to brain dynamics.
- Embody the Ricciardi-Umezawa conjecture.
- Provide a step for further refinement.

Physiological details are not held fast at this stage. For achieving these purposes, we will build up a quantum field theoretic model of brain dynamics based on the hypotheses:

- Only one kind of neurons exist.
- A neuron can be represented by the spin operator (called corticon).
- Neurons are capable of emitting or absorbing bosons (called stuartons).

By these hypotheses, a physical picture of an observed neuron can be illustrated as a bare corticon surrounded by stuartons. Such a situation is characteristic to quantum field theory with interaction, and familiar in relation with the notion of renormalization in the Lee model or quantum electrodynamics.

Just as quantum field theory describing the interaction between electrons and photons in the universe is called quantum electrodynamics and abbreviated as QED, quantum field theory describing the interaction between corticons and stuartons in the brain will be called quantum brain dynamics and abbreviated as QBD.

5. Quantum Brain Dynamics – QBD

We start QBD with the Hamiltonian

$$H \equiv \sum_{k=1}^{?} K_k \left\{A_k^*(t)A_k(t) + B_k^*(t)B_k(t)\right\}$$
$$+\sum_{i=1}^{N}\sum_{k=1}^{?} V_k^{(i)} \left\{\tau_+^{(i)}(t)\left(A_k(t) + B_k^*(t)\right) + \tau_-^{(i)}(t)\left(A_k^*(t) + B_k(t)\right)\right\} \quad (8)$$

where $i = 1, 2, \cdots, N \approx 10^{10}$ denotes the numeric suffices to identify each corticon, and $k = 1, 2, \cdots, ?$ denotes the numeric suffices to identify each mode of information transfer carried by a stuarton. A_k, B_k stand for the annihilation operators of a stuarton in the k-th mode, and conjugate operators A_k^*, B_k^* are creation operators. Those operators satisfy the canonical commutation relation for the Bose field, and for each $k = 1, 2, \cdots, ?$, the eigenvalues of $A_k^*A_k$ and $B_k^*B_k$ are found to be $0, 1, 2, \cdots, \infty$.

$\tau_+^{(i)}, \tau_-^{(i)}$ are the creation and annihilation operators for i-th corticon (i.e., the spin variable of the i-th neuron), and satisfy the spin commutation relations

$$\left[\tau_+^{(i)}, \tau_-^{(j)}\right] = 2\delta_{ij}\tau_3^{(i)} \quad (9)$$

$$\left[\tau_3^{(i)}, \tau_\pm^{(j)}\right] = \pm\delta_{ij}\tau_\pm^{(i)}. \quad (10)$$

Furthermore, $V_k^{(i)}$ is a coupling constant representing the strength of the coupling between the i-th corticon (i.e., neuron) and the k-th stuarton (i.e., information transfer mode). Just as the coupling constant between the electron and the photon is called 'electric charge', this coupling constant may be called 'neural charge'.

The Hamiltonian (8) happens to be invariant under the continuous transformation of the creation and annihilation operators for corticons and stuartons

$$\left.\begin{aligned} A_k &\to A_k e^{i\theta} \\ B_k &\to B_k e^{-i\theta} \end{aligned}\right\} \tag{11}$$

$$\left.\begin{aligned} \tau_+^{(i)} &\to \tau_+^{(i)} e^{-i\theta} \\ \tau_-^{(i)} &\to \tau_-^{(i)} e^{i\theta} \end{aligned}\right\}, \tag{12}$$

and the generator of the transformation

$$L_3 \equiv \sum_{k=1}^{?} \left(A_k^* A_k - B_k^* B_k\right) + \frac{1}{2} \sum_{i=1}^{N} \tau_3^{(i)} \tag{13}$$

becomes a conserved quantity.

5.1 Degenerate ground states and memory

In the Heisenberg picture of quantum field theory, the ground state (i.e., the minimum energy eigen state of the Hamiltonian) is given by the time-independent solutions to the Heisenberg equation. In QBD they are

$$A_k = -\frac{1}{K_k} \sum_{i=1}^{N} V_k^{(i)} \tau_-^{(i)} \equiv \alpha_k \tag{14}$$

$$B_k = -\frac{1}{K_k} \sum_{i=1}^{N} V_k^{(i)} \tau_+^{(i)} \equiv \beta_k \tag{15}$$

$$\tau_\pm^{(i)} = e^{\pm i\chi} \tag{16}$$

$$\tau_3^{(i)} = 0. \tag{17}$$

Since χ does not depend on the corticon suffix i, the spin variables of all the neurons are pointing in one and the same direction. The total number of stuartons in this ground state is given by

$$\begin{aligned} \overline{N} &= \sum_{k=1}^{?} \left(A_k^* A_k + B_k^* B_k\right) = \sum_{k=1}^{?} \left(\alpha_k^* \alpha_k + \beta_k^* \beta_k\right) \\ &= 2 \sum_{k=1}^{?} \sum_{i,j=1}^{N} \frac{1}{K_k^2} V_k^{(i)} V_k^{(j)}. \end{aligned} \tag{18}$$

As the number of stuartons surrounding the i-th corticon

$$n^{(i)} = 2 \sum_{k=1}^{?} \frac{1}{K_k^2} V_k^{(i)} V_k^{(j)} \tag{19}$$

denotes the number of information transfer modes available in the ground state for the i-th neuron, we interpret it as the number of active synapses.

5.2 Nambu-Goldstone bosons and memory retrieval

The ground state given by (14)–(17) is not invariant under the continuous transformations (11) and (12) which make the Hamiltonian (1) unchanged. This is the ordered state of neurons proposed by Ricciardi and Umezawa, and the uniform directionality of all the spin variables represents the memory storage (Ricciardi & Umezawa 1967).

The symmetry broken spontaneously in the ground state is restored by the emergence of Nambu-Goldstone bosons, that is, by turning the direction of each neuron's spin variable in the plane perpendicular to the third axis. Small oscillations of the spin variables around the ground state manifest a long-range correlation wave. Such a creation of long-range correlation waves in the ground state would correspond to the recall of memory in the Ricciardi-Umezawa conjecture.

Since the ground state is of spontaneous symmetry breaking type, the excitation energy levels of the system become double-structured in QBD described by spin variables and stuartons. In the first group the energy level is an eigen state of the radial oscillation in the plane perpendicular to the third axis with a finite eigen frequency Ω, and the excitation of all the neuron variables directed uniformly in the ordered state requires such a large amount of external energy as $N\Omega$. However, in the second group the energy level is an eigen state of the rotational oscillation along the valley of the interaction potential encircling around the origin in the plane perpendicular to the third axis, and its eigen frequency ω becomes infinitesimal. Therefore, the amount of external energy to excite all the neuron variables remains infinitesimal. Quanta of such oscillations modes distributing continuously around 0 are the Nambu-Goldstone bosons. In this way the memory recall is continuously activated by the incoming slight external energy (stimuli) in QBD.

6. Further problems

Several issues have been pointed out as to be investigated further on the prototype of QBD explained so far above:

- Refinement of the model.
- Determination of critical temperature of the ordered ground state.
- Identification of physiological entities with various quantities appearing in the Hamiltonian (8).
- Detailed investigation of mechanism of memory, learning, recall etc.
- Experimental verification.

Among those issues one of the authors (YT) could refine the QBD model in the intensive discussion with Umezawa and Stuart as is shown in the following subsection. As for the determination of critical temperature of the ordered ground state, important results have been obtained by Umezawa's coworkers in Italy (Del Giudice et al. 1986, 1988). It was the second author (MJ) who derived the Hamiltonian (8) within the fundamental framework of QED applied to the interaction between water and electromagnetic field inside and outside the cell membranes, and developed the QBD based theory of memory, learning and consciousness mechanism (Jibu & Yasue 1995; Jibu et al. 1996). Since those advancements are clearly exposed separately in this volume, we restrict ourselves to the refinement of the QBD prototype model.

6.1 Refined QBD model

Let $\vec{x}_j$ be the position of j-th neuron and $A^*_\alpha(j) = A^*_\alpha(\vec{x}_j), A_\alpha(j) = A_\alpha(\vec{x}_j)$ be the creation-annihilation operators representing the on-off of the α-th information channel connected to the j-th neuron. In the prototype model of QBD we have seen so far, $\alpha = 1, 2$ and the channel 1 has the creation-annihilation operator A^*, A and the channel 2 has B^*, B. For the general case of multi-channel, we derive the creation-annihilation operators for the quanta (i.e., stuartons) of the α-information transfer mode from those creation-annihilation operators representing the on-off of the α-th information channel connected to the j-th neuron by means of Fourier transformations

$$A^*_\alpha(j) = \int d^3k A^*_\alpha(\vec{k}) e^{-i\vec{k}\cdot\vec{x}_j} \tag{20}$$

$$A_\alpha(j) = \int d^3k A_\alpha(\vec{k}) e^{i\vec{k}\cdot\vec{x}_j} \tag{21}$$

As for the spin variable of the j-th neuron, $\tau^{(j)}_\pm = \tau_\pm(j) = \tau_\pm(\vec{x}_j)$ and $\tau^{(j)}_3 = \tau_3(j) = \tau_3(\vec{x}_j)$ are used just as in the prototype model.

Assuming the naturally generalized form of Hamiltonian from (8) the Heisenberg equations become

$$i\frac{d}{dt}A_\alpha(j) = KA_\alpha(j) + V_\alpha(-\nabla^2)\tau_-(j) \tag{22}$$

$$i\frac{d}{dt}\tau_+(j) = 2\tau_3(j)\sum_\alpha V_\alpha(-\nabla^2)A_\alpha(j) \tag{23}$$

$$i\frac{d}{dt}\tau_3(j) = \tau_+(j)\sum_\alpha V_\alpha(-\nabla^2)A_\alpha(j) - \tau_-(j)\sum_\alpha V_\alpha(-\nabla^2)A^*_\alpha(j) \tag{24}$$

where $\nabla^2 = \frac{\partial^2}{\partial x_1^2} + \frac{\partial^2}{\partial x_2^2} + \frac{\partial^2}{\partial x_3^2}$ denotes the Laplacian operator, and $V_\alpha(-\nabla^2)$ is given by the Fourier transformation:

$$V_\alpha(-\nabla^2)A_\alpha(j) = \int d^3k V_\alpha(\vec{k}^2)A^*_\alpha(\vec{k})e^{-i\vec{k}\cdot\vec{x}_j} \tag{25}$$

Here $\vec{k}$ is the wave vector of the "information wave" and $V_\alpha(\vec{k}^2)$ denotes the strength of coupling between the neuron and the information plane wave with the wave vector $\vec{k}$. For example, if the realistic situation would show that no information wave carries high $\vec{k}$, then $V_\alpha(\vec{k}^2)$ should be such that $V_\alpha(\vec{k}^2) \to 0$ for large $\vec{k}$.

6.2 Analysis by change of variables

By the change of variables

$$A(j) \equiv \sum_\alpha V_\alpha(-\nabla^2)A_\alpha(j) \tag{26}$$

$$V(-\nabla^2) \equiv \sum_\alpha V_\alpha(-\nabla^2)V_\alpha(-\nabla^2) \tag{27}$$

Eqs. (22)–(24) can be rewritten:

$$i\frac{d}{dt}A(j) = KA(j) + V(-\nabla^2)\tau_-(j) \tag{28}$$

$$i\frac{d}{dt}\tau_+(j) = 2\tau_3(j)A(j) \tag{29}$$

$$i\frac{d}{dt}\tau_3(j) = \tau_+(j)A(j) - \tau_-(j)A^*(j) \tag{30}$$

In the new variables the continuous transformation (11), (12) becomes

$$\left.\begin{array}{l} A(j) \to e^{i\theta}A(j) \\ A^*(j) \to e^{-i\theta}A^*(j) \end{array}\right\} \tag{31}$$

$$\left.\begin{aligned} \tau_+(j) &\rightarrow \tau_+(j)e^{-i\theta} \\ \tau_-(j) &\rightarrow \tau_-(j)e^{i\theta} \\ \tau_3(j) &\rightarrow \tau_3(j) \end{aligned}\right\} \tag{32}$$

and the Heisenberg equations (28)–(30) remain unchanged.

The variables $A(j), \tau_-(j)$ may be useful to describe the behavior of each neuron, but seem inadequate to describe the macroscopic (i.e., long-range) behavior of all the neurons in the gross. We will make use of certain collective variables. Since the macroscopic behavior of the whole ensemble of neurons manifest most apparently in the ground state with the spontaneous symmetry breaking, we focus on the collective variables in such a ground state.

Around the expectation values of the variables $\langle A(j)\rangle = U, \langle \tau_+(j)\rangle = v, \langle \tau_3(j)\rangle = 0$ in the ground state of spontaneous symmetry breaking type we may introduce the following fluctuation variables in the ground state.

$$A(j) = U + a_1(j) - ia_2(j) \tag{33}$$

$$\tau_+(j) = v + \rho_1(j) + i\rho_2(j) \tag{34}$$

$$\tau_3(j) = 0 + \rho_3(j) = \rho_3(j) \tag{35}$$

Within the Hartree approximation in which the products of fluctuation variables are ignored, the Heisenberg equations (28)–(30) can be rewritten:

$$U = -\frac{V(0)}{K}v \tag{36}$$

$$\frac{d}{dt}a_1(j) = -Ka_2(j) - V(-\nabla^2)\rho_2(j) \tag{37}$$

$$\frac{d}{dt}a_2(j) = Ka_2(j) + V(-\nabla^2)\rho_1(j) + KU + vV(0) \tag{38}$$

$$\frac{d}{dt}\rho_1(j) = 0 \tag{39}$$

$$\frac{d}{dt}\rho_2(j) = -2U\rho_3(j) \tag{40}$$

$$\frac{d}{dt}\rho_3(j) = 2U\rho_2(j) - 2va_2(j) \tag{41}$$

Furthermore, introducing $Q(k^2)$ by

$$Q(k^2) \equiv \sqrt{(K^2+4U^2)^2-16Uv\left\{V(k^2)-V(0)\right\}K(k^2)}$$

and $\omega_1(k^2), \omega_2(k^2), g_1(-\nabla^2), g_2(-\nabla^2)$ by

$$\omega_1(k^2) \equiv \frac{1}{2}\left\{K^2 + 4U^2 - Q(k^2)\right\} \tag{42}$$

$$\omega_2(k^2) \equiv \frac{1}{2}\left\{K^2 + 4U^2 + Q(k^2)\right\} \tag{43}$$

$$g_1(-\nabla^2) \equiv 4U^2 - \omega_1(-\nabla^2) \tag{44}$$

$$g_2(-\nabla^2) \equiv 4U^2 - \omega_2(-\nabla^2) \tag{45}$$

we finally obtain the collective variables.

$$\alpha_1(j) \equiv \frac{1}{v}\frac{-4Uva_1(j) + g_1(-\nabla_j^2)\rho_1(j)}{\omega_2^2(-\nabla_j^2) - \omega_1^2(-\nabla_j^2)} \tag{46}$$

$$\alpha_2(j) \equiv \frac{1}{v}\frac{-4Uva_2(j) + g_1(-\nabla_j^2)\rho_2(j)}{\omega_2^2(-\nabla_j^2) - \omega_1^2(-\nabla_j^2)} \tag{47}$$

The corresponding field variables $\alpha_1(\vec{x}, t), \alpha_2(\vec{x}, t)$ are subject to the field equations

$$\frac{\partial^2}{\partial t^2}\alpha_1 = -\omega_1(-\nabla^2)\alpha_1 \tag{48}$$

$$\frac{\partial^2}{\partial t^2}\alpha_2 = -\omega_2(-\nabla^2)\alpha_2 \tag{49}$$

and the quanta of the variable α_1 comes to be the Nambu-Goldstone bosons because $\omega_1(0) = 0$. Namely, the minimum value of energy spectrum is 0, and the long-range correlation wave is created only for lower frequencies. On the other hand, the variable α_2 is no longer the Nambu-Goldstone mode but a quasi particle mode, because $\omega_2(0)$ differs from 0.

We call the quanta of the variable α_2 quasi neurons. Then, the variables in the refined QBD model would be represented by the Nambu-Goldstone variable α_1 and the quasi neuron variable α_2.

$$A(j) = F_A\left[\alpha_2, \partial\alpha_1\right] e^{i\alpha_1(j)} \tag{50}$$

$$\tau_+(j) = F_+\left[\alpha_2, \partial\alpha_1\right] e^{-i\alpha_1(j)} \tag{51}$$

$$\tau_3(j) = F_3\left[\alpha_2, \partial\alpha_1\right] \tag{52}$$

Here, $\partial\alpha_1$ stands for any combination of the derivatives of the Nambu-Goldstone variable α_1, and $F_A\left[\alpha_2, \partial\alpha_1\right], F_+\left[\alpha_2, \partial\alpha_1\right], F_3\left[\alpha_2, \partial\alpha_1\right]$ are certain functions of $\alpha_2, \partial\alpha_1$.

6.3 Nambu-Goldstone bosons and memory

It is worthwhile to notice that the Nambu-Goldstone variable enters in the canonical transformation (50)–(52) only through the phase factor. The ground state of spontaneous symmetry breaking type is an ordered state in which the spin variables of all the neurons are pointing in one and the same direction, and

Nambu-Goldstone bosons are generated to form a condensation. This condensation of Nambu-Goldstone bosons is characterized by the fact that the expectation value of the total number of the Nambu-Goldstone bosons does not vanish.

$$\langle \alpha_1^* \alpha_1 \rangle = C^2 \neq 0 \tag{53}$$

C in the right hand side is an order parameter, and understood to manifest only tempered spatial and temporal changes when it describes the condensation as a classical field. Therefore, we can assume $\partial \alpha_1 = 0$ holds approximately, and the existence of the condensation of the Nambu-Goldstone bosons in the ordered state is found to appear as the phase modulation in the neuron variables $\tau_{\pm}$, τ_3 and the stuarton variables A. And Eq. (51) shows how the bare neurons are surrounded by a cloud of Nambu-Goldstone bosons.

In the refined QBD model, memory code is binary: 'yes' for all spin variables aligned, and 'no' for not aligned. When we have an uncoded system, then certain external stimulus, which is able to influence the Nambu-Goldstone mode, can print the yes-code. A similar external stimulus can also recall the memorized code. More precisely, a region in which all the spin variables of neurons are aligned in one and the same direction forming the ground state of spontaneous symmetry breaking type corresponds to a bit representing 'yes', and that in which the spin variables of neurons are not aligned forming the ground state with symmetry corresponds to a bit representing 'no'. A physical process which changes a region from the 'no' state into the 'yes' state triggered by certain external stimulus achieves the memory printing of the external stimulus. Similar external stimulus propagated to this coded region and the physical process of creation of the Nambu-Goldstone bosons due to this energy perturbation achieves the memory recall of the external stimulus already printed in this region as a 'yes' code.

6.4 Formulation as non-equilibrium states

Memory code is binary even in the refined QBD model: One 'bit' of information corresponds to one phase transition in the region of ground state. In this sense, the memory capacity in the QBD model would not be much larger in order than that in the classical model of neural networks with mutual synapse connections. Therefore, further theoretical development of QBD will be needed to cover the huge memory capacity in the real brain system. For example, such a development can be realized within the theoretical framework in which the phase transition is described in terms of a time-dependent non-equilibrium state. Recently, Vitiello and coworkers proposed the dissipative QBD model within the realm of non-equilibrium thermofield dynamics (TFD), and obtained an important result that the memory code in a region is no longer binary but manifests infinitely many variety (Pessa & Vitiello 1999).

In our workshop in 1975, the development of the QBD model in the direction to include the time-dependent dynamics of non-equilibrium states was discussed, too. We will show it briefly.

In terms of the creation-annihilation operators of stuartons with the wave vector $\vec{k}$ for the α-th information transfer mode, the free Hamiltonian becomes

$$H_0 = \sum_{\alpha} \int d^3k \varepsilon_\alpha(\vec{k}) A^*_\alpha(\vec{k}) A_\alpha(\vec{k}) \tag{54}$$

As the interaction Hamiltonian describing the action of external stimulus to the brain system, we assume the following form inspired by Hebb's hypothesis of self-organization through changing synaptic weight.

$$H_1 = \sum_{\alpha\beta} \int d^3k d^3q \delta V_{\alpha\beta}(\vec{k}, \vec{q}, t) \left\{ A^*_\alpha(\vec{k}) A_\beta(\vec{q}) + A^*_\beta(\vec{q}) A_\alpha(\vec{k}) \right\} \tag{55}$$

Response of the QBD non-equilibrium state to the time-dependent external stimulus $\delta V_{\alpha\beta}(\vec{k}, \vec{q}, t)$ can be described in the standard framework of quantum statistical mechanics by the density matrix

$$\rho(t) = e^{i(H_0+H_1)t} e^{-\frac{1}{k_B T} H_0} e^{-i(H_0+H_1)t} \tag{56}$$

Here, T denotes the temperature and k_B the Boltzmann constant. With this density matrix $\rho(t)$ non-equilibrium oscillations of the response of the system to the external stimulus can be represented as

$$\rho_\alpha(t) = \int d^3k \frac{Tr\left[A^*_\alpha(\vec{k}) A_\alpha(\vec{k}) \rho(t)\right]}{Tr\rho(t)}. \tag{57}$$

From the technical point of view of theoretical physics, quantum statistical mechanics of non-equilibrium states of such a system with spontaneous symmetry breaking as QBD requires the TFD formulation, not the conventional framework with the density matrix (Umezawa 1993). The reader can find a detailed exposition of such TFD formulation of QBD by Vitiello in this volume.

7. Outlook

We presented a brief exposition of the theoretical framework of QBD discussed in the monumental workshops in 1975. After the intensive discussions on the quantum field theoretical model over a few weeks, our colleague Kreuzer conducted a seminar in which he revisited the QBD model from the point of view of brain physiology, giving proper directions to develop further the QBD model and pointed out the necessity of experimental verifications. Fortunately, recent investigations on QBD seem to be quite along with the direction Kreuzer showed, and it may be worthwhile to close the present paper by playing his comments back.

The basic idea of QBD is that the formation of stable memory is associated with certain phase transition phenomena in the brain. Therefore, it will be impossible to verify QBD experimentally unless we might develop a concrete model of QBD based on physics of the real fundamental constituents of the brain tissue. For helping the future development in this direction, we point out here some of the physical properties of the brain tissue which might be tightly related to QBD.

As the fundamental constituents of the brain tissue, we have

- electric dipole layers: membranes
- permanent and induced electric dipoles: ions
- electric charges: ions
- electric currents: ionic currents
- visco-elastic medium with a degree of mechanical instability
- very complex molecular structure.

These are reminiscent of a ferro-electric medium. Ferro-electrica show a phase transition below Curie temperature T_c, where electric dipoles (i.e., spin variables of the neuron in QBD) are aligned within small domains.

Due to complex molecular structures, many ferro-electrica show many phase transitions showing up in the dielectric constant, for example. It may be possible, therefore, to verify the existence of those many phase transitions by measuring the dielectric constant of the brain tissue over small temperature region. Such phase transitions of many ferro-electrica in the brain imply the memory printing and recalling processes due to hysteresis:

Assume external stimulus results in local temperature drop in a given polarization domain sufficient to trigger a phase transition. This is the memory printing in QBD. If hysteresis is large no small temperature increase can bring the state back, and the polarization domain holds stable (long-term) memory.

If hysteresis is very narrow memory can be wiped out by any small temperature increase, and the polarization domain holds unstable (short-term) memory.

If those phase transition phenomena of the brain tissue as a ferro-electric medium were essential in the realistic QBD model, accommodation of large amounts of information would be possible, for example, in many phase transitions due to complex nature of brain matter, and in different polarization domains within the brain tissue. The latter possibility may force us to look for local phase transitions in the membrane structure that might be highly related to the yet unknown mechanism of anesthesia (Jibu et al. 1998). An important result in the theory of anesthesia has been obtained within the realm of QBD (Jibu 2001).

References

Del Giudice, E., Doglia, S., Milani, M., & Vitiello, G. (1986). Electromagnetic and spontaneous symmetry breaking in biological matter. *Nuclear Physics, B275 [FS17]*, 185–199.

Del Giudice, E., Preparata, E., & Vitiello, G. (1988). Water as a free electric dipole laser. *Physical Review Letters, 61*, 1085–1088.

Jibu M. (2001). Theory of cell membrane organizers and pressure reversal of anesthesia. *Medical Hyposethes, 56*, 26–32.

Jibu, M., Hagan, S., Hameroff, S. R., Pribram, K. H., & Yasue, K. (1994). Quantum optical coherence in cytoskeletal microtubules: Implications for brain function. *BioSystems, 32*, 195–209.

Jibu, M. & Yasue, K. (1995). *Quantum brain dynamics – An introduction.* Amsterdam: John Benjamins.

Jibu, M., Pribram, K. H., & Yasue, K. (1996). From conscious experience to memory storage and retrieval: the role of quantum brain dynamics and boson condensation of evanescent photons. *International Journal of Modern Physics, B10*, 1745–1754.

Jibu, M., Yamada, T., Yasue, K., & Hirakawa, M. (1998). Effect of anesthetics on the self-sustained oscillation in an artificial membrane induced by repetitive conformational change of DOPH molecules between hydrophilic and hydrophobic phases. *Journal of Anesthesia, 12*, 62–68.

Pessa, E. & Vitiello, G. (1999). Quantum dissipation and neural net dynamics. *Bioelectrochemistry and Bioenergetics, 48*, 339–342.

Pribram, K. H. (1991). *Brain and perception.* New Jersey: Lawrence Erlbaum.

Ricciardi, L. M. & Umezawa, H. (1967). Brain and physics of many-body problem. *Kybernetik, 4*, 44–48.

Stuart, C. I. J. M., Takahashi, Y., & Umezawa, H. (1978). On the stability and non-local properties of memory. *Journal of Theoretical Biology, 71*, 605–618.

Stuart, C. I. J. M., Takahashi, Y., & Umezawa, H. (1979). Mixed-system brain dynamics: neural memory as a macroscopic ordered state. *Foundation of Physics, 9*, 301–327.

Takahashi, Y. & Umezawa, H. (1975). Thermo field dynamics. *Collective Phenomena, 2*, 55–80.

Umezawa, H. (1993). *Advanced field theory: micro, macro and thermal physics.* New York: American Institute of Physics.

Chapter 17

The dissipative brain

Giuseppe Vitiello
Università di Salerno, Italy

I review the dissipative quantum model of brain and discuss its recent developments related with the rôle of entanglement, quantum noise and chaos. Some comments on consciousness in the frame of the dissipative model are also presented. Dissipation seems to account for the medial character of consciousness, for its being in the present (the Now), its un-dividable unity, its intrinsic subjectivity (autonomy). Finally, essential features of a conscious artificial device, if ever one can construct it, are briefly commented upon, also in relation to a device able to exhibit mistakes in its behavior. The name I give to such a hypothetical device is Spartacus.

1. Introduction

I have been always attracted by the unitary conception of knowledge, which is ever present in some streams of our cultural inheritance. Perhaps, one of the most vivid expressions of such a comprehensive view of the world has been provided by Titus Lucretius Carus in his *De rerum natura.* There are no territories forbidden to our search and there are no separate domains of knowledge,

> …we must not only give a correct account of celestial
> matter, explaining in what way the wandering of the sun
> and moon occur and by what power things happen on earth.
> We must also take special care and employ keen reasoning
> to see where the soul and the nature of the mind come from,…
>
> (Titus Lucretius Carus, 99–55 B.C.)

Therefore, searching in territories not traditionally explored by physicists should not require special justifications.

In this paper I would like to review briefly the extension of the Ricciardi and Umezawa (1967) quantum model of the brain to the dissipative dynamics. For sake of brevity, I will mostly present results rather than their derivations.

The interested reader may find formal details in the quoted literature. I will comment on consciousness in the dissipative quantum model frame and finally I will briefly discuss the essential features which an artificial conscious device should have, if ever it will be possible to construct it, and its relation with a device able to exhibit mistakes in its behavior. (For a recent review on quantum theory and consciousness see Atmanspacher 2004.)

The extension of the quantum model to the dissipative dynamics is required in order to solve the *overprinting* problem, namely the fact that in the Ricciardi and Umezawa model (Ricciardi & Umezawa 1967) the memory capacity is extremely small: any successive memory printing overwrites on the previously recorded memory.

The proposed solution (Vitiello 1995) relies on two facts. One is that the brain is a system permanently coupled with the environment (an "open" or "dissipative" system). The other one is a crucial property of quantum field theory (QFT), i.e. the existence of infinitely many states of minimal energy, the so called vacuum states or ground states. On each of these vacua there can be built a full set (a space) of other states of nonzero energy. We have thus infinitely many state spaces, which, in technical words, are called "representations of the canonical commutation relations". I will refer to the collection of all these spaces or representations as the "memory space". The vacuum of each of these spaces is characterized by a specific ordering and is identified by its *code*, which is the value of the "order parameter", the macroscopic observable characterizing indeed the ordering present in that vacuum. Vacua (or the corresponding spaces) identified by different codes are "distinct" vacua in the sense that one of them cannot be reduced (transformed) into another one of them. In technical words, they are "unitary inequivalent vacua" (or unitary inequivalent spaces or representations).

In the dissipative quantum model of brain the vacuum code is taken to be the memory code. A given memory is represented by a given degree of ordering. A huge number of memory records can be thus stored, each one in a vacuum of given code. In the original model by Ricciardi and Umezawa only one vacuum is available for memory printing. In the *dissipative* model all the vacua are available for memory printing.

2. Broken symmetry and order

In the quantum model a crucial role is played by the mechanism of "the spontaneous breakdown of symmetry" by which the invariance (the symmetry) of the field equations *manifests* itself into ordered patterns in the vacuum state. The symmetry is said to be broken since the vacuum state does not possess the full symmetry of the field equations (the dynamics). The "order" *is* indeed such a "lack of symmetry".

One can show that when symmetry is broken the invariance of the field equations implies the existence of quanta, the so called Nambu-Goldstone (NG) quanta, which, propagating through the whole system volume, are the carrier of the ordering information, they are the long range *correlation modes*: in the crystal, for example, the ordering information is the one specifying the lattice arrangement.

The presence (*the condensation*) in the vacuum state of NG quanta thus describes the ordering. When ordering is achieved, each of the elementary components of the system is "trapped" in a specific space-time behavior (e.g. a specific space position, a specific oscillation mode, etc.). Ordering implies thus the freezing of some of the degrees of freedom of the elementary components. Or, in other words, their *coherent* motion or behavior. It is such a coherent, collective behavior that macroscopically manifests itself as the ordered pattern: the microscopic quantum behavior thus provides macroscopic (collective) properties. We have a "change of scale", from microscopic to macroscopic, and the ordered state is called a *macroscopic quantum state*. NG quanta are therefore also called collective modes. In the dissipative model of brain these NG quanta are called "dipole wave quanta" (dwq) since they originate from the breakdown of the electrical dipole rotational symmetry (Vitiello 1995; Del Giudice et al. 1985, 1986; Jibu & Yasue 1995).

In order for the NG quanta to be able to span the full system volume and thus set up the ordered pattern, their mass has to be zero (or quasi-zero in realistic conditions of finite system sizes). In their lowest momentum state NG quanta thus do not carry energy. For this reason, the vacuum state where even a very large number of them is condensed is a state of minimal energy. This guaranties the stability of the ordered vacuum, namely of the memory record in the quantum model of brain.

Incidentally, I observe that in this model the variables are basic quantum field variables (the electrical dipole field). In the quantum model "we do not intend", Ricciardi and Umezawa say "to consider necessarily the neurons as the fundamental units of the brain". Moreover, Stuart, Takahashi and Umezawa (1978) have also remarked that "it is difficult to consider neurons as quantum objects".

In principle, any of the ordered patterns compatible with the invariance of the field equations can be realized in the process of symmetry breaking. This is why symmetry breaking is said to be "spontaneous". The point is that the ordered pattern which is actually realized is the output of the system *inner* dynamics. The process of symmetry breaking is triggered by some external input; the "choice" of the specific symmetry pattern which is actually realized is, on the contrary, "internal" to system. Therefore one speaks of *self-organizing* dynamics: ordering is an inner (spontaneous, indeed) dynamical process. This feature of spontaneous symmetry breaking is common to solid state physics and high energy physics, and it is of particular interest when modelling the brain: in the brain, contrary to the computer case, ordering is not imported from the outside, it is the outgrowth of an "internal" dynamical process of the system.

The generation of ordered, coherent patterns is thus the dynamical result of the system elementary component interactions.

3. The brain is a dissipative system

In the quantum model of brain a specific memory is associated to a specific degree of ordering (a specific value of the vacuum *code*). The overprinting problem then reduces to the problem of making available all possible vacua, or, in other words, to attach a specific label (the code value) to a given vacuum under the trigger of a specific external input.

On the other hand, it is evident that in its continual interaction with the environment the brain's time evolution is irreversible, i.e., technically speaking, it is *non*-unitary. Getting information from the outside world, which is a feature characterizing the brain, makes the brain dynamics intrinsically irreversible. I have elsewhere depicted such a situation by mentioning the way of saying ...*Now you know it!*..., which indeed means that once one gets some information, he/she is nevermore the same person as before. Getting information introduces the *Now* in our experience, or, in different words, the feeling of the past and of the future, of the arrow of time pointing forward in time. Without getting information there would be neither forward nor backward in time. However, we *cannot avoid* getting information, being opened on the world (including our inner world, ourself). The brain is an open, dissipative system. The brain closed on the world is a dead brain, physiology tells us. Isolation of the brain (closure to the world) produces serious pathologies. Thus, the extension of the quantum model of brain to the dissipative dynamics appears to be a necessity (Vitiello 1995).

Then, the mathematical formalism for quantum dissipation (Celeghini, Rasetti, & Vitiello 1992) *requires* the doubling of the brain degrees of freedom. The doubled degrees of freedom, say $\tilde{A}$ (the tilde quanta; the non-tilde quanta A denoting the brain degrees of freedom), are meant to represent the environment to which the brain is coupled. The physical meaning of the doubling is the one of ensuring the balance of the energy flux between the system and the environment.

The environment thus represented by the doubled degrees of freedom appears described as the "time-reversed copy" (the *Double*) of the brain. The environment is "modelled" on the brain. Time-reversed since the energy flux outgoing from the brain is incoming into the environment, and vice versa.

In addition, the quantum dissipation formalism implies that the full operator describing the system time evolution includes the operator describing the coupling between the non-tilde and the tilde quanta. At the same time, such a coupling term acts as the mathematical tool to attach the label to the vacua (and thus to distinguish among different memories). This label is time-dependent: the system states are thus time-dependent states.

In this new light, the time evolution operator is readily recognized to be the "free energy" operator, not just the Hamiltonian operator, as it would be in the absence of dissipation. The free energy operator is made indeed by the Hamiltonian operator, which controls the reversible (unitary) part of time evolution, plus the non-tilde/tilde coupling term, which is recognized to be proportional to the *entropy* operator and controls indeed the non-unitary, irreversible part of time evolution. In a thermodynamical language this last term describes the heat term in the system energy.

The doubling of the degrees of freedom in the dissipative model thus arises as a consequence of the irreversible time evolution.

Once thermal aspects in the dissipative model have been also recognized, the memory state is found (Vitiello 1995) to be a *non-equilibrium* Thermo Field Dynamics (TFD) state. TFD is the QFT formalism for thermal systems introduced by Takahashi and Umezawa (Takahashi & Umezawa 1975; Umezawa 1993) which provides an explicit representation of the so called Gelfand-Naimark-Segal (GNS) construction in the C^*-algebra formalism (Ojima 1981). TFD was not devised for the study of the brain, but for the study of solid state physics, to which it has been successfully applied.

In equilibrium TFD the system time evolution is fully controlled by the Hamiltonian. The operator necessary to attach the label to the thermal states (the label is temperature in that case) is not a term of the time evolution operator (as, on the contrary, in the dissipative model). Non-equilibrium transitions (non-unitary time evolution) in thermal systems have been considered later on in the time-dependent TFD formalism (Umezawa 1993; and references therein quoted). The non-equilibrium character of the brain dynamics makes the dissipative model substantially different from equilibrium TFD.

One can show that the dynamics now includes, when the system volume is large but finite, the possibility of transitions *through* inequivalent (labelled by different codes) vacua: in this way, at once, familiar phenomena such as memory associations, memory confusion, even the possibility to forget some memories, or else difficulties in recovering memory, are described by the dissipative model. The dissipative character of the dynamics thus accounts for many features of the brain behavior and for its huge memory capacity: now, indeed, all the differently coded vacua become accessible to the memory printing process. Their unitary inequivalence at large volume guaranties protection from overprinting, not excluding, however, due to realistic boundary effects, the processes of association, confusion, etc. just mentioned.

The general scheme of the dissipative quantum model can be summarized as follows. The starting point is that the brain is permanently coupled to the environment. Of course, the specific details of such a coupling may be very intricate and changeable so that they are difficult to be measured and known. One possible strategy is to average the effects of the coupling and represent them, at some degree of accuracy, by means of some "effective" interaction. Another possibility is to take into account the environmental influence on the brain by a suitable *choice* of the brain vacuum state. Such a choice is triggered by the external input (breakdown of the symmetry), and it actually is the end point of the internal (spontaneous) dynamical process of the brain (self-organization). The chosen vacuum thus carries the *signature* (memory) of the reciprocal brain-environment influence at a given time under given boundary conditions. A change in the brain-environment reciprocal influence then would correspond to a change in the choice of the brain vacuum: the brain state evolution or "story" is thus the story of the trade of the brain with the surrounding world. The theory should then provide the equations describing the brain evolution "through the vacua", each vacuum for each instant of time of its history.

The brain evolution is thus similar to a time-ordered sequence of photograms: each photogram represents the "picture" of the brain at a given instant of time. Putting together these photograms in "temporal order" one gets a movie, i.e. the story (the evolution) of open brain, which includes the brain-environment interaction effects.

Pasquale Vitiello (1912–1962), Composizione con figure, ink 1962

The evolution of a memory specified by a given code value, say $\mathcal{N}$, can be then represented as a trajectory of given initial condition running over time-dependent vacumm states, denoted by $|0(t)>_{\mathcal{N}}$, each one minimizing the free energy functional (Pessa & Vitiello 2003, 2004; Vitiello 2004). These trajectories are known (Manka, Kuczynski, & Vitiello 1986; Del Giudice et al. 1988; Vitiello 2004) to be *classical* trajectories in the infinite volume limit: transition from one representation to another inequivalent one would be strictly forbidden in a quantum dynamics.

4. Entanglement, chaos and coherence

Since we have now two-modes (i.e. non-tilde and tilde modes), the memory state $|0(t)>_{\mathcal{N}}$ turns out to be a two-mode coherent state. This is known (Perelomov 1986; Vitiello 2004; Pessa & Vitiello 2003, 2004) to be an entangled state, i.e. it cannot be factorized into two single-mode states, the non-tilde and the tilde one. The physical meaning of such an entanglement between non-tilde and tilde modes is in the fact that the brain dynamics is permanently a dissipative dynamics. The entanglement, which is an unavoidable mathematical result of dissipation, represents the impossibility of cutting the links between the brain and the external world.

I remark that the entanglement is permanent in the large volume limit. Due to boundary effects, however, a unitary transformation could disentangle the tilde and non-tilde sectors: this may result in a pathological state for the brain. It is known that forced isolation of a subject produces pathological states of various kinds.

I also observe that the tilde mode is not just a mathematical fiction. It corresponds to a real excitation mode (quasiparticle) of the brain arising as an effect of its interaction with the environment: the couples of non-tilde/tilde dwq quanta represent the correlation modes dynamically created in the brain as a response to the brain-environment reciprocal influence. It is the interaction between tilde and non-tilde modes that controls the irreversible time evolution of the brain: these collective modes are confined to live *in* the brain. They vanish as soon as the links between the brain and the environment are cut.

Here, it is interesting to recall (Vitiello 1998, 2001) that structure and function constitute an un-dividable unity in the frame of QFT: the dipole wave quanta are at the same time structure (they are *real particles* confined to live inside the system) and function, since they *are* the collective, macroscopic correlations characterizing the brain functioning.

The structure/function unity in the dissipative model thus accounts for the observed strong "reciprocal dependence" existing between the formation of neuronal correlates and nets and the functional activity of the brain, including the brain's *plasticity* and *adaptiveness*. The dissipative model implies that the insurgence of some structural (physiological) pathologies of the brain may be caused by the reduction and/or inhibition of its functions due to externally imposed constraints in some severe conditions.

As mentioned, transitions among unitary inequivalent vacua may occur (phase transitions) for large but finite volume, due to coupling with the environment. Due to dissipation the brain appears as "living over many vacuum states" (continuously undergoing phase transitions). Even very weak (although above a certain threshold) perturbations may drive the system through its macroscopic configurations. In this way, occasional (random) weak perturbations play an important rôle in the complex behavior of the brain activity. In a recent paper (Pessa & Vitiello 2003, 2004) the tilde modes have been shown to be strictly related to the quantum noise in the fluctuating random forces coupling the brain with the environment.

It has been also found (Pessa & Vitiello 2003, 2004) that, under convenient conditions, in the infinite volume limit, trajectories over the memory space are classical chaotic trajectories (Hilborn 1994), namely: (i) they are bounded and each trajectory does not intersect itself (trajectories are not periodic); (ii) there are no intersections between trajectories specified by different initial conditions; (iii) trajectories of different initial conditions are diverging trajectories.

In this connection, it is interesting to mention that some experimental observations by Freeman (1990, 1996, 2000) show that noisy fluctuations at the neuronal level may have a stabilizing effect on brain activity, noise preventing to fall into some unwanted state (attractor) and being an essential ingredient for the neural chaotic perceptual apparatus (especially in neural aggregates of the olfactory system of laboratory animals).

In the dissipative model noise and chaos turn out to be natural ingredients of the model. In particular, in the infinite volume limit the chaotic behavior of the trajectories in memory space may account for the high perceptive resolution in the recognition of the perceptual inputs. Indeed, small differences in the codes associated to external inputs may lead to diverging differences in the corresponding memory paths. On the other side, it also happens that codes differing only in a finite number of their components (in the momentum space) may easily be recognized as being the "same" code, which makes possible that "almost similar" inputs are recognized by the brain as "equal" inputs (as in pattern recognition).

Summarizing, the brain may be viewed as a complex system with (infinitely) many macroscopic configurations (the memory states). Dissipation is recognized to be the root of such a complexity.

The brain's many structural and dynamical levels (the basic level of coherent condensation of dwq, the cellular cytoskeleton level, the neuronal dendritic level, and so on) coexist, interact among themselves and influence each other's functioning. The crucial point is that the different levels of organization are not simply structural features of the brain; their reciprocal interaction and their evolution is intrinsically related to the basic quantum dissipative dynamics.

On the other hand, the brain's functional stability is ensured by the system's "coherent response" to the multiplicity of external stimuli. Thus dissipation also seems to suggest a solution to the so called *binding problem*, namely the understanding of the unitary response and behavior of apparently separated units and physiological structures of the brain. In this connection see also the holonomic theory by Pribram (1971, 1991).

The coherence properties of the memory states also explain how memory remains stable and well protected within a highly excited system, as indeed the brain is. Such a "stability" is realized in spite of the permanent electrochemical activity and the continual response to external stimulation. The electrochemical activity must also, of course, be coupled to the correlation modes which are triggered by external stimuli. It is indeed the electrochemical activity observed by neurophysiology that provides a first response to external stimuli.

This has suggested (Stuart, Takahashi, & Umezawa 1978, 1979) to model the memory mechanism as a separate mechanism from the electrochemical processes of neuro-synaptic dynamics: the brain is then a "mixed" system involving two separate but interacting levels. The memory level is a quantum dynamical level, the electrochemical activity is at a classical level. The interaction between the two dynamical levels is possible because the memory state is a *macroscopic quantum state* due, indeed, to the *coherence* of the correlation modes. The coupling between the quantum dynamical level and the classical electrochemical level is then the coupling between two macroscopic entities. This is analogous to the coupling between classical acoustic waves and phonons in crystals (phonons are the crystal NG quanta). Such a coupling is possible since the macroscopic behavior of the crystal "resides" in the phonon modes, so that the coupling acoustic-waves/phonon is nothing but the coupling acoustic-wave/crystal.

Finally, let me observe that, considering time-dependent frequency for the dwq, modes with higher momentum are found to possess longer life-time. Since the momentum is proportional to the reciprocal of the distance over which the mode can propagate, this means that modes with shorter range of propagation will survive longer. On the contrary, modes with longer range of propagation will decay sooner. This mechanism may produce the formation of ordered domains of finite different sizes with different degree of stability: smaller domains would be the more stable ones. Thus we arrive at the dynamic formation of a hierarchy (according to their life-time or equivalently to their sizes) of ordered domains (Alfinito & Vitiello 2000). On the other hand, since any value of the momentum is in principle allowed to the dwq, we also see that a scaling law is present in the domain formation (any domain size is possible in view of the momentum/size relation).

5. The trade with the Double: A route to consciousness?

We have seen that the tilde modes are a representation of the environment "modelled" on the (non-tilde) system: they constitute the time-reversed copy of it. And, we have seen, they are "necessary", they cannot be eliminated from the game. The mathematical operation of doubling the system degrees of freedom, required by dissipation, thus turns out to produce the system's *Double*. I have then suggested that consciousness mechanisms might be involved in the continual "trade" (interaction) between the non-tilde and the tilde modes (Vitiello 1995, 2001).

Here I would be tempted to say: trade "between the subject and his Double". However, the word "subject" may be evocative of rich but intricate philosophical scenarios, which here are absolutely out of my considerations. I have experienced indeed that using in a simple minded way that word in connection with the brain (as I did in my book (Vitiello 2001)) may be highly misleading, pushing the reader far from the much more modest, but concrete, mathematical and physical features of the dissipative model.

My attention is rather on the dynamics, the "inter-action", the trade, the "between" (as Gordon Globus 2003 would say), l'"entre-deux" (as Nadia Prete 2004 would prefer). The use of the word "subject" could instead evoke the idea of something, the "one" (the non-tilde), pre-existing the relation with the other "one", his Double (tilde). However, this would correspond neither to the physics, nor to the mathematics, both of which are my fixed starting points.

The physics of the problem is, technically speaking, a non-perturbative physics, the one of the open systems, and it can be shown that in such a case the system-environment "inter-action" cannot be switched off (Celeghini, Rasetti, & Vitiello 1992). This means that we cannot even think of the system deprived of its physical essence which is its openness (even the physiology tells us that an isolated brain is a dead brain; namely, if "closed", it does not exist *as a brain*). The physics does not allow the existence of the "one" (non-tilde) *independently* of the existence of the "other one" (tilde), and vice versa.

The mathematics, on the other hand, imposes a strong limit on the description (the "language") we have to use for a quantum dissipative system: we cannot avoid starting from two reciprocal (in the mirror of time) images. This "un-divided two", mathematics tells us, is more elementary than "the one". The non-tilde one "cannot" be the subject.

The temptation could be to think that the Double is the subject. But this simply means being captured by the Narcissus self-mirroring fatal trap (Vitiello 2001): it is equivalent to think of the brain as the subject, and vice versa in an endless loop. The Double is not trivially the system image. It is the environment representation *modelled* on the system. The Double cannot be the subject.

Tilde and non-tilde cannot *individually* pre-exist prior to their being each other's images, an "un-dividable two". They are actors *forced* (without alternative choice) to be on the stage. The "one", the subject, is the action, the play, their entre-deux. This is the meaning of the entanglement: the entangled state cannot be factorized (is un-dividable) into two single-mode states. Non-tilde and tilde modes share a common, entangled vacuum at each instant of time.

In some sense, here we face the root of the (ontological) prejudice that some "being" might exist as a "closed", i.e. non-interacting, system, and therefore, capable to exist *by itself*, *independently* of the existence of any other system, *complete* in its own individuality (Vitiello 1997, 2001). If so, it might also happen that such a system could be, in absolute, the *only* existent system ("being").

Such a prejudice seems to be intrinsic to our same language, where any action presupposes pre-existing actors having the possibility of being fully non-interacting, and thus each one *independently* existing from the other one (fully disentangled) *before* the action started. Notable exceptions (Stamenov 2001) might be those actions, such as *to exchange*, *to trade*, indeed, which exclude the *separate* (disentangled) pre-existence of the actors: to be possible that such actions could occur, the *joint* existence of (at least) a couple of actors is necessary (even if not sufficient, of course). Thus, those actions are special ones in that they presuppose entangled existences of the actors. Each one of these cannot exist by himself. And also, actors cannot be *separated* from their action and vice versa. Without exchange there are no exchangers, and vice versa. Such a situation also reminds me of the rheomode language of Bohm, whose structure is aimed to allow "the verb rather than the noun to play a primary role" (Bohm 1980; cf. Stamenov 2004 for a discussion on Bohm's rheomode language).

The unavoidable dialog with the Double is the continual, changeable and reciprocal (non-linear) interaction with the environment. If the consciousness phenomenon basically resides in such a permanent dialog, one of its characterizations seems to be the relational (medial) one, which agrees with Desideri's standpoint (Desideri 2004). Consciousness seems thus to be rooted and diffused in the large brain-environment world, in the dissipative brain dynamics. There is no conflict between the *subjectiveness* of the first person experience of consciousness and the *objectiveness* of the external world.[1] Without such an *objectiveness* there would be no possibility of "openness" (openness on what?), no dissipation out of which consciousness could arise. Objectiveness of the external world is the primary, necessary condition for consciousness to exist.

On the other hand, the question Desideri poses, namely "whether it is possible to reverse also the relationship between structure and function and then if it is possible to consider brain as a function of consciousness" (Desideri 2004) also finds a positive answer in the dissipative quantum model. The answer is positive in a true physical sense, since the brain cannot avoid to be an active/passive system, and promoting or inhibiting its activity (summing up in the consciousness) would produce the creation or destruction, respectively, of structural features of the brain, such as, e.g., long range correlations, pattern structures. As observed in Section 4, the different levels of organization are not simply structural features of the brain, their reciprocal interaction and their evolution is intrinsically related to the brain-environment entanglement, namely to that medial "one" which is the dialog with the Double. In this sense, the adaptiveness, the plasticity of the brain *is* the function of consciousness.

It is also interesting to observe that the dialog with the Double is "evolutive" and never repeats itself in the same form: from one side, it carries the memory, the story of the past; from the other side, the permanent openness on the world implies its continual updating. Recurrent resolutions into "new synthesis" of the non-tilde/tilde reciprocal presence are thus reached. The mentioned process of minimizing the free energy, namely of reaching the equilibrium between the numbers of non-tilde and tilde modes, is indeed the process by which such synthesis are recurrently reached by permanently tuning the constantly renewed brain-environment "relation". The actors are never engaged in a boring reply. And such a *truly dialectic* relation with the Double is

Pasquale Vitiello (1912–1962), Gli attori, ink 1962

inserted in the unidirectional flow of time, it is itself a "witness" of the flow of time. This depends on the fact that its mathematical description is provided by the coupling term in the time evolution operator and this is proportional to the entropy operator. It is possible to talk of *unidirectional* flow of time because time-reversal symmetry is broken due to dissipation (cf. also the beginning of Section 2). Then, the time axis gets divided by a *singular* point: *the origin*, which divides the past from the future. The singularity of this point consists in the fact that *it cannot be translated*, it is the *Now*.

Without dissipation, *any* point, any time, can be *arbitrarily* taken to be the origin of the time axis, which means that the origin (and any other point on the time axis) can be freely *translated* without inducing any observable change in the system (time translational invariance): thus there is no *singular* origin of the time. There is no *Now*. All the origins are alike. There is no a *true* origin.

In the absence of dissipation, we could say that time, in its flowing, swallows those fictitious Nows we might assign as (non-singular) origins on its axis, as *Κρόνος* eats his sons. This destructive property of time (oblivion) is, paradoxically, eluded, avoided by dissipation: dissipation introduces a life-time, a time scale which carries the *memory* of "when" (the origin) the dissipative system "has started". From the observation of a (radiative) decay process (typically with carbon fourteen) we "can trace back" the time, reach the origin and say "how old" is the object we are interested in. So we know where the true, non-forgettable origin (the truth), not a fictitious, false one easily eaten by *Κρόνος*, sits on the time axis. Memory (non-oblivion) and truth are the same thing, which the ancient Greeks denoted, indeed, with the same word, *αλήθεια* (Tagliagambe 1995; Vitiello 1997, 2001).

The Now is that point on the time-mirror where the non-tilde and the tilde, reciprocal time-reversed images, join together, in the *present* (Vitiello 1997, 2001). The non-tilde *unveils* its Double and they conjugate in a circular (non-linear) recognition, each being "exposed" to the other's eyes. Perhaps this is *intuition*, the *instantaneous apprehension* (Webster Dictionary 1968) of the "between". Literally, *intueri* is such a looking inside "without the conscious use of reasoning" (Webster Dictionary 1968), an immediate, out of time, not in the past, not in the future, act of *unconscious* knowledge, an "unknowable act" (Plotnitsky 2002) of knowledge. An act which repeats itself continuously, not translating the Now (dissipation forbids it!), but re-creating another independent, but equally true, Now, in an endless, dense sequence of Nows, all different, singular origins of different paths in the future, all starting points of chaotic memory paths in the memory space, which then we recognize as the "identity" or the "self" space. Identity, dynamically living in the memory space through the dissipative Nows, thus escapes the destructive fury of *Κρόνος*.

Perhaps, in these Nows is realized the primary property of consciousness, the one of self-questioning (Desideri 2004), i.e. the unveiling the Double, and the photographer's "sur-prise"..."when at the precise instant an image suddenly stands out and the eye stops" forcing "the time to stop his course" (Prete 2004): "and suddenly, all at once, the veil is torn away, I have understood, I have *seen*" (Sartre 1990; see also the related discussion by Prete 2004).

Unveiling the Double is then to see and to be seen, the *συνειδώς*, the being conscious of the ancient Greeks, which literally is to "see together", indeed; or, as in the lifting the veil in the Prete's photobjects, "more precisely, to have a perception of this togetherness as a whole and to understand that it was made of two images in strong relation" (Prete 2004); or else Bohm's self-recursive mirroring loops of the spontaneous and unrestricted act of "lifting into attention" (Bohm 1980; Stamenov 2004): *συνειδώς* then comes to be confidants, secret friends (Bandini 2002), to be each other "witness".

Such a sudden act of knowledge remains, however, an intuitive knowledge, an *unum*, not susceptible to be "divided" into rational steps, thinkable but "non-computational", not "translatable" into a language (i.e. logical) frame, which would require its breaking up (analysis) into linguistic fragments (cf. the traditional language fragmentation discussed in Bhom 1980, and the related discussion by Stamenov 2004). (It is interesting that the *ειδώς* in the word *συνειδώς* (being conscious) denotes the act of immediate vision; the word *οράω* is used instead for the act of lasting vision (Bonazzi 1936).)

In conclusion, from the sequence of these acts inserted into the "objective" time flow a sequence of independent, subjective Nows is generated, which constitute the multi-time dimensions of the *self*, its *own* time space, the *dynamic* archive of chaotic trajectories in the memory space which depicts its identity; that spring of time-lines through which the self can move "freely", apparently unconstrained by the external time-ordering.

Without such an internal freedom there could be neither the "pleasure" of the perception (the *αίσθησις*), the aesthetical dimension, that erotic charge of the *unveiling*, which continuously renews itself in the dialogic relation with the Double, nor the "active response" to the world. Neither pleasure, nor intentionality could be allowed in a rigidly constrained system. Active responses imply responsibility and thus they become moral, ethical responses through which the self and its Double become part of the larger social dialog. Aesthetical pleasure unavoidably implies disclosure, to *manifest* "signs", artistic *communication*. An interpersonal, collective level of consciousness then arises, a larger stage where again the actors are mutually dependent, each one bounded (entangled) in his very existence (including any sort of physical needs) to the other ones, simply non-existing without the others.

6. Doubts and mistakes. Toward the construction of an erratic device

I finally observe that the strong influence of even slight changes in the initial conditions on the memory paths (their chaotic behavior) leads us to consider the rôle of the "doubt" in consciousness mechanisms (Desideri 1998). In this connection I will also very briefly comment on a provocative proposal of mine: to construct an artificial device able to make "mistakes", namely able of taking a step, in its behavior, not logically consequent from the previous ones, or not belonging to any pre-ordered chain of steps or events, an erratic step. For shortness, and in a provisional way, I will refer to it as to the "erratic device". Such a device is perhaps in strict relation with an artificial conscious device (if ever it will be possible to construct an artificial conscious device!).

My erratic device is not a machine "out of order", not properly functioning. It *cannot* be a machine at all, since a machine, in the usual sense, is *by definition* (and by construction) something which must work *properly*, in a strictly predictive way, producing processes of sequentially ordered steps according to some functional logic. Also, the erratic device is not meant to be a device exhibiting chaotic behavior: the *value* (!) of the mistake is in its infrequent occurrence, an exceptional "novelty" with respect to an otherwise "normal" (correct) behavior.

But let me go back to the dissipative quantum model of brain. There, tilde modes also account for the quantum noise in the fluctuating forces coupling the brain with the environment. The dialog with the Double lives therefore on a noisy background of quantum fluctuations. "Listening" sometimes at such noisy background in the continual dialog (self-questioning) might slightly perturb the initial conditions of the memory paths and manifest in their drastic differences. This might be sometimes a welcome event, pushing the brain activity out of unwanted loops or fixations (attractors), (which also suggests a possible relation with Freeman's (1990, 1996, 2000) observations on neuronal noisy activity). Doubt might well be such a kind of self-questioning in a noisy background, being tempted by new perspectives, testing new standpoints by more or less slightly perturbing old certainties, leaving room for erratic fluctuations, listening to them; in a word, allowing fuzziness in the initial conditions, the starting assumptions of our travelling in the memory space (our archive of certainties); the consequences of the doubt will be then chaotically diverging trajectories in such a space. Consciousness modes then acquire their uncertain (doubtful) predictability with their precious unfaithfulness, their secret flavor of subjectivity, their full *autonomy*.

I suspect that the great privilege of being able of making mistakes finds its roots in these consciousness features. And perhaps here is the bridge between the program of constructing the erratic device and the one of constructing an artificial conscious device.

Perhaps, if ever it will be possible to construct a conscious artificial device, it will not be indeed a "machine", i.e. its behavior cannot be like a chain of logically predetermined steps, it must be an artificial being taking upon itself the best of the human model: unpredictably erratic, able to learn, but unfaithful, full of doubts, fully entangled to the world, but irreducibly free. We might name it Spartacus.

Note

1. Although it might sound philosophically unpleasant, I adopt the physicist's working hypothesis that the external world is objectively existing. In rough words, this amounts to adopt the working hypothesis that we do exchange energy with some other system. For example, we do need to eat. Without eating we cannot think. Of course, this does not mean that thinking is less important than eating, but simply that neglecting to eat leads to weak (or null) thinking.

References

Alfinito, E. & G. Vitiello (2000). Formation and life-time of memory domains in the dissipative quantum model of brain. *Int. J. Mod. Phys., B14*, 853–868.

Atmanspacher, H. (2004). Quantum theory and consciousness. *Stanford Encyclopedia of Philosophy*.

Bandini, F. (2002). La Coscienza e il Tempo. In *Exploring consciousness* (pp. 147–153). Milano: Fondazione Carlo Erba.

Bonazzi, B. (1936). *Dizionario Greco-Italiano*. Napoli: A. Morano.

Bohm, D. (1980). *Wholeness and the implicate order*. London: Routledge and Kegan Paul.

Celeghini, E., M. Rasetti, & G. Vitiello (1992). Quantum Dissipation. *Ann. Phys., 215*, 156–170.

Del Giudice, E., S. Doglia, M. Milani, & G. Vitiello (1985). A quantum field theoretical approach to the collective behavior of biological systems. *Nucl. Phys., B251* [*FS 13*], 375–400.

Del Giudice, E., S. Doglia, M. Milani, & G. Vitiello (1986). Electromagnetic field and spontaneous symmetry breakdown in biological matter. *Nucl. Phys., B275 [FS 17]*, 185–199.

Del Giudice, E., R. Manka, M. Milani, & G. Vitiello (1988). Non-constant order parameter and vacuum evolution. *Phys. Lett., A 206*, 661–664.

Desideri, F. (1998). *L'ascolto della coscienza*. Milano: Feltrinelli.

Desideri, F. (2004). The self-transcendence of consciousness towards its models (This Volume).

Freeman, W. J. (1990). On the the problem of anomalous dispersion in chaotic phase transitions of neural masses, and its significance for the management of perceptual information in brains. In H. Haken & M. Stadler (Eds.), *Synergetics of cognition, 45* (pp. 126–143). Berlin: Springer Verlag.

Freeman, W. J. (1996). Random activity at the microscopic neural level in cortex ("noise") sustains and is regulated by low dimensional dynamics of macroscopic cortical activity ("chaos"). *Intern. J. of Neural Systems, 7*, 473–480.

Freeman, W. J. (2000). *Neurodynamics: An exploration of mesoscopic brain dynamics*. Berlin: Springer.

Globus, G. (2003). *Quantum closures and disclosures: Thinking-togheter postphenomenology and quantum brain dynamics*. Amsterdam: John Benjamins.

Hilborn, R. (1994). *Chaos and nonlinear Dynamics*. Oxford: University Press.

Jibu, M., & K. Yasue (1995). *Quantum brain dynamics and consciousness*. Amsterdam: John Benjamins.

Manka, R., J. Kuczynski, & G. Vitiello (1986). Vacuum structure and temperature effects. *Nucl. Phys., B 276*, 533–548.

Ojima, I. (1981). Gauge fields at finite temperature – "Thermo Field Dynamics" and the KMS condition and their extension to gauge theories. *Annals of Physics, 137*, 1–37.

Perelomov, A. (1986). *Generalized Coherent States and Their Applications*. Berlin: Springer.

Pessa, E. & G. Vitiello (2003). Quantum noise, entanglement and chaos in the quantum field theory of mind/brain states. *Mind and Matter, 1*, 59–79. arXiv:q-bio.OT/0309009.

Pessa, E. & G. Vitiello (2004). Quantum noise induced entanglement and chaos in the dissipative quantum model of brain. *Int. J. Mod. Phys., B18*, 841–858.

Plotnitsky, A. (2002). *The knowable and the unknowable*. Ann Arbor: The University of Michigan.

Prete, N. (2004). Doubling image to face the obscenity of photography (This Volume).

Pribram, K. H. (1971). *Languages of the brain*. Englewood Cliffs, NJ: Prentice-Hall.

Pribram, K. H. (1991). *Brain and perception*. Hillsdale, NJ: Lawrence Erlbaum.

Ricciardi, L. M. & H. Umezawa (1967). Brain and physics of many-body problems. *Kybernetik, 4*, 44–48. Reprint in this Volume.

Sartre, J. P. (1990). *La nausea* (p. 171). Torino: Giulio Einaudi.

Stamenov, M. I. (2001). Nouns, Verbs, and Consciousness. Pre-Conference Workshop, *Toward a science of consciousness. Consciousness and its place in Nature*. Skövde, Sweden: unpublished.

Stamenov, M. I. (2004). The rheomode of language of David Bohm as a way to re-construct the access to physical reality (This Volume).

Stuart, C. I. J., Y. Takahashi, & H. Umezawa (1978). On the stability and non-local properties of memory. *J. Theor. Biol., 71*, 605–618.
Stuart, C. I. J., Y. Takahashi, & H. Umezawa (1979). Mixed system brain dynamics: neural memory as a macroscopic ordered state. *Found. Phys., 9*, 301–327.
Tagliagambe, S. (1995). Creatività. *Atque, 12*, 25–46.
Takahashi, Y. & H. Umezawa (1975). Thermo field dynamics. *Collective Phenomena, 2*, 55–80.
Titus Lucretius Carus (99–55 B.C.). *De rerum natura.* Translation by Walter Englert, 2003. Newbury Port, MA: Focus.
Umezawa, H. (1993). *Advanced field theory: micro, macro and thermal concepts.* New York: American Institute of Physics.
Vitiello, G. (1995). Dissipation and memory capacity in the quantum brain model. *Int. J. Mod. Phys., B9*, 973–989.
Vitiello, G. (1997). Dissipazione e coscienza. *Atque, 16*, 171–198.
Vitiello, G. (1998). Structure and function. An open letter to Patricia Churchland. In S. R. Hameroff, A. W. Kaszniak, & A. C. Scott (Eds.), *Toward a science of consciousness II. The second Tucson Discussions and debates* (pp. 231–236). Cambridge: MIT Press.
Vitiello, G. (2001). *My Double Unveiled.* Amsterdam: John Benjamins.
Vitiello, G. (2004). Classical chaotic trajectories in quantum field theory. *Int. J. Mod. Phys., B18*, 785–792. arXiv:hep-th/0309197.
Webster's New World Dictionary (1968). College Edition, Cleveland: The World Pu. Co.

Name index

U

V

W

Y

Z

Subject index

D

List of the contributors

Emilio Del Giudice
INFN – Via Celoria 16,
I-20133 Milano, Italy
delgiudice@mi.infn.it

Fabrizio Desideri
Dipartimento di Filosofia
Università di Firenze
Via Bolognese 52
50139 Firenze, Italy
fabdesideri@tin.it.

Martin Fleischmann
Bury Lodge, Duck Street,
Tisbury, Salisbury, Wilts., SP3 6LJ, UK
Martinsp36lj@aol.com

Georg Franck
Vienna University of Technology
Department of Computer Aided
Architecture and Planning
Treitelstrasse 3/E259.1
A-1040 Vienna, Austria
franck@iemar.tuwien.ac.at

Gordon Globus
2990 Zurich Ct.
Laguna Beach, CA 92651, USA
ggglobus@aol.com

B. J. Hiley
Theoretical Physics Research Unit,
Birkbeck College, Malet Street,
London WC1E 7HX, UK
b.hiley@bbk.ac.uk

Mari Jibu
Research Institute for Informatics
and Science
Notredame Seishin University
Okayama 700-8516, Japan
mariej@ndsu.ac.jp

Teruaki Nakagomi
Department of Information Science
Faculty of Science
Kochi University
2-5-1 Akebono-cho
Kochi 780-8520, Japan
nakagomi@is.kochi-u.ac.jp

Eliano Pessa
Dipartimento di Psicologia
Università degli Studi di Pavia
Piazza Botta 6, 27100 Pavia, Italy
eliano.pessa@unipv.it

Arkady Plotnitsky
Professor of English
Theory and Cultural Studies Program
Department of English
Purdue University
West Lafayette, IN 47907
aplotnit@sla.purdue.edu

Nadia Prete
I.U.T. Lens,
Artois University, France
9, rue Saulnier
75009 Paris, France
pretenadia@hotmail.com

Karl H. Pribram
Cognitive Neuroscience,
Department of Psychology
Georgetown University, Georgetown, USA
&
School of Computational Sciences
George Mason University
Professor Emeritus Stanford
and Radford Universities
pribramk@georgetown.edu

Paavo Pylkkänen
Consciousness Studies Programme
School of Humanities and Informatics
University of Skövde
Box 408, SE – 541 28 Skövde, Sweden
paavo@ihu.his.se

Maxim Stamenov
Seminar für Slavische Philologie
Georg-August-Universität Göttingen
Humboldtallee 19
37073 Göttingen, Germany
mstamen@gwdg.de

Yasushi Takahashi
Department of Physic
University of Alberta
Edmonton, T6G 2J1, Canada
ashi@phys.ualberta.ca

Giuseppe Vitiello
Dipartimento di Fisica "E. R. Caianiello"
Universita' di Salerno
84100 Salerno, Italy
vitiello@sa.infn.it

Kunio Yasue
Research Institute for Informatics
and Science
Notre Dame Seishin University
2-16-9 Ifuku-cho
Okayama 700-8516, Japan
kunio@ndsu.ac.jp
&
Okayama Institute for Quantum Physics
1-9-1 Kyoyama, Okayama 700-0015, Japan
kunio_yasue@pref.okayama.jp

In the series *Advances in Consciousness Research* the following titles have been published thus far or are scheduled for publication:

1 **GLOBUS, Gordon G.:** The Postmodern Brain. 1995. xii, 188 pp.
2 **ELLIS, Ralph D.:** Questioning Consciousness. The interplay of imagery, cognition, and emotion in the human brain. 1995. viii, 262 pp.
3 **JIBU, Mari and Kunio YASUE:** Quantum Brain Dynamics and Consciousness. An introduction. 1995. xvi, 244 pp.
4 **HARDCASTLE, Valerie Gray:** Locating Consciousness. 1995. xviii, 266 pp.
5 **STUBENBERG, Leopold:** Consciousness and Qualia. 1998. x, 368 pp.
6 **GENNARO, Rocco J.:** Consciousness and Self-Consciousness. A defense of the higher-order thought theory of consciousness. 1996. x, 220 pp.
7 **MAC CORMAC, Earl and Maxim I. STAMENOV (eds.):** Fractals of Brain, Fractals of Mind. In search of a symmetry bond. 1996. x, 359 pp.
8 **GROSSENBACHER, Peter G. (ed.):** Finding Consciousness in the Brain. A neurocognitive approach. 2001. xvi, 326 pp.
9 **Ó NUALLÁIN, Seán, Paul Mc KEVITT and Eoghan Mac AOGÁIN (eds.):** Two Sciences of Mind. Readings in cognitive science and consciousness. 1997. xii, 490 pp.
10 **NEWTON, Natika:** Foundations of Understanding. 1996. x, 211 pp.
11 **PYLKKÖ, Pauli:** The Aconceptual Mind. Heideggerian themes in holistic naturalism. 1998. xxvi, 297 pp.
12 **STAMENOV, Maxim I. (ed.):** Language Structure, Discourse and the Access to Consciousness. 1997. xii, 364 pp.
13 **VELMANS, Max (ed.):** Investigating Phenomenal Consciousness. New methodologies and maps. 2000. xii, 381 pp.
14 **SHEETS-JOHNSTONE, Maxine:** The Primacy of Movement. 1999. xxxiv, 584 pp.
15 **CHALLIS, Bradford H. and Boris M. VELICHKOVSKY (eds.):** Stratification in Cognition and Consciousness. 1999. viii, 293 pp.
16 **ELLIS, Ralph D. and Natika NEWTON (eds.):** The Caldron of Consciousness. Motivation, affect and self-organization — An anthology. 2000. xxii, 276 pp.
17 **HUTTO, Daniel D.:** The Presence of Mind. 1999. xiv, 252 pp.
18 **PALMER, Gary B. and Debra J. OCCHI (eds.):** Languages of Sentiment. Cultural constructions of emotional substrates. 1999. vi, 272 pp.
19 **DAUTENHAHN, Kerstin (ed.):** Human Cognition and Social Agent Technology. 2000. xxiv, 448 pp.
20 **KUNZENDORF, Robert G. and Benjamin WALLACE (eds.):** Individual Differences in Conscious Experience. 2000. xii, 412 pp.
21 **HUTTO, Daniel D.:** Beyond Physicalism. 2000. xvi, 306 pp.
22 **ROSSETTI, Yves and Antti REVONSUO (eds.):** Beyond Dissociation. Interaction between dissociated implicit and explicit processing. 2000. x, 372 pp.
23 **ZAHAVI, Dan (ed.):** Exploring the Self. Philosophical and psychopathological perspectives on self-experience. 2000. viii, 301 pp.
24 **ROVEE-COLLIER, Carolyn, Harlene HAYNE and Michael COLOMBO:** The Development of Implicit and Explicit Memory. 2000. x, 324 pp.
25 **BACHMANN, Talis:** Microgenetic Approach to the Conscious Mind. 2000. xiv, 300 pp.
26 **Ó NUALLÁIN, Seán (ed.):** Spatial Cognition. Foundations and applications. 2000. xvi, 366 pp.
27 **GILLETT, Grant R. and John McMILLAN:** Consciousness and Intentionality. 2001. x, 265 pp.
28 **ZACHAR, Peter:** Psychological Concepts and Biological Psychiatry. A philosophical analysis. 2000. xx, 342 pp.
29 **VAN LOOCKE, Philip (ed.):** The Physical Nature of Consciousness. 2001. viii, 321 pp.
30 **BROOK, Andrew and Richard C. DEVIDI (eds.):** Self-Reference and Self-Awareness. 2001. viii, 277 pp.

31 **RAKOVER, Sam S. and Baruch CAHLON:** Face Recognition. Cognitive and computational processes. 2001. x, 306 pp.

32 **VITIELLO, Giuseppe:** My Double Unveiled. The dissipative quantum model of brain. 2001. xvi, 163 pp.

33 **YASUE, Kunio, Mari JIBU and Tarcisio DELLA SENTA (eds.):** No Matter, Never Mind. Proceedings of Toward a Science of Consciousness: Fundamental approaches, Tokyo 1999. 2002. xvi, 391 pp.

34 **FETZER, James H. (ed.):** Consciousness Evolving. 2002. xx, 253 pp.

35 **Mc KEVITT, Paul, Seán Ó NUALLÁIN and Conn MULVIHILL (eds.):** Language, Vision and Music. Selected papers from the 8th International Workshop on the Cognitive Science of Natural Language Processing, Galway, 1999. 2002. xii, 433 pp.

36 **PERRY, Elaine, Heather ASHTON and Allan H. YOUNG (eds.):** Neurochemistry of Consciousness. Neurotransmitters in mind. With a foreword by Susan Greenfield. 2002. xii, 344 pp.

37 **PYLKKÄNEN, Paavo and Tere VADÉN (eds.):** Dimensions of Conscious Experience. 2001. xiv, 209 pp.

38 **SALZARULO, Piero and Gianluca FICCA (eds.):** Awakening and Sleep–Wake Cycle Across Development. 2002. vi, 283 pp.

39 **BARTSCH, Renate:** Consciousness Emerging. The dynamics of perception, imagination, action, memory, thought, and language. 2002. x, 258 pp.

40 **MANDLER, George:** Consciousness Recovered. Psychological functions and origins of conscious thought. 2002. xii, 142 pp.

41 **ALBERTAZZI, Liliana (ed.):** Unfolding Perceptual Continua. 2002. vi, 296 pp.

42 **STAMENOV, Maxim I. and Vittorio GALLESE (eds.):** Mirror Neurons and the Evolution of Brain and Language. 2002. viii, 392 pp.

43 **DEPRAZ, Nathalie, Francisco J. VARELA and Pierre VERMERSCH:** On Becoming Aware. A pragmatics of experiencing. 2003. viii, 283 pp.

44 **MOORE, Simon C. and Mike OAKSFORD (eds.):** Emotional Cognition. From brain to behaviour. 2002. vi, 350 pp.

45 **DOKIC, Jérôme and Joëlle PROUST (eds.):** Simulation and Knowledge of Action. 2002. xxii, 271 pp.

46 **MATEAS, Michael and Phoebe SENGERS (eds.):** Narrative Intelligence. 2003. viii, 342 pp.

47 **COOK, Norman D.:** Tone of Voice and Mind. The connections between intonation, emotion, cognition and consciousness. 2002. x, 293 pp.

48 **JIMÉNEZ, Luis (ed.):** Attention and Implicit Learning. 2003. x, 385 pp.

49 **OSAKA, Naoyuki (ed.):** Neural Basis of Consciousness. 2003. viii, 227 pp.

50 **GLOBUS, Gordon G.:** Quantum Closures and Disclosures. Thinking-together postphenomenology and quantum brain dynamics. 2003. xxii, 200 pp.

51 **DROEGE, Paula:** Caging the Beast. A theory of sensory consciousness. 2003. x, 183 pp.

52 **NORTHOFF, Georg:** Philosophy of the Brain. The brain problem. 2004. x, 433 pp.

53 **HATWELL, Yvette, Arlette STRERI and Edouard GENTAZ (eds.):** Touching for Knowing. Cognitive psychology of haptic manual perception. 2003. x, 322 pp.

54 **BEAUREGARD, Mario (ed.):** Consciousness, Emotional Self-Regulation and the Brain. 2004. xii, 294 pp.

55 **PERUZZI, Alberto (ed.):** Mind and Causality. 2004. xiv, 235 pp.

56 **GENNARO, Rocco J. (ed.):** Higher-Order Theories of Consciousness. An Anthology. 2004. xii, 371 pp.

57 **WILDGEN, Wolfgang:** The Evolution of Human Language. Scenarios, principles, and cultural dynamics. 2004. x, 237 pp.

58 **GLOBUS, Gordon G., Karl H. PRIBRAM and Giuseppe VITIELLO (eds.):** Brain and Being. At the boundary between science, philosophy, language and arts. xii, 350 pp.